Dimensions of the Meal
The Science, Culture, Business, and Art of Eating

Editor

Herbert L. Meiselman, PhD
Senior Research Scientist
U.S. Army Natick Research
Development and Engineering Center
Natick, Massachusetts

AN ASPEN PUBLICATION®
Aspen Publishers, Inc.
Gaithersburg, Maryland
2000

Library of Congress Cataloging-in-Publication Data

Dimensions of the meal: the science, culture, business, and art of eating / editor,
Herbert L. Meiselman.
p. cm.
Includes bibliographical references and index.
ISBN 0-8342-1641-8
1. Gastronomy. 2. Food habits. I. Meiselman, Herbert L.
TX631 .D49 2000
641'.01'3—dc21
99-044079

Orders: (800) 638-8437
Customer Service: (800) 234-1660

About Aspen Publishers • For more than 40 years, Aspen has been a leading professional
publisher in a variety of disciplines. Aspen's vast information resources are available in both
print and electronic formats. We are committed to providing the highest quality information available in the most appropriate format for our customers. Visit Aspen's Internet site
for more information resources, directories, articles, and a searchable version of Aspen's full
catalog, including the most recent publications: **www.aspenpublishers.com**
Aspen Publishers, Inc. • The hallmark of quality in publishing
Member of the worldwide Wolters Kluwer group.

Editorial Services: Denise Hawkins Coursey
Library of Congress Catalog Card Number: 99-044079
ISBN: 0-8342-1641-8

Printed in the United States of America

1 2 3 4 5

*To my parents, Leo and Mollie Meiselman,
who taught me to appreciate good meals*

Table of Contents

Contributors

John S.A. Edwards, PhD
Granada Professor of Food Service
    Management
Director
The Worshipful Company of Cooks
Centre for Culinary Research
Bournemouth University
Poole, Dorset, England

Cees de Graaf, MSc, PhD
Division of Human Nutrition and
    Epidemiology
Wageningen University
Wageningen, The Netherlands

Harry R. Kissileff, PhD
Associate Professor of Clinical Psychology
    in Psychiatry and Medicine
New York Obesity Research Center
St. Luke's/Roosevelt Hospital and
    Columbia University
New York, New York

Harry T. Lawless, PhD
Professor
Food Science Department
Cornell University
Ithaca, New York

Lucy M. Long, MA, PhD
Assistant Professor
Department of Popular Culture
Bowling Green State University
Bowling Green, Ohio

Johanna Mäkelä, PhD
Research Associate
Department of Sociology
University of Helsinki
Helsinki, Finland

D.W. Marshall, PhD
Senior Lecturer
The University of Edinburgh Management
    School
Edinburgh, Scotland

Herbert L. Meiselman, PhD
Senior Research Scientist
U.S. Army Natick Research Development
    and Engineering Center
Natick, Massachusetts

Howard R. Moskowitz, PhD
President
Moskowitz Jacobs Inc.
White Plains, New York

**Jacqueline M. Newman, PhD, RD, CFCS,
    CDN, FADA, FACN**
Professor
Family, Nutrition, and Exercise Sciences
    Department
Queens College
Flushing, New York
Editor
Flavor & Fortune
Kings Park, New York

Shigeru Otsuka, PhD
Mukogawa Women's University
School of Life and Environmental
    Sciences
Department of Food and Nutrition
Nishinomiya-shi, Hyogo-ken, Japan

Patricia Pliner, PhD
Professor of Psychology
Department of Psychology
University of Toronto
Toronto, Ontario, Canada

Ritva Prättälä, PhD
Senior Researcher
Department of Epidemiology and Health
    Promotion
National Public Health Institute (KTL)
Helsinki, Finland

Barbara J. Rolls, PhD
Professor and Guthrie Chair in Nutrition
Nutrition Department
The Pennsylvania State University
University Park, Pennsylvania

Elisabeth Rozin
Havertown, Pennsylvania

Paul Rozin, PhD
Department of Psychology
University of Pennsylvania
Philadelphia, Pennsylvania

J.M. Schafheitle, MPhil, Res Dip, Cert Ed
Senior Lecturer
School of Service Industries
Bournemouth University
Poole, Dorset, England
Academicien, Academy of Culinary Art
Master Craftsman, Craft Guild of Chefs
Christchurch, Dorset, England

Jeffery Sobal, PhD
Associate Professor
Division of Nutritional Sciences
Cornell University
Ithaca, New York

Acknowledgments

This volume represents the encouragement and effort of many people. The idea for the book came from my long association with my colleagues at Natick Laboratories, USA, and my colleagues at Bournemouth University, UK. I wish to acknowledge their tireless discussions, challenges, and collaborations over the years. My experiences with both Natick and Bournemouth have taught me the importance of meals in considering any food topic. Both Natick and Bournemouth are interested in feeding people, not just in developing food products. My collaborations with the visiting scientists at Natick, who have come from all over the world, have also helped. My idea for a book on meals was further refined during several of the Food Choice meetings arranged by Dr. David Booth of Birmingham University, who has successfully brought together many diverse groups within the food research community. The original idea for the book was discussed with and encouraged by Dr. Jon Walmsley, who published my last book and supported the concept for this book. I would like to acknowledge the work of the authors who contributed special perspectives and material that are not collected elsewhere. I am very proud of the uniqueness of this volume. In the task of organizing the material, editing some of the papers, and communicating with authors, I was aided by my editorial assistant, Ms. Jane Johnson. Publication of the book was aided by the friendly and professional staff at Aspen.

On a personal level, I acknowledge the fundamental interest in food that was instilled in me by my parents to whom I dedicate this book. My father and grandfather were both in the food business, and my childhood home was filled with talk of food and examples of good meals. Finally, I acknowledge the support and encouragement of my wife, Deborah.

Introduction

Herbert L. Meiselman

Your food consists of a hard roll and coffee with milk. You are probably eating breakfast in Italy. Or perhaps the food is baked fish with boiled potatoes and cooked green vegetables. You are probably eating the main meal of the day in Scandinavia. Or maybe the food is a steak, baked potato, and large vegetable salad. You are probably eating in the United States. If the steak were served with rice instead of potato, you would probably be in Japan. All these food examples represent popular meals in different cultures. Some foods, like hamburger, pizza, and steak, are becoming worldwide foods, although even for these foods, the way they are served and the foods that accompany them might vary from country to country. What makes foods into a meal? What is a meal? What is it about us as people, about our psychology and physiology, that forms the basis for eating meals? What roles do meals play in how people interact within groups and within cultures? What do typical meals from different cultures look like? And how does understanding meals lead to providing the highest level of meal preparation?

This book is intended to address long-standing gaps in our understanding of meals. The science and technology of food products has focused on individual foods. The science and technology dealing with what we eat and how much we eat has dealt with individual foods. The business areas dealing with hospitality, food service, and catering have also dealt with individual foods. Yet despite this focus on individual foods, most food is eaten as part of a meal. Therefore most food products are consumed with other food products. The factors controlling what we eat and how much we eat must include the fact that foods are selected and consumed in combination. And the business of providing prepared food must consider meals, not just individual food items.

The reasons for these gaps in our approach to food science, product development, and food service lie in the complexity of meals. Defining and understanding what a meal is turns out to be a very difficult task. This is because the word "meal" means different things to different people. To the nutritionist, the meal is an opportunity to eat the wide variety of nutrients that the body needs. To the psychologist, the meal is an important phenomenon dealing with an individual's food preferences and food attitudes. To the sociologist, the meal is a prime daily opportunity for social interaction or family interaction. To the caterer or food service professional, the meal can be anything from a simple breakfast to a five-course formal dinner. And what about snacks: Are they meals? And so on.

Turning to the dictionary for simpler definitions of the "meal" does not resolve the confusion. The *Oxford English Dictionary* of 1933 ("Being a corrected re-issue

1

with an introduction, supplement, and bibliography of *A New English Dictionary*") lists eight completely different meanings for "meal." The second listing is what concerns us: "Meal (mil), sb2A measure. Any of the occasions of taking food that occur by custom or habit at more or less fixed times of the day, as a breakfast, dinner, supper, etc. Without reference to time: An occasion of taking food, a repast. Also, the material of a repast; the food eaten at or provided for a repast." (p. 267). The 1997 *Supplement to the Oxford English Dictionary* adds some more recent, specific uses of meal (such as "meals-on-wheels") but does not alter the basic definition. Thus the dictionary definition of meals emphasizes *when we eat* and *what we eat*. We shall see in this volume that the word "meal" contains an enormous richness well beyond when we eat and what we eat.

Another level of complexity concerning the meal is encountered when we consider what constitutes a meal in different cultures and different countries and even in different regions within the same country. A breakfast of bacon and eggs in the United States would seem very foreign in many parts of the world. Further, a breakfast in the United States that contained grits would identify the meal as being in the southern part of the country. Bacon and eggs in England would be accompanied by a grilled tomato and in Japan might be accompanied by a salad. The accompaniments can identify location, as well as the main dish itself. An open-faced sandwich lunch in Scandinavia would seem just as foreign in southern Europe. Pasta, which is eaten as a meal in the United States, is usually eaten as only one course of a meal in Italy.

Thus, meals have physiological, psychological, nutritional, anthropological, sociological, culinary, economic, business, and many other dimensions. One question that arises is whether we should examine these dimensions separately or together. That is where, in my opinion, this book is unique. There are two main points to this volume: the first point is the complexity of meals. I hope that every reader sees meals as far more complex when he or she has finished the book. The second point is the interdisciplinary nature of understanding meals and dealing with meals in practical situations. Nutrition, acceptability, cost, and other factors vary together and must be considered together. This book attempts to contribute to that process by bringing the different aspects of meals into one volume. Another question is whether the basic aspects of eating (physiology and nutrition) should be dealt with separately from the social science aspects (psychology, sociology, anthropology) and from the applied (product development) and business aspects (food service). Again, my answer is that we will obtain better answers to all these questions if we take a more interdisciplinary approach.

To deal with the enormous complexity of meals, I have asked the world's foremost experts to address various topics. The first chapters of the book in Part I seek to define what a meal is, although in some sense, the entire book addresses that question. The questions addressed in this section include the following: What distinguishes a meal from just eating? Does the meal have real psychological significance, or has it evolved as merely a convenient means of consuming adequate nutrition? Is mealtime just an eating occasion, or is it also a social and family occasion? Are there regular patterns of eating that define meals? Which meals engender the most or the least consumption? To answer these questions Dr. Johanna Mäkelä, Dr. Patricia Pliner, and Dr. Paul Rozin provide a social science perspective to defining a meal, addressing both the cultural and psychological dimensions. Dr. Cees de Graaf adds the nutritional component to the definition of the meal.

Part II of the book deals with the biological bases of meals, although some of this material was introduced previously by Pliner and Rozin. Although it is perhaps more intuitive to acknowledge cultural differences of meals and the important sociology of meals, much can also be made of the biological foundations that produce meal-eating in both animals and in man. The types of questions addressed in this section of the book include the following: What is the basis for treating meals as a basic biological phenomenon common to man and other animals? What biological events signal mealtime (other than the clock!)? What biological variables control meal spacing? Can we identify basic sensory events that form the basis of food appreciation in meals? And how does food appreciation extend from simple foods to complex meals, with their almost infinite variety and complexity? Are there biological mechanisms that account for man's appreciation of meal variety?

Thus, in this section, Dr. Harry Kissileff addresses the role of the meal in the mechanisms of hunger, satiety, and thirst, using data from both animals and man. Dr. Harry Lawless examines the role of sensory factors in meal appreciation. And Dr. Barbara Rolls expands the dimension of meals by examining food variety in meals and the underlying mechanism of sensory specific satiety.

The very first chapter in the book deals with a cultural definition of the meal. Part III deals with many issues related to culture and meals. This section tries to define some of the cultural variables and phenomena that are critical at mealtimes. From a cultural perspective, a meal is as much a time of interpersonal activity as a time of eating. Culture not only controls interaction at mealtime, but actually defines what a meal is and how mealtime is conducted. And culture even defines the very basis of food preparation, using culturally accepted flavor principles. Finally, culture achieves its greatest impact at festive meals, which certainly are as much cultural as nutritional.

Dr. Jeffrey Sobal begins by examining meals and social contact, emphasizing the social phenomena that are critical within meals. Elizabeth Rozin introduces the important concept of flavor principles to help us understand our basis for distinguishing foods and meals from culture to culture. And finally, Lucy Long addresses the meals that to many people exemplify what meals are: special holiday meals.

Part IV of the book deals with four representative cultures and their meals. Two European regions and two Asian regions are represented. The purpose of this section is to begin to appreciate the enormous differences, and some similarities, among meals of different cultures. Which meals appear in each culture? Which meal is the largest and which the smallest? Which meals involve similar foods from day to day, and which involve more variety? What are the basic patterns of foods within meals in different cultures (e.g., meat and two vegetables).

In this section, Dr. Ritva Prättälä discusses the differences and similarities among Scandinavian meals. Dr. Jacqueline Newman presents an authentic view of Chinese meals, and Professor Shigeru Otsuka presents Japanese meals. And finally, Dr. David Marshall looks at meal structures and social structures in Britain, also addressing the notion of what a meal is in British culture. Clearly, this section of the book could be expanded to an entire volume, or even multiple volumes, to cover more of the world's cultures. Other volumes could explore regional differences in meals within cultures. How many meal patterns exist in China or in India, for example? The reader should note that material on meals in different cultures appears in other chapters throughout the book. English meals are described in the chapters by Edwards, Marshall, and Schafheitle. Scandinavian meals are described by

Mäkelä, as well as by Prättälä. And numerous chapters use examples of American meals, including those by Pliner, Rozin, and Long.

Part V of the book deals with the business and artistic aspects of meals, that is, meals within catering and food service and food product development. Again, an entire book or many books could be devoted to meal planning and meal composition. How do we develop food products for our contemporary style of eating? How do chefs design meals? What is the influence of available food ingredients? What rules govern the sequence of courses? Dr. John Edwards begins by putting meals within both an historical perspective and within a food service/catering perspective. What is the history of meals; have we always eaten as we do now? Dr. Howard Moskowitz follows with a business orientation, dealing with food product development from a meal perspective. With the help of Joachim Schafheitle, we will hear from four of Britain's most acclaimed chefs on how they design meals, and we will see examples of meals that they have served in their restaurants.

The historical perspective, which is addressed by Dr. David Marshall and Mr. Joachim Schafheitle, as well as by Dr. John Edwards, is interesting because meals as we know them are of fairly recent origin in the history of man. This adds another source of variation in meals. Previously, I noted the complexity of meals when viewed from the perspective of different disciplines, technologies, and businesses, as well as from the perspective of different cultures. However, meals differ when viewed across history. The meal of the eighteenth century was quite different from that of the nineteenth and twentieth centuries, not just because different foods were available but also because meal customs have changed and continue to change. Several chapters within this volume address the issues of changing meals (see Marshall, Long, Prättälä, and Otsuka).

As you read through the book, keep in mind our objectives: To appreciate the complexity of meals; to see the psychological, physiological, cultural, nutritional, biological, sensory, food service/catering, and other business aspects of meals; and to see the interdisciplinary nature of understanding meals. Meals *are* complex, but understanding meals and addressing meals in the practical world requires a more complex view of the meal.

Part I

Definitions
of the Meal

Chapter 1

Cultural Definitions of the Meal

Johanna Mäkelä

INTRODUCTION

People eat meals and people talk about meals, but the question is: what is meant by a meal? Defining a meal is an intriguing task. On the one hand, a meal is a self-evident and common word in the everyday life vocabulary of the Western world. On the other hand, it is a concept in which scholars try to pinpoint some features of our eating habits and our essence as social animals and members of a certain culture.

The choice of food is a process in which nutrition produced by nature is transformed into food, a product of culture. People do not accept all possible substances as edible but make choices. Culture defines how possible nutrition is coded into acceptable food. (See Lévi-Strauss, 1966; MacClancy, 1992; & Mäkelä, 1990). Ecological, biological, and economic conditions affect our choice of food, but it is the cultural understanding and categorization that structure food as a part of the world and as edible or inedible. (Abrahams, 1984; Falk, 1994; & Ilmonen, 1993). According to the French anthropologist Claude Lévi-Strauss (1966), no culture is without language and cooking skills. However, every culture has its own definitions of a meal (MacClancy, 1992, p. 54).

To approach the constitution of meals three relevant dimensions of meals can be distinguished: meal format, eating pattern, and the social organization of eating (see Mäkelä et al., 1999). The meal format takes both the composition of the main course and the sequence of the whole meal (i.e., the starter, the main course, the dessert) into account. The types of meals vary from simple cold meals to elaborate meals with many courses. Second, the eating pattern is defined by three different elements: time (the rhythm of eating events), the number of eating events, and the alternation of hot and cold meals and snacks. Third, the aspects of where and with whom people eat and who did the cooking form the social organization of eating.

MEAL FORMAT

A classical approach to defining meals was worked out by the British anthropologist Mary Douglas, who asked, "What defines the category of a meal in our home?" (Douglas, 1975a, p. 250). She started to analyze the structure of daily meals by using classifications and binary oppositions. To determine the categories of food she classified meals, courses, helpings, and mouthfuls from the standpoint of a linguistic analogy. To Douglas, meals and drinks are two important opposite categories, as in the relation between solids and liquids. A meal has both solid and liquid elements, and it has to have a dimension of bland, sweet, and sour. For Douglas the meaning of a meal is based on a system of repeated analogies (Douglas, 1975a).

Before this general overview, Douglas together with Michael Nicod had ana-lyzed the structure of British working class meals. First, Douglas and Nicod drew attention to the structural and sensory qualities embedded in the binary opposi-tions savory/sweet, hot/cold, liquid/dry. This is essential in defining different types of meals. The other criteria for the classification are complexity, copiousness, and ceremoniousness. A meal is not, however, the only type of eating. Nicod has pre-sented four terms that describe the different forms of eating: (1) A food event is an occasion when food is eaten. (2) A structured event is a social occasion organized by rules concerning time, place, and sequence of action. (3) Food eaten as a part of a structured event is a meal. A meal is connected to the rules of combination and sequence. (4) A snack is an unstructured food event without any rules of combina-tion and sequence. The meal system consists of three types of meals: (A) a major meal/the main meal, (B) a minor meal/the second meal, and (C) an even less signifi-cant meal/the third meal (biscuits and a hot drink). (Douglas, 1983, p. 83; Douglas & Gross, 1981, p. 6–7; Douglas & Nicod, 1974.) Thus, the model of Douglas and Nicod consists of complementary classifications.

The first course of meal A has, both on Sundays and weekdays, the same basic structure based on a staple (potato), a center (meat, fish, or egg), trimmings (veg-etables), and dressing (gravy). Everything is savory and hot. The second course has the same structure, except that everything is sweet. The staple is cereal, the center is fruit, and the dressing is liquid custard or cream. Meal B follows the structure of meal A, but the staple is cereal (bread) and not potato (Douglas & Nicod, 1974).

The British sociologist Anne Murcott has studied the concept of a proper meal in South Wales. A proper meal is always a cooked dinner, very similar to what Dou-glas and Nicod found. According to Murcott, proper meals are always of a certain type. Creating a proper meal means transforming food items into a meal by cooking and combining ingredients in the right way. A proper meal consists of one course only, a plateful, which is always a variation on meat and two vegetables. "Meat" must be (fresh) meat; sausages or offal cannot be used. In an emergency, poultry qualifies, but fish is ruled out, unlike in Douglas' construction. Potatoes and other vegetables are necessary, and at least one of them must be green. Finally, the gravy gives the finishing touch to the plateful and combines all the ingredients into a proper meal (Murcott, 1982, 1983, 1986). Nickie Charles and Marion Kerr (1988, p. 19–20) found a similar proper meal as Murcott: a cooked meal with meat (fish), potatoes, and vegetables as opposed to a snack, which is bread based and merely prepared instead of cooked.

Interestingly, a recent study conducted by Kemmer, Anderson, and Marshall (1998) in Scotland shows that the notions of a proper British meal have changed over time. Although the proper meal as a concept was easily understood by young couples, their ideas about a proper meal were more flexible than the traditional "meat and two veg," which was considered to be the proper meal *par excellence* but at the same time a bit dull. For young Scottish couples a proper meal could be a salad, pasta, chili or curry, or other non-British dishes that have become a natural part of the younger generation's food habits. The so-called fusion or crossover kitchen popular right now all over the world is a concrete example of the melting pot of different food cultures.

According to Marianne Ekström (1990), a cooked meal in Sweden has four ele-ments: a main course, which gives the name to the dish and is usually animal pro-tein; trimming n. 1, which is the starchy base of the meal; trimming n. 2, which consists of vegetables; and extra trimmings, which are vegetables or different condiments. In Finland, a proper meal is often defined as "a hot dish" (i.e., cooked food) accompanied by a salad (i.e., raw vegetables) (Mäkelä, 1996). The basic structure of a Finnish meal has not changed much over the past few decades. However, the transition from a rural to an urban eating pattern can be observed on the level of foodstuffs or dishes. The central role of potatoes, bread, and porridge has been overtaken by meat or fish (Prättälä, Pelto, Pelto, Ahola, & Räsänen, 1993).

As we can see, meals are often defined by "counting" the components: the foodstuffs chosen, the way they are prepared and combined, the number of courses served, and the number of dishes within each course (see Abrahams, 1984). This point of view is perhaps taken furthest by Michael P. Carroll (1982). According to Carroll, meals are characterized by a variety of tripartite classification schemes. A typical American eats three meals a day, uses three table utensils, eats three courses, and so forth. Carroll gives this as an example of tabular thinking and making lists. This kind of thinking reflected the rise of literacy, which meant that people started to list different features of meals. Slowly these lists changed so that every list had the same number of elements and thereby created a table.

Following the linguistic analogy of meals as a language with a certain grammar further, we see that each meal follows both syntagmatic and paradigmatic rules. The syntagmatic rules define the order of dishes. The basic structure of a Western meal would be three courses: a starter, a main course, and a dessert. Of course, there are variations on this structure from a simple main course and dessert version at home to a 12-course formal dinner at a restaurant. The paradigmatic rules define what kind of dish can be eaten in each group. For example, for a dessert it is possible to choose ice cream, apple pie, or soufflé, whereas chicken soup would never qualify as a dessert.

A clarifying example of how deeply these structures are actually rooted in our everyday life is a history of a dreadful dinner party told by Sally Cline (1990). A lady outraged and confused her friends completely by serving an exclusive six-course dinner in reverse order. Even though the trimmings and drinks with each course followed the conventions, the meal started with a cheese board followed by icecream with chocolate sauce, and the final course was avocado vinaigrette—the guests were freaked out by this obvious backward order. The hilarious gimmick planned by the hostess to entertain her guests ended in catastrophe because everybody felt offended.

Yet, this habit of eating meals in precise helpings and a certain order was established in the late nineteenth century. The current modern method of serving, *service à la russe*, in which a helping of each dish is prepared in the kitchen and then served, was adapted in France, England, and the United States during the second half of the nineteenth century. These meals have fewer dishes but more courses in sequence than the older *service à la française* from the Middle Ages. *Service à la française* resembles a modern buffet. The guest can choose from a large number of dishes served at the same time in each course (Levenstein, 1988:61; Mennell, 1985, p. 150).

The real meal (*un vrai repas*) with a three-course structure found by Annick Sjögren-de Beauchaine (1988) in her study of French bourgeoisie meals is a good example of how *service à la russe* is established in everyday life routines. However, the choice of meal components and level of elaboration and variety vary according to budget and the time constraints (Messer, 1991; Sharman, 1991).

EATING PATTERN

Besides the content and order of the courses, a meal is structured by time and space (Sjögren-de Beauchaine, 1988). Moreover, the structure and rhythm of meals also organize and affect the temporal pattern of the day and the year (Nordström, 1988). Meals are our clocks and calendars (MacClancy, 1992, p. 53). In agrarian society meals were part of the daily rhythm, and even today people seem to like to tie together their daily timetables by sharing a meal (Nordström, 1991).

In a Nordic comparative study the eating pattern was defined by three elements: time (rhythm of eating), the number of snacks and meals, and the alternation and the sequence of hot and cold snacks and meals (Gronow et al., 1998). According to this study, most people had at least a couple of meals daily, and they were concentrated into certain peak eating hours. In Finland two clear peaks exist: the first one ("breakfast") is eaten between 6 AM and 10 AM, the second one ("lunch") is eaten between 11 AM and 1 PM, whereas afternoon and evening eating is spread out more evenly.

This confirms the historical change in the number of hot meals especially; earlier, people ate more meals. Nowadays, the tendency is toward one hot meal, which on weekdays is quite often the lunch consumed in the work canteen. In the 1990s 44% of Finns followed the conventional meal pattern (i.e., three meals per day); 49% had two meals per day, and 7% had one meal or less per day (Roos & Prättälä, 1997).

The number of meals and snacks has varied between social groups and over time. A tendency to eat fewer meals can be detected all over Europe during the twentieth century because of changing working and living conditions. Robert Rotenberg (1981) shows how industrialization changed the allocation of time and the organization of work and housing in Austria. This influenced eating patterns and transformed the traditional five-meal pattern into the modern three-meal pattern within which meals are shorter and less elaborate and are less often eaten with the family or among friends.

In many countries in which the number of meals is decreasing, the role of different meals has changed. This is especially true of the breakfast, which used to be quite a substantial meal with hot dishes like meat sauce and boiled potatoes in Finland or bacon and eggs in Britain (MacClancy, 1992, p. 61–67; Prättälä & Helminen, 1990). Nowadays, the only possible hot dish of a Finnish breakfast is porridge, which is frequently replaced by cereals, bread, or yogurt; whereas in Britain the traditional "English breakfast" seems to be a special treat in hotels rather than the everyday breakfast.

SOCIABILITY OF MEALS

Consuming meals does not only amount to eating certain ingredients cooked in a certain way in a certain order. It is significant with whom and under what conditions we eat. The sociability of eating, the fact that a meal is shared with other people, is often considered as a necessary feature of meal definition.

Many scholars exploit this social nature of a meal to build more sophisticated analyses and definitions. For example, Mary Douglas and Michael Nicod not only analyze the construction of a meal, they are also especially interested in food as a marker of social relations and its role in everyday life and on special occasions. They also seek the regularities between social behavior and eating (Douglas, 1975b; Douglas & Nicod, 1974). The idea is to connect meals to a wider social system in which a meal is one ordered system related to the other ordered systems (Douglas, 1975a). For Anne Murcott (1982), a proper meal is also essentially a social affair in the same spirit. Annick Sjögren-de Beauchaine (1988) describes a meal as ordered social action that includes ritualized sharing of food. To Georg Simmel (1910; Gronow 1997), the sociability of eating is related to the refinement of social forms of interaction. Robert Rotenberg (1981) argues that a meal is essentially a social affair, "a planned social interaction centered on food" (Rotenberg, 1981, p. 26). Eating together not only means sharing the same food but also sharing the ideas related to food.

This section explores the social nature of meal eating. First, the importance of family meals to social relationships will be described. Second, the division of labor of meal preparation will characterize the social organization related to meals. Third, the differences between social groups show the role of food habits as a marker of social stratification.

Family Meals

A meal is essential for the unity of the family and a form of socialization and social reproduction. Meals are both formal and informal and intimate at the same time. Meals play an important role in marking closeness and distance (Sjögren-de Beauchaine, 1988). The sequence, rules, and ranking of the meals are vital to social life because they are part of a system of intimacy and distance. In British culture, drinks are for strangers, whereas meals are for family and friends (Douglas, 1983, p. 88; 1975a, p. 256.)

A meal is a symbol of social coherence (Brown, 1984). This is especially clear in the case of family meals, which are an important part of the social reproduction in the family. Eating together means staying together. The properness of meals is crystallized in Sunday and Christmas dinners, which are key symbols of the family (Charles & Kerr, 1988). For example, during our Christmas and Thanksgiving dinners we honor the past generations while preparing the same traditional dishes that our grandmothers used to cook. Each year these dinners celebrate the chain of family ties.

Marjorie DeVault (1991) furthers this argument by pointing out that family meals actually construct the family. Johanna Mäkelä (1996) found that, in the metropolitan area of Finland, working mothers' ideas of a proper meal consist of three factors: a hot dish, a salad, and company. Here the notion of a proper meal is a combination of components and context that illustrates how the idea of sharing food with others is essential for the concept of a meal. Sidney Mintz (1986) has stressed that meals must be consumed by everyone at the same time. They consist of the same items for all participants, served and eaten in a fixed order. Before the age of the microwave, it was laborious to realize everyone's individual preference during a meal. Yet, organizing a shared meal is not always easy. On the one hand, the different timetables of family members can make it difficult to get all to the table at the same time (Mäkelä, 1996). On the other hand, lack of space is an obstacle to joint meals in some households (Charles & Kerr, 1988, p. 184).

The fact that the free irregular eating habits of younger people tend to give way to a more precise mode when they start to share their everyday life with a partner emphasizes the role of the family. The birth of a family with children often marks a certain shift in eating habits. Then the concern for healthy eating increases (Valentin & Granzin, 1987) and food becomes a more important part of everyday life. Many studies show that one essential feature of living together is eating together and gradually learning about the partner's likes and dislikes. In the long run spouses will usually make compromises to overcome different food preferences (Kemmer et al. 1998). A Finnish study (Laitinen, Högström, & Räsänen, 1997) also reports a high resemblance in food choices among married and cohabiting couples who had children and who had meals at home.

The family environment acts as an arena for socialization. Children learn about food and how to behave at family meals (McIntosh, 1996, p. 63). At the table information is shared. Furthermore, meals civilize the children in a certain culture. Parents usually have a project of promoting and requiring table manners. Good table manners are considered an integral part of good social behavior, especially among middle-class families. Children learn the limits of edible and inedible food, but, in addition, they learn to manage their bodies and to appreciate the proper way to both cook and eat food in general. (Bell & Valentine, 1997, p. 64; Fischler, 1986; Grieshaber, 1997; Nordström, 1991.)

Division of Labor

A lot of work is required before a meal is hot and on the table. The ingredients have to be bought, prepared, and cooked and the table set. In the previously mentioned classic studies by Murcott (1982; 1983; 1986) and Charles and Kerr (1988) the division of labor was found to be one-dimensional: cooking proper meals was a female chore, as was clearing up after a meal.

Interestingly, the division of labor seems to have changed in Britain since the beginning of the 1980s. The study on changes in food choice and eating habits during the transition from single to married or cohabiting in Scotland (Kemmer et al., 1998) shows that although in more than half the couples the main responsibility for cooking was assumed by the woman, in some marriages the man was the main cook or the food preparation was shared. The reasons for this change are manifold. First, attitudes to the division of labor have changed, and, second, the participants in this particular study were partly from higher social groups. Their life phase was different, too (i.e., they did not have children and the women were working outside the home) (Kemmer et al., 1998; see also Warde & Hetherington, 1994).

Even if women are responsible for food preparation and maybe exercise control over cooking, this does not mean they also have power over it. It has been hypothesized that a woman's power increases when her status and income equals or exceeds her husband's (McIntosh 1996, p. 75). A Finnish study (Laitinen et al., 1997) supports this hypothesis; one of the findings was that if a woman's educational level is high, the home chores are shared more evenly. Interestingly, the division of labor of couples with no children was more even than that of couples with children. One reason for this might be the rather long Finnish maternity leave, which could establish certain routines so that taking care of the children and the household come to be regarded mainly as female tasks.

It also seems that women's and men's criteria for food preparation vary because their reports of their food preparation responsibilities differ significantly. The fact that women often emphasize cooked meals as proper could be interpreted as an indication that, for them, the preparation of food "boils down" to cooking hot meals, whereas men might consider making sandwiches just as important. Moreover, gender differences may exist in the views on the size and types of household chores that count as food preparation. Many studies show that women are responsible for both buying and cooking food, and they may consequently see the chain of food preparation as a longer and more time-consuming exercise than men. By the same token, they may report the time they spend more thoroughly. Marianne Ekström (1990) has aptly described this as "having the cupboard in their heads," meaning that women ensure that the stock of certain basic foodstuffs like flour or sugar is replenished.

The division of labor has changed over the past few decades. Now would be a good time to incorporate more observations on men's ideas of meals, eating habits, and cooking in contemporary research to get a wider picture of our everyday life. However, a study by Susan Grieshaber (1997) shows that girls are still socialized in preparing, serving, and clearing up after meals at an early age, whereas such skills are not required from boys.

The allocation of different food items has also varied within a family. Many earlier studies report a difference between men, women, and children. Working men got the best bits of meat, whereas the children and women managed with the lesser parts (Charles & Kerr, 1988). Men's preferences were respected more than children's and women's. This tendency is definitely weakening. Everyday experience seems to indicate that in the families of the 90s the role of children is growing. The child in a highchair has claimed the traditional prestigious place at the head of the table and is served first, while the others wait. Yet, women and mothers are willing to take their husband's and children's needs into account: while they favor dishes they know will be accepted and appreciated by the family, they also carry the responsibility for serving healthy, nutritious meals (see Mäkelä 1996; Valentin & Granzin, 1987). Some studies actually indicate that getting married often makes men's eating habits more healthy, and single men are considered to have the most unhealthy eating habits (see Kemmer et al., 1998; Karisto, Berg, & Prättälä, 1993).

Meals and Social Hierarchy

The structure of meals in a certain society is like a microcosm that reflects that society at the historic moment in question. During this century, the differences in food use between social groups have diminished in Western countries (Beardsworth & Keil, 1997; Levenstein, 1988; Mennell, 1985). The problems of diet have changed from scarcity to excess. Luxuries like white flour and sugar have been transformed into health hazards. On the level of nutrients, different social groups eat quite similarly. However, some differences still exist on the food item level. The cultural preferences and choices vary (see Roos, 1998).

The birth of gastronomy is related to the rise of the bourgeoisie after the French Revolution (Mennell, 1985, p. 266). First, the meal was democratized and politicized, but soon it became a symbol of social status and economic success. Food is a powerful marker of certain historical and societal moments even in works of art (see Brown, 1984) in relation to both other cultures and different social groups in the same society. As Stephen Mennell (1985) has put it, nowadays we live in a world of diminishing contrasts and increasing varieties. Yet, one's food habits still have distinctive class features.

For example, the distinctive lifestyles created by certain cultural practices are at the core of Pierre Bourdieu's (1984) concept of class. Taste is an important part of the distinctive lifestyle. According to Bourdieu, the right taste assures that one's choices are in harmony with the lifestyle. However, taste means not only the ability to make aesthetic or cultural judgments but it also refers to the taste of food. The double meaning of taste sits well in the Bourdieuan way of thinking because on each level it is easy to understand taste as intrinsic. You cannot learn good taste and you cannot dispute over matters of taste.

According to Bourdieu, taste is a product of living conditions, and differences between social groups are differences in consumption founded in the opposition between the tastes of luxury and the tastes of necessity. The distance from necessity describes the taste of the upper class, whereas the working class has no other choice than to choose the necessary. People do not have similar economic or cultural possibilities to make choices. Yet, Bourdieu sees that the working class resists the legitimate upper class way of life with its own eating and drinking habits. Working class meals are informal and easy. In the working class food is plentiful and it is nutritious, whereas bourgeois meals are more concerned with the form of the meal. The emphasis is on certain manners and forms, not on eating as consumption and nutrition.

Some of the features related to the lifestyles of the upper social classes trickle down to the lower social classes. Many foodstuffs that were once reserved for or accessible to only the upper classes have lost their role as prestige food and are available to all (Ilmonen, 1993, p. 255). A good example of the rotation of foodstuffs is the story of white bread, which until the end of the last century was a delicacy of the rich. During the twentieth century the new milling techniques made white bread available to all. Now prestigious bread is made of whole meal and spiked with seeds and nuts and is more expensive than the once luxurious white bread.

Today, the differences in the patterns of food consumption between social groups are not as sharp as they used to be. Yet, differences exist in attitudes toward the healthiness of eating and the percentage of income spent on food (Beardsworth & Keil, 1997). Nowadays, differences between groups are not only defined by social stratification. The style of certain groups is imitated selectively. It would be interesting to speculate whether the field of eating is one of the ways through which changes in social or cultural positions are manifested. It is relatively easy to learn the "language" of new culinary cultures (e.g., different ethnic cuisines) that seem to flow in one after the other. Cooking can be a hobby with a certain amount of rivalry and competition with other people interested in gastronomy. Dinner parties as part of the reproduction of social relationships are also a good shop window for one's lifestyle.

The role of food in the reproduction of social relationships is crystallized in food exchange and reciprocity. For example, Janet Thephano and Karen Curtis (1991) have studied food exchange among women in an Italian-American community. Marcella, an acknowledged cook, had guests in her household for meals or other food events more than 100 times and food was exchanged in her network 58 times during a 2-month period. The forms of exchange include hospitality, sharing eating activities other than meals, exchanging foods, specialty foods as gifts or payments for services, and cooperative cooking (Thephano & Curtis, 1991, p. 165). The idea of entertaining at dinner parties became part of American middle-class life in the late nineteenth century by following the example of the upper class (Levenstein, 1988, p. 61).

Even though the discussion of shared meals is often related to family meals, eating together creates a feeling of community and solidarity among people who do not share family ties. For example, people also have a tendency to share their meals at work or school canteens. Presumably, a kind of togetherness resembling the family meal may emerge in such fairly stable groups.

The idea of the social nature of meals is often clarified when a snack is mentioned as a counterpart for a meal. Snacking is usually described as the spontaneous and irregular consumption of food and a meal as a ritual regulated by rules concerning behavior and sharing food at the table. Meals are taken at routine times and they are social events, whereas random snacks are not (Whitehead, 1984). However, it might be fruitful to see meals and snacks as a continuum rather than counterparts to better grasp the richness and variety of different types of eating events in relation to aspects like complexity, structure, and sociability (see Mäkelä, 1991).

THE MEAL IN TRANSFORMATION

Meal eating patterns are changing. This has raised concern at the demise of the family meal as an essence of sociality (see Mennell, Murcott, & van Otterloo, 1992, p. 116; Mintz, 1982; Whit, 1995, p. 146). This view is based on the diminishing number of shared family meals and the growing tendency to eat snacks and fast food frequently. These irregular eating patterns have been described as "grazing" (see Caplan, 1997). The disappearance of seasonal variations is considered to be ruining the traditional annual rhythm of eating (see Grignon, 1992). Claude Fischler (1990) has described the situation as a state of "gastroanomy" in which people no longer have a clear idea of how to eat, when to eat, and how much to eat. Pasi Falk (1994) anticipates that the eating community will not be the principle structuring social life anymore . Meals are nowadays communication rather than communion. Although meals are marginalized, nonritual oral consumption like chewing gum, eating snacks and sweets, and smoking is increasing.

The new technologies of food storage and production have had an impact on households' food management. During the twentieth century refrigerators, freezers, electric stoves, and microwave ovens, not to mention smaller kitchen gadgets, have revolutionized our home cooking. The range of convenience foods available in the Western world is steadily widening. These developments seem to evoke different reactions. Some see them as a blessing for busy parents, whereas others see convenience foods as a threat to traditional home cooking and the microwave oven as a medium for individualizing meals.

The past few decades have also witnessed the birth of broad debate on food-related health issues. Growing health consciousness has put some foodstuffs on a black list. The once luxurious butter and sugar are seen as health hazards. Many prefer white meat to red meat or abandon all meat products. These changing food habits are creating problems for the home cook. In the same family the father might prefer red meat, the mother white, while one of the children loves fast food and the other is a strict vegan. Composing a meal acceptable to all requires a certain amount of knowledge and creativeness.

Although people strive for "good" and "healthy" eating, everyday practices are the result of a bargain with a trade-off between "good" meals and "bad" junk food (Anderson, Milburn, & Lean, 1995). If, by and large, you believe that you eat properly, a controlled relapse or an occasional treat is not harmful. Furthermore, weekend meals are nowadays important because then people are more willing to cook elaborate meals from scratch, partly to compensate for the scattered meals of weekdays.

Nevertheless, as Alex W. McIntosh (1996, p. 93) puts it: "despite warnings and some evidence, it is apparent that the family meal has not disappeared or lost its importance for family." The family meal is still a symbol of a family (DeVault, 1991; Lupton, 1996). Even though it is evident from many empirical studies that eating patterns are changing (see, for example, Bell & Valentine, 1997, p. 82), it seems that the change is not as drastic as some have anticipated. Proper meals and the breakfast, lunch, dinner pattern are still appreciated and snacking or grazing have not—yet—overtaken meal eating (Mäkelä, 1996; Roos & Prättälä, 1997; Warde, 1997, p. 149; Wood, 1995, p. 63–66).

However, the importance of the family meal and concern for its decline are part of a powerful myth. As Anne Murcott (1997) points out, the interesting feature of this idea of wane is the very debate itself. If we look at meals from the point of view of gender, age, and class, the idealistic picture of family meals starts to fall apart. The children of upper-class families did not share their meals with their parents but ate in their own rooms, and working-class wives merely played the role of waiter instead of sharing the meal with their husbands.

The new possibilities either to eat outside the home or to buy convenience foods and take-away foods are influencing the division of labor at home (see McIntosh, 1996, p. 74). Family meals are not the only way to feed the family. It is important to bear in mind that family meals are actually only one type of meal. A lovers' dinner at a restaurant, a teenagers' hamburger meal at a fast food joint, and a lunch with colleagues at a work canteen are all meals, but outside the home. And last, but not least, the debate on waning family meals has the tendency to forget that the traditional family is no longer the representative household pattern in real life. In 1992, 24% of households in the United Kingdom were classic nuclear families with married parents and children. At the same time 27% of households were single-person households (Beardsworth & Keil, 1997, p. 98).

Research on meals has been criticized for using a very traditional or even stereotypical notion of the nuclear family with both parents and children. To overcome this the term "postmodern family" has been suggested to describe the variety of households today (Bell & Valentine, 1997). The meaning of different life phases and courses has also been emphasized lately. As Bell and Valentine point out, "changes in individual identity are articulated on individual's plates" (Bell & Valentine, 1997, p. 77). The history of one's food habits is closely related to life history. Patterns of eating are influenced by childhood, residential history, school and employment history, relationships within the household, composition of the household, and skills and resources related to cooking (see Sharman, 1991).

One reason for the importance of meals is that structured meals help to structure one's life in general. Nevertheless, the role, structure, and content of a meal is gradually changing. Our grandparents' meals were different from those of today. Our grandchildren will definitely experience pleasures of food unknown to us. Yet it will take a long time before a shared meal eaten at table is replaced by an individually eaten meal pill.

REFERENCES

Abrahams, R. (1984). Equal opportunity eating: A structural excursus on things of the mouth. In K.L. Brown & K. Mussell (Eds.), *Ethnic and regional foodways in the United States: The performance of group identity* (pp. 19–36). Knoxville, TN: The University of Tennessee Press.

Anderson, A.S., Milburn, K., & Lean, M. (1995). Food and nutrition: Helping the consumer understand. In D. Marshall (Ed.), *Food choice and the consumer* (pp. 105–128). London: Blackie Academic & Professional.

Beardsworth, A., & Keil, T. (1997). *Sociology on the menu: An invitation to the study of food and society.* London: Routledge.

Bell, D., & Valentine, G. (1997). *Consuming geographies: We are where we eat.* London: Routledge.

Bourdieu, P. (1984). *Distinction: A social critique of the judgement of taste.* London: Routledge & Kegan Paul.

Brown, J.W. (1984). *Fictional meals and their function in the French novel.* University of Toronto. Romance Series 48. Toronto, Canada: University of Toronto Press.

Caplan, P. (1997). Approaches to the study of food, health and identity. In P. Caplan (Ed.), *Food, health and identity* (pp. 1–31). London: Routledge.

Carroll, M.P. (1982). The logic of Anglo-American meals. *Journal of American Culture 5*(3), 36–45.

Charles, N., & Kerr, M. (1988). *Women, food and families*. Manchester, England: Manchester University Press.

Cline, S. (1990). *Just desserts: Women and food*. London: Andre Deutsch.

DeVault, M.L. (1991). *Feeding the family: The social organization of caring as gendered work*. Chicago: The University of Chicago Press.

Douglas, M. (1975a). *Implicit meanings: Essays in anthropology*. London: Routledge & Kegan Paul.

Douglas, M. (1975b). The sociology of bread. In A. Spicer (Ed.), *Bread: Social, nutritional and agricultural aspects of wheaten bread* (pp. 7–24). London: Applied Science Publishers.

Douglas, M. (1983). Culture and food. In M. Freilich (Ed.), *The pleasures of anthropology* (pp. 74–101). New York: New American Library.

Douglas, M., & Gross, J. (1981). Food and culture: Measuring the intricacy of rule systems. *Social Science Information 20*(1), 1–35.

Douglas, M., & Nicod, M. (1974). Taking the biscuit: The structure of British meals. *New Society 30*(637), 744–747.

Ekström, M. (1990). *Kost, klass och kön* (in Swedish, Diet, class, and gender). Umeå Studies in Sociology No 98. Umeå, Sweden: Umeå Universitet.

Falk, P. (1994). *The consuming body*. London: Sage Publications.

Fischler, C. (1986). Learned versus "spontaneous" dietetics: French mothers' views of what children should eat. *Social Science Information 25*(4), 945–965.

Fischler, C. (1990). *L'Homnivore. La goût, la cuisine et le corps*. Paris: Editions Odile Jacob.

Grieshaber, S. (1997). Mealtime rituals: Power and resistance in the construction of mealtime rules. *The British Journal of Sociology 48*(4), 649–666.

Grignon, C. (1992). Manger en temps et en heure: La popularisation d'une discipline dominante. *Social Science Information 31*(4), 643–668.

Gronow, J. (1997). *The sociology of taste*. London: Routledge.

Gronow, J., Mäkelä, J., Kjærnes, U., Pipping Ekström, M., Holm, L., & Bjœrkum, E. (1998). A comparative study of the Nordic meal. In J.S.A. Edwards & D. Lee-Ross (Eds.), *Culinary arts and sciences II: Global and national perspectives* (pp. 377–382). Bournemouth, England: Worshipful Company of Cooks and Centre for Culinary Research at Bournemouth University.

Ilmonen, K. (1993): *Tavaroiden taikamaailma: Sosiologinen avaus kulutukseen* (in Finnish, The magic world of things). Tampere, Finland: Vastapaino.

Karisto, A., Berg, M.-A., & Prättälä, R. (1993): The good, the bad, and the ugly: Differences and changes in health related lifestyles. In U. Kjærnes, L. Holm, E. Fürst L'Orange & R. Prättälä (Eds.), *Regulating markets, regulating people: On food and nutrition policy*. Oslo, Norway: Novum Press.

Kemmer, D., Anderson, A.S., & Marshall, D.W. (1998). Living together and eating together: Changes in food choice and eating habits during the transition from single to married/cohabiting. *The Sociological Review 46*(1), 48–72.

Laitinen, S., Högström, P., & Räsänen, L. (1997). Similarity of food choices among young Finnish couples. *Journal of Human Nutrition and Dietetics 10*, 353–360.

Levenstein, H.A. (1988). *Revolution at the table: The transformation of the American diet*. New York: Oxford University Press.

Lévi-Strauss, C. (1966). The culinary triangle. *New Society*, 937–940.

Lupton, D. (1996). *Food, the body and the self*. London: Sage Publications.

MacClancy, J. (1992). *Consuming culture*. London: Chapman & Hall.

Mäkelä, J. (1990). *Luonnosta kulttuuriksi, ravinnosta ruoaksi: Neljä näkökulmaa ruoan sosiologiaan* (in Finnish, From nature to culture, from nourishment to food). Jyväskylän yliopisto: Nykykulttuurin tutkimusyksikön julkaisuja 21. Jyväskylä, Finland: University of Jyväskylä.

Mäkelä, J. (1991). Defining a meal. In E.L. Fürst, R. Prättälä, M. Ekström, L. Holm, & U. Kjærnes (Eds.), *Palatable worlds: Sociocultural food studies* (pp. 87–95). Oslo, Norway: Solum.

Mäkelä, J. (1996). Kunnon ateria: Pääkaupunkiseudun perheellisten naisten käsityksiä (in Finnish, A proper meal: exploring the views of women with families). *Sosiologia 33*(1), 12–22.

Mäkelä, J., Kjærnes, U., Pipping Ekström, M., Fürst L'Orange, E., Gronow, J., & Holm, L. (1999). Nordic meals: Methodological notes on a comparative survey. *Appetite 32,* 73–79.

McIntosh, W.A. (1996). *Sociologies of food and nutrition.* New York: Plenum Press.

Mennell, S. (1985). *All manners of food: Eating and taste in England and France from the Middle Ages to the present.* Oxford, England: Basil Blackwell.

Mennell, S., Murcott, A., & van Otterloo, A.H. (1992). *The sociology of food: Eating, diet and culture.* London: Sage Publications.

Messer, E. (1991). Getting through (three) meals a day: Diet, domesticity, and cash income in a Mexican community. In A. Sharman, J. Theophano, K. Curtis, & E. Messer (Eds.), *Diet and domestic life in society* (pp. 33–60). Philadelphia: Temple University Press.

Mintz, S.W. (1982). Choice and occasion: Sweet moments. In L.M. Barker (Ed.), *The psychobiology of human food selection* (pp. 157–169). Chichester, England: Ellis Horwood Limited.

Mintz, S.W. (1986). *Sweetness and power.* New York: Penguin Books.

Murcott, A. (1982). On the social significance of "cooked dinner" in South Wales. *Social Science Information 21*(4/5), 677–696.

Murcott, A. (1983). It's a pleasure to cook for him: Food, mealtimes and gender in some South Wales households. In E. Gamarnikow (Ed.), *The public and the private* (pp. 78–90). Portsmouth, NH: Heinemann.

Murcott, A. (1986). Opening the "black box": Food, eating and household relationships. *Sosiaalilääketieteellinen Aikakauslehti 23*(2), 85–92.

Murcott, A. (1997). Family meals: A thing of the past? In P. Caplan (Ed.), *Food, health and identity* (pp. 32–49). London: Routledge.

Nordström, I. (1988). *Till bords: Vardagsmoral och festprestige i det sydsvenska bondesamhället* (in Swedish, To table). Stockholm: Carlssons.

Nordström, I. (1991). Går det an? Skick och oskick vid bordet (in Swedish, Table manners good and bad) (pp. 115–135). In J. Frykman & O. Löfgren (Eds.), *Svenska vanor och ovanor.* Stockholm: Natur och kultur.

Prättälä, R., & Helminen, P. (1990). Finnish meal patterns. *Bibliotheca Nutritio et Dieta (45),* 80–91.

Prättälä, R., Pelto, G., Pelto, P., Ahola, M., & Räsänen, L. (1993). Continuity and change in meal patterns: The case of urban Finland. *Ecology of Food and Nutrition 31,* 87–100.

Roos, E. (1998). *Social patterning of food behaviour among Finnish men and women.* Helsinki, Finland: Publications of the National Public Health Institute A 6/1998.

Roos, E., & Prättälä, R. (1997). Meal pattern and nutrient intake among adult Finns. *Appetite 29*(1), 11–24.

Rotenberg, R. (1981). The impact of industrialization on meal patterns in Vienna, Austria. *Ecology of Food and Nutrition 11*(1), 25–35.

Sharman, A. (1991). From generation to generation: Resources, experience, and orientation in the dietary patterns of selected urban American households. In: A. Sharman, J. Theophano, K. Curtis, & E. Messer (Eds.), *Diet and domestic life in society* (pp. 173–203). Philadelphia: Temple University Press.

Simmel, G. (1910). Soziologie der Mahlzeit. *Berliner Tageblatt* (10 Oktober 1910).

Sjögren-de Beauchaine, A. (1988). *The bourgeoisie in the dining-room: Meal ritual and cultural process in Parisian families of today.* Stockholm: Institutet för folklivsforskning vid Nordiska museet och Stockholms universitet.

Thephano, J., & Curtis, K. (1991). Sisters, mother, and daughters: Food exchange and reciprocity in an Italian-American community. In A. Sharman, J. Theophano, K. Curtis, & E. Messer (Eds.), *Diet and domestic life in society* (pp. 147–171). Philadelphia: Temple University Press.

Valentin, E.K., & Granzin, K.L. (1987). Food attribute importance, situational effects, and homemaker values. *Home Economics Research Journal 16*(1), 57–64.

Warde, A. (1997). *Consumption, food and taste: Culinary antinomies and commodity culture.* London: Sage Publications.

Warde, A., & Hetherington, K. (1994). English households and routine food practices: research note. *The Sociological Review 42*(4), 758–778.

Whit, W.C. (1995). *Food and society: A sociological approach.* Dix Hills, NY: General Hall, Inc.

Whitehead, T.L. (1984). Sociocultural dynamics and food habits in a Southern community. In M. Douglas (Ed.), *Food in the social order: Studies of food festivities in three American communities* (pp. 97–142). New York: Russell Sage Foundation.

Wood, R.C. (1995). *The sociology of the meal.* Edinburgh, England: Edinburgh University Press.

Chapter 2

The Psychology of the Meal

Patricia Pliner and Paul Rozin

The meal is a very real psychological entity. It is a virtually universal physical and behavioral feature of human life. It does not depend on any particular theories or point of view. Any reasonable criterion for some kind of clustering of eating will show that humans binge; they do most of their eating in relatively short periods of time, separated by periods of minimal if any consumption. Finns do it, Russians do it, Malays do it, Chinese do it, Zambians do it, Brazilians do it. And even hunter-gatherers do it. Not only do all adults eat meals, but all children do too, and they do so from the first day of postnatal life. Their constant placental food infusion ends abruptly, and they seek and receive milk periodically, that is, in meals. The meal is a meaningful unit of life for most animals: virtually all nongrazing mammals, including most primates and that best investigated of all eaters, the laboratory rat.

Of course, other important or "physical/behavioral" units of eating exist: bites, dishes or courses, daily intakes, long-term patterns. But we think it is fair to say, that from both the lay and scientific perspective, the meal is a privileged unit. We will provide evidence for this claim throughout this chapter.

Why is meal eating so common? Most likely, this results from an ecological influence. Animals have many things to do, including sleep, watch for predators, and find mates. Eating is critical, but it competes with other activities; surely, one's vigilance is not maximum while gobbling up food. And sleep, that peculiarly essential activity for many animals, would be maladaptively segmented into short bouts were food intake to be accomplished semicontinuously or in a great many short bouts or "snacks." Short periods of rapid ingestion minimize the time devoted to feeding. In addition, animals, most of whom live in a world of spotty resources (the exceptions being some grazing animals and perhaps some predators of relatively small prey, such as insect-eating birds or frogs), often come upon rather large food offerings, and they had best be consumed before they disappear. For the carnivorous animals, this is most obvious. A snake or a lion meal begins with the capture of prey, a large and perishable amount of food. Capture is followed by a long and plentiful eating episode, a literal gorging, a meal of meals.

For many animals, including humans, the normal food diet does not provide sufficient water. Water ingestion, or ingestion of liquids that consist almost entirely of water, itself occurs in bouts. In both rats and humans, these drinking bouts are associated with meals. Although this association may be partly determined by the same set of ecological factors that promote meals, some additional factors promote the incorporation of drinking bouts into meals. Hunger and thirst tend to co-occur, and drinking may facilitate ingestion of solid food. In one study of drinking and eating by humans in natural situations over a period of days, drinking without eating occurred only rarely, and the best predictors of the occurrence of a drinking bout were food and meal-related factors, rather than self-rated thirst (de Castro, 1988).

A major force would favor a rather continuous, even eating pattern. Regulation of the internal environment, Claude Bernard's "milieu interne," and Walter Cannon's homeostasis, is a fundamental condition of life. A great deal of the physiological and behavioral machinery of animals is devoted to buffering them against stresses that threaten to dysregulate the internal environment. The meal is one of the great threats to homeostasis. A large load of nutrients is dropped into the system at one time, causing a flood of glucose and other nutrients to enter the system. Regulated parameters, including temperature and blood glucose level, are challenged. Meal-taking accounts for many anatomical and physiological adaptations: large stomachs, gradual delivery of food from stomach to intestine, insulin secretion, glycogen as a short-term energy store (see Smith, 1982, for a general discussion of the meal from the physiological point of view).

There are a surprising number of behavioral/psychological adaptations to the stress of meals, as described by Woods and Strubbe (1994). They point out that under conditions where meals are predicted by either specific environmental events or time, the organism has the opportunity to anticipate and prepare for a nutrient onslaught. Based primarily on data from rats, premeal adaptations include elevated body temperature (presumably to facilitate the energy-expending processes of digestion and assimilation) and a slow decline in blood glucose and metabolic rate, both in the service of compensating for substantial increases in both after ingestion. These "cephalic phase" responses operate, in some cases (e.g., increases in body temperature), to prepare the system to handle the load, and in others (e.g., decreases in glucose and metabolic rate), to compensate for some of the consequences of ingestion. In either event, they operate to smooth out the perturbation produced by the meal nutrient load. In the case of blood glucose, it is known that a component of the anticipatory glucose drop is secretion of insulin.

Just as the discussion of the meal, in some cases, pits physiological (e.g., metabolic) against psychological factors (e.g., learning), the study of premeal physiological changes has two competing accounts: the more traditional account sees at least some of these changes (e.g., the drop in blood glucose and metabolic rate) as purely metabolic signals of a decreasing fuel supply (e.g., see Campfield & Smith, 1990), whereas others interpret these same events as conditioned compensations in anticipation of the meal load (Woods & Strubbe, 1994). It is, of course, possible that both processes are at work as are both physiological and psychological determinants of meal onset, size, and offset.

THE MEAL AS A FOOD INTAKE PATTERN

The Rat Meal

Dating from Richter's (1927) classical analysis of rat feeding patterns, the rat meal has been a focus of research efforts—the focus of analysis of how rats regulate their food intake. Richter identified the diurnal pattern, including predominant meal eating during the dark phase of the light cycle. Under laboratory conditions, in individual cages, rats consume somewhere between 5 and 15 meals per day. There are many schemes for parsing meals; a common one is to define a meal as a period of ingestion bounded by a period of at least 10 to 20 minutes without eating (Panskepp, 1978). Typically, rats become active before a feeding bout and usually drink some (Fitzsimons & LeMagnen, 1969); after a meal, usually a thorough bout of grooming, a short period of investigatory activity, and then sleep occur. Somnolence may be a good criterion for meal termination.

A great deal of research has been done on the role of metabolic events, status of the stomach, palatability, and conditioning factors in determining the onset and termination of rat meals. These factors have been shown to influence meal size. However, there is another major determinant: the availability of competing activities. Nicolaidis, Danguir, & Mather (1979) made an important observation about the rat meal pattern. When rats were provided with a "chambre de coucher," a niche expansion of the typical small cage in which they could curl up and sleep, the number of meals they ate over 10 to 15 consecutive days dropped from a mean of 9.3 to a mean of 6.8. There is good reason to believe that the sleep cycle may drive the meal cycle.

The Nicolaidis et al. study reminds us that the meal pattern of a domesticated rat in a cage smaller than 1 foot on a side may be a function of the cage as much as the rat. We do not have extensive data on meal patterns of either domestic or wild rats in larger, more natural enclosures or in real world settings (for wild rats).

The Human Meal

The Human Infant

The pattern of infant food ingestion is well reviewed by Hammer (1992) and Birch, Fisher & Grimm-Thomas (1996). From these accounts, we extract some of the basic descriptive features appropriate for this review. From birth, infants adopt a meal eating/nursing pattern that is at least in part internally generated, perhaps in relation to the sleep cycle. Over the first week of life, the number of daily meals increases, on average from about 5 to about 7, with both meal size and frequency increasing over the early days of life. Meal size is probably limited by gastric capacity. Young infants (8 weeks) tend to take the largest meal on arising after their longest (overnight) fast, whereas by 6 months of age, there is a shift to the adult pattern of a largest meal at night. The shift to solid foods may well engage a somewhat different control system. Birch and others have tracked the development of basic aspects of adult regulation in human infants. As with rats, aspects of physiological regulation come in separately over time. Opportunities for the child to develop conditioned anticipations of meals and so forth vary in terms of the predictability of food availability, as in the demand-feeding option. In some cultures, demand feeding is the only option.

The Adult Human Meal

The three-meal-a-day pattern is widespread among adults around the world. This is suggested by the existence of words for precisely three distinct meals in many (but not all) languages (Rozin, Pliner, & Berman, 1999). This general human characteristic could be a result of ecological pressures and compromises with competing behavioral/motivational systems, or it could represent a natural feeding rhythm in humans. Booth and Mather's (1978) mathematical model of human eating assumes that the onset and offset of eating are determined by particular values of the rate of flow of energy into body cells; feeding begins when energy flow falls below a certain value and ends when it rises above another. When several parameters are entered into this model (e.g., daily activity pattern, body weight, energy density, gastric emptying time), the model predicts three meals per day. Most likely, humans are predisposed to a pattern of about three meals a day and are not optimally served by delivery of energy and nutrients in fewer than three occasions. However, the prevalence of three as opposed to four or five meals per day may result from competing pressures. We have only minimal data on the free-running meal patterns of humans, and none, of course, from humans not already enculturated to the three-meal-a-day pattern.

Perhaps the most natural yet well-controlled study of spontaneous human meal eating comes from research by Green, Pollak, & Smith (1986). Eight adults lived individually for several weeks under continuous illumination in a one- or two-room apartment without windows, clocks, or other means of telling time of day. Subjects could call for breakfast, lunch, dinner, or snacks from a computer-displayed menu. Subjects had a definite tendency to eat 3 meals for each sleep-wake period (to some extent, perhaps, prompted by the availability of meals labeled as breakfast, lunch, and dinner). These varied from an average of 25 hours early in the study to a mean of 33 hours later in the sessions. This work, along with the rat studies of Nicolaidis et al. (1979), links the eating cycle to the sleep cycle.

As with the rat research, serious questions need to be asked about the meaningfulness of meal consumption by humans in laboratory settings. Green et al. came a long way in producing a naturalistic setting in the laboratory, but, of course, many normal meal-related activities, such as company for eating and activities competing with eating, were missing. Kissileff, Guss, & Nolan (1996) report some comparability of meal sizes in laboratory or cafeteria settings; also, in both cases, the presence of others increased intake.

Microstructure of the Meal

The meal is a highly structured event, in terms of an orchestrated presentation of feeding opportunities and a variety of food choices and detailed patterns of ingestion. A number of studies have sought to describe these events in humans, using either videotape technology or direct measurements of chewing and swallowing. A major focus of this work, not relevant to this review, is the comparison of eating patterns of obese and normal-weight people.

Hill (1974) pioneered in this area, using analysis of videotaped meals eaten by normal subjects in the laboratory. His main focus was changes in meal structure (bite and chew sequence) as a function of palatability and hunger and will be discussed later under palatability as a determinant of meals. Rogers and Blundell (1979), using video records, reported a steady decline in eating rate across each quarter of the meal. Surprisingly, especially in light of major advances in video technology, this useful technique has not been applied much in subsequent work on the events in human meals.

Stunkard, Coll, Lundquist, and Meyers (1980) used an even more basic approach, direct observation of eating under natural conditions. Observations were made on obese and normal-weight women eating in a fast food restaurant. The women were given a coupon entitling them to a free meal of either 985 or 1800 calories. They reported lower eating rates and fewer chews per mouthful for the larger meal. In contrast to results from some laboratory studies, they found the highest rate of intake in the middle third of the meal and the lowest in the last third.

From 1980 onward, much of the work on meal structure was carried out with direct measurement of chewing and swallowing by means of sensors attached to the body. Some of the early studies, following up on prior work in LeMagnen's laboratory, reported the general pattern of chewing and swallowing in French adult subjects, using foods of varying palatability (Bellisle & LeMagnen, 1980; Bellisle, Lucas, Amrani, & LeMagnen, 1984). Results indicated a decrease in eating rate from the beginning to the end of the meal, as manifested by gradual increases in chewing time per food unit and in the interval between food units. Water sips increased as the meal progressed (Bellisle & LeMagnen, 1981). In addition, there was evidence that eating rate was a stable individual difference variable.

Stellar and Shrager (1985) used a different oral sensor system, mounted like a retainer against the palate. Using small, 15-kcal sandwich rolls (their SFU or solid food unit), they reported a characteristic pattern of 14 to 16 chews per bite, followed by a swallow. This cycle averaged 12 to 13 seconds in length. A large bite can be followed by several smaller chew-swallow cycles before another bite.

Kissileff and Thornton (1982a) used Kissileff's universal eating monitor to obtain a fine-grained record of intake of a liquid diet of yogurt and fruit consumed through a straw. They found that a quadratic equation accounts for almost all the variation in the cumulative food intake curve in single-course meals in nonobese men and women. The linear coefficient is the initial rate of eating, and the quadratic is half the rate of deceleration (in absolute value). Men exhibited a higher initial rate of eating but also decelerated faster than women. As with all studies, these results no doubt depend on the particular food being offered, as well as other aspects of the setting. Kissileff and Thornton (1982a) reported some correspondence between these results and real world intake patterns; subjects kept food diaries, and intake from the diaries correlated .60 with intake in the laboratory.

THE MEAL AS A NATURAL UNIT

Semantic Recognition of the Meal Category

The meal is a recognized semantic category, and across languages, there is a modest set of words designated to describe meals or their subdivisions. Interviews about meal terminology with native speakers (in English) of 18 languages were carried out (Rozin et al., 1999). The languages covered a wide geographical area and included many different language groups: English, French, Spanish, Swedish, Icelandic, Polish, Russian, Finnish, Turkish, Arabic, Ibo, Hindi, Vietnamese, Thai, Tagalog, Cantonese, Mandarin, and Japanese. Seventeen of the 18 languages have a term for meal, 16 have unique words for breakfast and for lunch, all have a word for a main meal eaten later in the day, and 16 have a word corresponding to snack. Approximately half of the languages have single words for parts of the major meal, or courses (none had corresponding words for lunch or breakfast). The words, when present, correspond generally to the English appetizer, entree, and dessert. As might be expected, the French seem to have the largest meal- (and food-) related vocabulary. The French identify a specific fromage (cheese) course, and, at least for elaborate meals, include "amuse gueule" as a preappetizer, a (typically sorbet) palate cleanser between two main courses, and a mignardise, a delicacy after dessert. And then, of course, there is that French monument to variety seeking, the degustation.

This is typically a sampling of small servings of the highlights of foods from various courses at a restaurant.

The Meal as a Unit of Day Segmentation

We typically punctuate or segment our days in terms of stable key events, such as waking up, breakfast, lunch, supper, and going to bed or sleep. That is, meals are important in organizing and remembering our days. We suspect that when people remember their intake of the prior day, they spontaneously organize it by meals. We also suspect that in recounting the day's events, people will be inclined to use meals as time markers, as in reporting that "after lunch, I went shopping." The prominence of meals in thinking about food is also illustrated by research on free associations to either specific foods or arrangements of food sequences into meals (Rappaport, Peters, Downey, McCann, & Huff-Corzine, 1993; Rozin, Kurzer, & Cohen, 1999). Meal terms occur prominently in both types of free associations.

The Meal as a Unit of Social Interaction

Meals, especially the dinner meal, are usually taken in a social setting. For many families, the evening meal is the main time for the whole family to interact. Meals often serve, as well, as occasions for meetings, for dating, business, or general social interaction. They offer a period of time long enough to accomplish something but short enough to provide a fixed time commitment, and they offer a ready subject for discussion, even as the main topic for discussion. Major life occasions, such as marriages, and important holidays, such as Christmas, are typically associated with an elaborate meal or banquet. This association is surely not a peculiar feature of Western cultures; rather, the banquet or feast is an important part of celebrations in many cultures.

Co-presence at a meal establishes a minimal bond. Typically, there is some sharing of food. The great majority of meals (virtually all, in some traditional cultures) are shared with relatives. The meal is an opportunity for exchange of information and experiences and, at the same time, a much more substantial sharing of substance, as represented in the food. Sharing food is a homogenizing act, just as not sharing it is a form of social distancing. This contrast is particularly clear in Hindu India, where food-sharing rules are more explicit (Appadurai, 1981). In this and some other traditional contexts, the sharing of food at meals establishes and reaffirms closeness, and the ritualized practices in the meals express and affirm particular relationships. Who eats first, who eats the choice foods, and who can eat foods already sampled by others; all of these affirm hierarchical family relations. A central concern is "consubstantiation," actual sharing of substance, as when two individuals eat from the same pot, trade food, or even when one person makes the food that another person eats. Such matters figure heavily in Hindu Indian food exchanges and, as well, in the elaborate contagion-dominated food experiences of the Hua of Papua New Guinea (Meigs, 1984).

In Western-developed cultures, some of these explicit functions of meals are muted. However, they are still present, just as an invitation to share a family dinner with a person is an indication of liking and ease in a relationship. Even among Americans, modest acts of food sharing during a meal carry major implications for type of relationship between the eaters (Miller, Rozin, & Fiske, 1998). Sharing of food in a restaurant implies a close personal relationship; feeding by one or the other eater implies (if their ages and genders are appropriate) a romantic relationship. Consubstantiation seems to enhance the perceived closeness of eating partners as viewed by others.

The Meal as a Memory: Retrospective Meals

At any given point in time, we are likely not to be eating a meal. But, at the same time, we are frequently thinking about meals, past or anticipated. The average person surely has a repertoire of many meal memories. Studies on diet recall suggest a rather accurate recall of the previous day's meals, with accuracy fading rapidly beyond that. But we all remember special meals, either because of their very high or low culinary quality, their unusual nature, or important nonculinary events that occurred at the meal (first dates, proposals of marriage, and the like).

Because we live with so many meals in our minds, it would seem reasonable to ask about the relation between the memorial representation of the meal and the actual experience of the meal. This is of particular importance because when we make a food choice now, it is based on our memories of relevant past experiences with the same or similar foods, not our actual experience with the foods. Insofar as the memories differ from the actual experience, it is the memories that are relevant.

Of particular interest are our memories for the quality of the meal, that is, the pleasure we experienced. Recent work by Daniel Kahneman and his colleagues (reviewed in Kahneman, Wakker, & Sarin, 1997) on experienced and remembered hedonic episodes makes this issue more compelling and interesting. Kahneman distinguishes between remembered and experienced pleasure and pain and asks what principles relate the experience with the memory; the relationship is quite complex but lawful. The empirical work centers on experienced and remembered pain. Kahneman and colleagues report two aspects of a pain experience that powerfully determine memories for it: the peak level (disproportionately large contribution to the remembered pain) and the rapidity of the offset (sharp offsets are remembered as more painful). On the other hand, they report one important aspect of a pain episode that is not well represented in memory: duration. Five versus ten minutes of pain are represented in memory as about equally unpleasant.

Kahneman's work suggests a series of questions about memories for meals.

1. Are certain parts of the meal (e.g., first, last, main course) disproportionately determinative of our evaluation of the meal? (Note that a postmeal evaluation, even if made right after a meal, is based on memory of experience, not direct experience.)
2. Is the peak (low, high, or both, depending on the meal) disproportionately represented in the hedonic representation carried in memory?
3. Is there duration neglect? Do we rate meals that had, say, 4 versus 8 ounces of our favorite food (certainly a big difference in experienced pleasure) as equally pleasurable?

Only a few studies shed light on these questions. One study obtained a primacy effect ("onset dominance"). Subjects provided ratings of many foods, from which a meal sequence of food was constructed. The two "halves" of the hypothetical meal each consisted of three items, each previously rated high (H) or low (L). Subjects were read a list of the six foods (e.g., HHH-LLL, LLL-HHH) and provided a rating of how much they would like this meal. Primacy effects were found; that is, subjects generally rated meals beginning with three highly rated foods (followed by three low-rated foods) as better than those presented in the reverse order (Anderson & Norman, 1964).

Rogozenski & Moskowitz (1982) developed a regression model for predicting meal liking from a weighted linear combination of liking for the components. Subjects first rated 140 foods and then rated meals (menus) composed of sequences of these foods. The meals consisted of a main dish, starch, vegetable, salad, and dessert. The model was applied to average ratings of components and meals across many subjects. Under these conditions, the main dish (entree) had a disproportionately strong contribution to the total meal rating. Turner and Collison (1988) used a similar procedure but used ratings of actual meal components and meals in a student training restaurant. Nine different meals were tested. The entree had the largest coefficient in the multiple regression for seven of the nine meals, displaying an average correlation of .74 with total meal rating. The second most important component was the sweet/dessert (correlation of .64).

Hedderley & Meiselman (1995), in keeping with Meiselman's general emphasis on studying food choice and intake in natural settings, tried to model meal acceptability in a student cafeteria in the United Kingdom. After completing a self-selected meal, students provided ratings of the components and of the overall meal. More than previous investigators, these researchers are aware of the problem of combining multiple regressions from meals with different components and of aggregating data across subjects. In this study, multiple regression equations were computed for individual subjects (at least three meals/subject) and combined only across meals with similar structures. Concordant with prior results, the main dish had the major predictive value (highest for pizza, then sandwich, and then other main dish). As with Turner and Collison, the next highest regression coefficient went with dessert.

The disproportionate effect of the main dish in these studies could be accounted for in a number of ways. It is usually the largest amount of food, and hence consumes the most time. It is often the most palatable part of the meal. It is usually the most expensive part of the meal. It is located in the temporal center of the meal; according to Kahneman this might decrease its contribution to meal memory, but for meals, the central position might be most salient.

With respect to Kahneman's findings, these studies provide some general support in that total meal liking is not a simple unweighted linear function of components. Furthermore, there is evidence for dominance of the entree, which may, in many cases, also represent the peak of liking. Finally, the finding that dessert is the second most predictive meal component is compatible with the idea that the end of a sequence contributes disproportionately to memory for that sequence. Duration neglect was not tested in these studies. However, we have recent data on this subject from actual mini-meals we constructed from three flavors of jelly beans in sequence. Under these circumstances, rating of the overall jelly bean meal was best predicted by the average liking for the component jelly beans, a finding that is not consistent with Kahneman's results. On the other hand, doubling the number of one of the jelly bean flavors (from the standard two to four jelly beans) did *not* increase the contribution of this bean to the total sequence rating, a finding arguing in favor of duration neglect (Rozin, Rode, & Ostovich, 1999).

As Meiselman has indicated, it is difficult to do this type of study. Besides measurement problems (including the subject's memory for ratings for the components when rating the whole meal), problems are due to the complexity of the meal as a stimulus. Compatibility of items (Schutz's [1989] appropriateness) and context effects and interactions (as described by Rozin and Tuorila [1993] and Meiselman [1996]) make rating of components a poor substitute for the on-line experience of pleasure. Perhaps more appropriate, in line with Kahneman's studies on pain, would be a moment-to-moment pleasure rating during the meal, although this could be quite disruptive of the meal itself.

PSYCHOLOGICAL DETERMINANTS OF MEAL INITIATION, TERMINATION, AND SIZE

Visibility/Availability of Food

It seems likely that, given an appropriate situation, palatable food, and at least a moderate degree of hunger, a major determinant of amount eaten in a meal would be the amount of food served. Supporting this idea, Edelman, Engell, Bronstein, and Hirsch (1986) provided participants with small (225 g), medium (426 g), or large (1000 g) portions of lasagna, finding greater intake of the large portion than of the other two, which did not differ. However, the increases in intake were modest (about 20% from the medium to the large portion) compared with the differences in portion size (more than 100%). Booth, Fuller, and Lewis (1981) increased the size of the sandwiches they served by a factor of 1.5 and found increases in caloric intake on the order of 1.2 or 1.3.

Most of the other evidence relevant to the effect of food availability on amount consumed has been collected in the context of evaluating the relative importance of food presence or salience for obese versus normal individuals in accordance with the predictions of Schachter's (1971) externality hypothesis. This hypothesis predicted that overweight individuals would eat more than those of normal weight in the presence of salient or prominent external cues. Accordingly, a number of studies was conducted to test that notion. In those studies, salience was operationalized in two rather different ways: (1) by varying conditions so that the physical visibility or availability of food was altered, and (2) by varying the extent to which subjects were thinking about food. With both kinds of manipulations, overweight individuals ate more under conditions of high salience, whereas the manipulation had little or no effect on their leaner peers (Johnson, 1974; Nisbett, 1968; Ross, 1974; Tom & Rucker, 1975). For example, Ross (1974) manipulated the salience of food by brightly illuminating it in one condition by means of an unshaded 40-watt bulb and dimly illuminating it in another by means of a shaded 7½-watt bulb and permitted obese and normal weight subjects to eat *ad lib*. The salience manipulation had a large and dramatic effect on the amount eaten by overweight subjects and virtually no effect on those of normal weight. Thus, when food was prominent, overweight subjects increased their meal size considerably. Other studies have shown an effect of salient food cues independent of body weight. Wooley and Wooley (1973), measuring salivation as a stand-in for quantity consumed, found that both obese and normal subjects salivated more when thinking about or looking at food than during control periods.

Salience can also affect food choice within a meal. Stunkard and Levitz (1975) examined dessert choices in a hospital cafeteria under control conditions with high- and low-calorie desserts equally visible and available; then, on one test day they placed high-calorie choices in front and low-calorie choices in the rear, reversing these placements on a second test day. The normal-weight subjects, for whom there were the most observations, consistently selected the desserts that were the most visible and available, increasing their choices of high-calorie desserts on the first test day and increasing their choices of low-calorie desserts on the second day.

Recently, a series of studies appeared that directly tested in normal-weight individuals the role of amount of food available on self-determined portion size. Wansink's (1996) general interest was in the relation between package size of all kinds of consumer goods and amount purchased. In one set of studies, he explored the relationship between package size and the amount selected as an individual portion. When adult subjects were asked to pour out M&Ms into a bowl for watching TV by themselves, they poured an average of 63 from a package of 114 but 103 from a package that was twice as large. An increase to a package three times the size of the original produced only a small and nonsignificant increased increment, to a mean of 122 candies. In another study, the tasks included pouring out Crisco oil for frying chicken and putting a portion of dried spaghetti from a commercial package into a pot for cooking. Wansink reported increases of more than 20% in the portions of oil and spaghetti selected when the source container volumes were doubled. The experiment eliminated the account that this increase is due to the amount of the supply (as opposed to the package size) because it was arranged that the double-sized package was only half full. Participants also estimated the cost of the product they selected as lower when it came from the larger package. Another study found, using bottled or tap water, that the increase in portion size with package size (1-liter vs. 2-liter bottles) occurred only for bottled water. Thus, the results of this study support perceived cost as an explanation for the package-portion effect because it did not appear for cost-free tap water. Yet another study provided more evidence for a determining role for perceived cost; people selected larger portions of Crisco oil for frying from products that were advertised as on sale.

Effort

A number of studies, originally done in the context of Schachter's externality theory of obesity, examined the effects of the effort required to obtain or consume food on meal size. Several studies showed that overweight subjects ate more when food was easily available or easily eaten than when it was difficult to obtain or consume (McArthur & Burstein, 1975; Nisbett & Gurwitz, 1970; Schachter & Friedman, 1974; Singh & Sikes, 1974); the results of the Nisbett (1968) study, cited previously, can also be interpreted in terms of effort. Other studies have found that even normal-weight subjects eat less and change their food choices when effort is high (Lappalainen & Epstein, 1990; Meiselman, Hedderley, Staddon, & Pierson, 1994; Rodin, 1975). For example, in a student cafeteria Meiselman et al. (1994) increased the effort required to obtain candy bars in one study and potato chips in another. Consumption of both decreased, and the students substituted other dessert or starch items, respectively, in the two studies. The Stunkard and Levitz (1975) study described previously can also be interpreted in terms of effort. The U.S. Army has extensively tested a general-purpose ration, the meal, ready to eat (MRE), intended for consumption when hot food is not available. In each of nine studies, in which the MRE was the sole source of subsistence for periods ranging from 5 to 34 days, conducted in a wide variety of environments, soldiers failed to consume enough to maintain body weight. Although many obvious explanations exist for these findings, one possibility is that the effort required to eat these rations (e.g., extracting them from the packages, rehydrating and heating them, cleaning up) can account for at least some of the effect. Consistent with this speculation, in an experiment in which troops received identical MREs in the field or prepared for them in the garrison dining room, the former consumed about 1000 calories per day less than the latter (Hirsch & Kramer, 1993). When Lester and Kramer (1991) provided troops with one of three means of heating their rations in the field, which varied in their perceived convenience of use, group differences in energy intake were positively related to convenience. Engell and Hirsch (1991) varied the relative ease with which subjects could obtain water during a lunch meal, finding that twice as much was consumed when it was on the table than when it was at a more remote location.

Palatability and Preference

It will come as no surprise to learn that people eat larger meals when they are eating food they like. This has been shown to be true whether "liking" is defined in terms of individual preferences (Hill, 1974; Hill & McCutcheon, 1975; Spiegel, Shrager, & Stellar, 1989) or manipulated by "doctoring" the food to alter its palatability (Nisbett, 1968; Yeomans, 1996); it is true for meals consisting of several courses (Guy-Grand, Lehnert, Doassans, & Bellisle, 1994) or a single course (Yeomans, Gray, Mitchell, & True, 1997), as well as sandwich meals (Bellisle & LeMagnen, 1980) and snacks (McKenna, 1972; Decke, 1971); it is true for neonates (Milstein, 1980; Nisbett & Gurwitz, 1970) and children (Ballard, Gipson, Guttenberg, & Ramsey, 1980), as well as for adults; it occurs in the laboratory and when individuals are observed in natural environments (Ballard et al., 1980) or report on their behavior in such environments (Feunekes, de Graaf, & van Staveren, 1995). Some research has suggested that people with certain "aberrant" nutritional statuses (i.e., those who are overweight, hungry, or disinhibited dieters) are more sensitive to variations in palatability than are appropriate controls (see Pliner, Herman, & Polivy, 1990, for a review).

In addition to its effects on meal size, palatability has also been shown to have effects on how people eat meals—on their microstructure. More palatable/preferred meals are eaten faster (i.e., at a higher rate) because people take bigger bites and chew their food for less time and/or more quickly (Bellisle & LeMagnen, 1980; Hill, 1974; Hill & McCutcheon, 1975; Spiegel et al., 1989; Yeomans, 1996). Many studies show that these effects of palatability on rate of intake are confined to the beginning of meals (Bellisle et al., 1984; Bobroff & Kissileff, 1986; Kissileff & Thornton, 1982b; Spiegel et al., 1989). Just as infants show effects analogous to those for adults in terms of the effect of palatability on meal size, so they show similar effects in terms of the microstructure of their eating. In their research on newborn infants, Lipsitt and his colleagues have shown differences in the patterns of sucking responses to solutions varying in palatability (as defined by sweetness). With increasing palatability, the infant increases its overall rate of sucking; although the rate of sucking within sucking bursts decreases, the length of the bursts increases and the infant takes fewer and smaller pauses between them (see Lipsitt & Behl, 1990, for a review).

Mood

There is no lack of theoretical positions positing a relationship between mood and eating. It has been proposed both that moods affect eating and that eating affects mood; in the likely event that there is some truth in both proposals, the meal presents a condition in which both causal arrows are manifested. For this section, with its focus on the psychological determinants of meals, only the effect of mood on meals will be explored. In many theoretical accounts, such a relationship is posited only for a particular subgroup of individuals. A central tenet of many clinical theories of obesity is that overweight people eat in response to emotional distress, especially anxiety and depression (Bruch, 1957; Kaplan & Kaplan, 1957), whereas a more recent rendition of this view suggests that it is only diffuse (i.e., unlabeled) and/or uncontrollable emotional distress that leads to overeating in the obese (Slochower, 1983). Schachter's (1971) theorizing leads to the expectation that anxiety should, by its effects on gastric motility and blood sugar levels, depress eating in individuals of normal weight while having no effect on the obese, who are hypothesized to be unresponsive to such internal cues. A prediction derived from Herman and Polivy's (1975) distinction between restrained and unrestrained eaters is that emotional states such as anxiety (particularly that produced by ego threats) and depression should disrupt the usual control displayed by the restrained eaters, thereby increasing eating.

Whether the hypothesized increases or decreases in eating predicted by these theories should occur in the form of changes in meal size, meal frequency, or both is not addressed in any of them, although nearly all the experimental work examines quantity consumed in single test meals scheduled by the experimenter. In that context, all the positions described previously have received some empirical support. Slochower (1976) produced "diffuse" arousal by exposing subjects to what they believed were the sounds of their own heartbeats; some heard fast (84–92 bpm) and others heard slow (66–74 bpm) heart rates, and to reinforce this aural feedback, she provided written feedback indicating that their heart rates were either "very high" or "slightly slow." This manipulation affected both actual heart rate and reported anxiety. Subsequently, half of each group received information that could provide a neutral explanation for any heart rate effects, whereas the remainder did not. Afterwards, the subjects, obese and normal, were given the opportunity to eat. The results indicated that overweight subjects increased their eating significantly when they were both aroused and had no label for their arousal, whereas normal-weight subjects were unaffected by these manipulations.

Schachter, Goldman, and Gordon (1968) threatened obese and normal-weight subjects with either strong or mild electric shock and found that high fear markedly decreased the amount eaten by normal subjects but had no effect on the amount eaten by the obese. Heatherton, Herman, & Polivy (1991) produced ego-related anxiety in restrained and unrestrained eaters by informing some they would be required to make a speech and others that they had failed an easy task. These manipulations had large effects on restrained eaters, both groups of high-anxiety subjects eating significantly more than those in a low-anxiety condition; unrestrained eaters were not affected by the manipulations. In another study, Polivy and Herman (1976) demonstrated that, although a period of clinical depression produced weight loss (the "typical" effect of depression) in unrestrained eaters, it produced weight gain in restrained eaters.

Yet another theoretical view predicts a change in the macronutrient composition of meals (or snacks) as a function of mood and is based on the notion that ingestion of carbohydrate increases brain serotonin and may, therefore, elevate mood. Wurtman and Wurtman have suggested that among obese carbohydrate cravers and individuals with seasonal affective disorder, consumption of sweet or starchy snacks may serve as a form of self-medication, reducing depression by increasing brain serotonin levels (J.J. Wurtman, 1987; R.J. Wurtman & J.J. Wurtman, 1986). A study by de Castro (1987b), in which subjects recorded all food ingested for 9 consecutive days, as well as rating three mood dimensions (tired-energetic, anxious-tranquil, elated-depressed) before each meal, enables us to examine mood-meal relationships. When correlations between premeal moods and proportions of the three macronutrients ingested were computed, no relationships were found between any of the three mood dimensions and intake of any of the macronutrients. These findings do not provide any support for the Wurtman and Wurtman position, although they do not contradict it either because the mood-macronutrient relationship is posited to exist only for certain individuals (obese carbohydrate cravers and individuals with seasonal affective disorder) and perhaps for snacks only (Wurtman et al., 1985).

Variety and Sensory-Specific Satiety

LeMagnen (1956) demonstrated that if rats successively received, for 30 minutes each, four distinctively flavored versions of their diet, they ate 72% again as much as they ate in a comparable period with only one flavor available. Rolls and her colleagues (1981) and others (Pliner, Herman, Polivy, & Zakalusny, 1980; Spiegel & Stellar, 1990) have shown that presenting humans with a variety of foods within a meal can increase meal size. The magnitude of this effect varies with the distinctiveness of the foods; for example, varying the flavor of cream cheese–based sandwich fillings enhanced intake by 15% over intake of the favorite food only (Rolls, Rowe, & Rolls, 1982b), whereas presenting four very different foods (sausages, bread and butter, chocolate desert, and bananas), which obviously varied on many dimensions, increased intake by 60% (Rolls, van Duijvenvoorde, & Rolls, 1984). The variety effect occurs whether the foods are presented sequentially, as in a many-course meal (Rolls, Hetherington, Burley, and van Duijvenvoorde, 1986), or simultaneously, as at a smorgasbord (Bellisle & LeMagnen, 1981; Pliner et al., 1980; Spiegel & Stellar, 1990).

This enhancement of meal size by a variety of foods suggests that satiety is to some degree "sensory specific" (LeMagnen, 1971). Rolls and her colleagues (1986) have shown that the variety effect is accompanied by decreases in the palatability of foods that are eaten, and Birch and Deysher (1986) have replicated the decline in palatability in young children. Furthermore, foods that are similar to the eaten foods on sensory dimensions such as sweetness and savoriness also decline somewhat in pleasantness, whereas dissimilar foods do not, although similarity in macronutrient composition produces no such effect (Rolls et al., 1984). It seems likely that these changes in palatability mediate the variety effect. That is, if it is assumed that an individual will eat less of a food as its palatability declines (and the section on palatability indicates that amount consumed is certainly affected by palatability), then the provision of a variety of foods should delay the occurrence of such decreases in palatability, thereby increasing intake.

The decrease in palatability that follows ingestion of a food should be distinguished from the negative alliesthesia described by Cabanac (1971; 1979), to which it appears, at first glance, to be similar. In Cabanac's work, previously fasted subjects who ingest a glucose or sucrose load report a decline in the pleasantness of sweet taste. In Cabanac's view, the pleasantness or palatability of a food depends on its physiological usefulness; when people are in a fasted state, glucose is more useful than it is after a glucose load, and for that reason, sweet taste is more pleasant in the former state than in the latter. Thus, for Cabanac, the nutritional consequences of ingestion are responsible for the changes in hedonic responses. However, the time courses of the two phenomena (sensory-specific satiety and negative alliesthesia) appear to be different; sensory-specific satiety is greatest within 2 minutes after ingestion of a food and decreases gradually over the hour after eating (Rolls, 1990), whereas in the Cabanac studies, the largest changes in pleasantness are seen 45 to 60 minutes after sucrose ingestion begins (Cabanac, 1979).

The data on sensory-specific satiety suggest that a decrease in the hedonic value of food may play a role in meal termination, and a small literature exists that is relevant to individuals' reported reasons for ending meals, which examines this notion. Mook and Votaw (1992) had subjects respond to the item, "I usually stop eating a meal when...," including as options: "everyone else is finished" (social alternative), "I've had all I'm allowed" (restrained alternative), "the food stops tasting good" (hedonic alternative), "the food is all gone" (external alternative), and "I feel full" (internal alternative). Of the reasons given, the feeling of fullness was the overwhelming favorite (61% of subjects), and subjects rarely indicated that they stopped eating when the food stopped tasting good (9%). A replication with a larger sample (Zylan, 1996), which also examined sex differences, again showed fullness to be the most common response for both men and women (44% and 47%, respectively) with a decrease in taste far behind (10% and 17% for men and women, respectively).

If we take these reports at face value, it appears that hedonic factors are relatively unimportant in meal termination. However, some data suggesting that these studies may have underestimated the importance of hedonic factors come from a study by Hetherington (1996), who provided subjects with a two-course meal and assessed reasons for stopping after each of the courses. In addition, she added a second "hedonic" alternative, "I got tired of eating that food." The two hedonic alternatives combined were cited more frequently as the most important reason for terminating the first course ("...got tired of that food": 40%; "...food tasted less pleasant": 11%) than was feeling full (25%). More than half of the subjects did not have a second course; among those who did, feeling full was cited more frequently (48%) than were the two hedonic alternatives combined (15% and 18%) as reasons for terminating the second course. In this study, hedonic reasons assumed more importance than in the previous ones. This may be attributable to the addition of the second hedonic alternative; it is possible that changes in the hedonic value of a food are perceived as getting tired of it rather than finding its taste less pleasant. It is also possible that having subjects provide reasons immediately after they stop eating promotes recall of the more subtle and transient hedonic changes that may be forgotten more quickly than the more obvious and longer lasting feelings of fullness. Finally, it is likely that the reasons for stopping the first course might not be the same as those for stopping the second, with fullness a more reasonable account for the second course.

Learning, Experience, and Expectations

Meal Initiation

Given that many, if not most, meals eaten by humans are eaten in accordance with relatively fixed schedules, it is unlikely that a threshold level of energy depletion is the principal cause of eating. This would require that all meal sizes be precisely predicted by the size of the prior meal, the interval since the last meal, and energy expenditure during this interval. In fact, in free-living animals (rat data from LeMagnen & Tallon, 1966; human data from Bernstein, Zimmerman, Czeisler, & Weitzman, 1981), meal size is a better predictor of time to next meal as opposed to time since last meal. But with a fixed meal schedule, the latter prediction cannot hold.

Under these conditions, it has been argued, people may eat because hunger signals become conditioned to cues (including temporal ones) predicting the imminence of a meal, and eating is a response to this conditioned hunger (Bellisle, 1979). In a series of elegant studies, Weingarten (1983; 1984; 1985) taught rats an association between a specific exteroceptive cue and food availability by signaling meals with a tone/buzzer compound (CS+) and nonmeals with a different pure tone (CS-). Subsequent presentation of the conditioned stimulus reliably induced meal initiation, even when the animals were tested under sated conditions. Birch, McPhee, Sullivan, and Johnson (1989) conducted a pair of similar studies with preschool children as subjects. The children received 10 pairs of conditioning trials in which a distinctive stimulus compound was presented just before access to snacks (CS+), whereas a different one was presented in the absence of access to snacks (CS-). During testing, which occurred in a sated state (children had just finished eating a snack), the children showed a quicker latency to eat and ate more after presentation of the CS+ than of the CS-. Consumption at the snack cued by the CS+ averaged 10 to 15% of the recommended daily allowance for calories. Of course, another interpretation of reliable meal eating at specific times by humans is that it has little to do with hunger of any sort and has rather to do with custom and the availability and appetizing quality of the food.

Meal Termination

Conditioning also appears to have an effect on meal termination. Many years ago LeMagnen (1955) noted that, by the end of a meal, absorption is not great enough to explain the cessation of feeding and invoked a conditioning explanation to account for the paradox. The notion is that conditioning arises from the delayed aftereffects of eating a particular food on earlier occasions; thus, intake can be adjusted to the nutritive value of a familiar food. In this conditioning, the postingestive effects of a food serve as an unconditioned stimulus with which the sensory aspects of the food become associated as a conditioned stimulus. Booth has demonstrated this phenomenon empirically in rats (Booth, 1972; Booth & Davis, 1973) and in humans (Booth, Lee, & McAleavey, 1976; Booth, Mather, & Fuller, 1982). In the last-named study, after one experience with a meal beginning with a distinctively flavored soup of high caloric density, on subsequent days when the same soup was served, subjects decreased their intake of food later in the meal; with one experience, they increased intake of later courses after ingestion of a soup with low caloric density. Again, a comparable study with young children as subjects has come from Birch's laboratory (Birch & Deysher, 1985). The children showed evidence of conditioned satiety, eating less after the flavor that was paired with the high-calorie (vs. low) preload after experience with the flavor-calorie pairing.

This anticipatory control may be unconscious, but it need not be so. Certainly, individuals make quite deliberate decisions about what to eat later in a meal based on what they have eaten earlier, or make decisions about what to eat earlier based on what they would like to eat later (e.g., "I'll skip an appetizer because the desserts look really good"). What has been learned in the past about the postingestional consequences of various foods can be conscious and it can operate symbolically. One need not actually see or even taste a particular food to adjust intake to its postingestional effects—its name alone will suffice. Booth (1977) asked subjects, both before and after they ate a meal, to indicate how pleasant it would be to ingest various foods. High-calorie foods showed the greatest decline from before to after the meal. However, some very rich foods, usually served at the end of the meal, remained highly palatable even for replete subjects (so maybe it is not necessary to forgo a nice appetizer in order to "have room" for dessert).

Several studies have shown that meal size is regulated in part by individuals' beliefs about the caloric content of what they have previously eaten, what they are currently eating, or what they will be eating. S.C. Wooley (1972) preloaded obese and normal-weight subjects with drinks that were actually high or low in caloric density (containing 600 vs. 200 kcal) and, crosscutting actual caloric content, that appeared to be high or low in calories. Subsequent meal intake for both weight groups was unaffected by actual caloric density, but subjects ate significantly more (about 10%) when they believed they had consumed a drink low (vs. high) in calories. Nisbett and Storms (1974) obtained a similar effect of perceived calories. O.W. Wooley, S. C. Wooley, and Dunham (1972) found that subjects' judgments about the caloric content of their liquid meals were better predictors of the amounts they consumed than was the actual caloric content.

A more complicated prediction about the effects of caloric beliefs on consumption comes from Herman and Polivy's (1980) research on dieting. In their view, eating a large amount should increase subsequent eating in dieters. Although dieters ordinarily restrict their eating, a "forced" high-calorie preload should cause a dieter to throw in the towel, the diet being blown for the day, and allow him or her to succumb to chronic hunger and overeat. A low-calorie preload would leave the dieter's restraint intact; thus, paradoxically, restrained eaters should "counterregulate," eating less after a preload low (vs. high) in calories. This prediction has been confirmed in many studies (see Herman & Polivy, 1980, for a review). More interesting in the present context are the results of several studies in which the *perceived* caloric content of a preload was manipulated; dieters responded to the manipulation by eating less when they thought they had eaten a low-calorie than when they thought they had eaten a high-calorie preload (Polivy, 1976; Spencer & Fremouw, 1979; Woody, Costanzo, Leifer, & Conger 1981).

It might also be expected that meal size should be regulated by individuals' beliefs about what they will be eating in the future. Plans concerning subsequent meals later in the day should affect intake in the here and now. For example, someone expecting to eat a particularly large dinner might purposely eat a light lunch, compensating in advance for the large meal. Similarly, someone anticipating a late dinner might have a late afternoon snack, again compensating in advance—this time for the extended deprivation. When Lowe (1982) told some of his subjects, before they ate, that they would have to wait 4 hours before eating again, they ate more than those who were not expecting a period of deprivation (but see Tomarken and Kirshenbaum [1984] for puzzling evidence that both dieters and nondieters increase intake when told of a high-calorie "postload").

Much of the research described in this section was motivated by LeMagnen's observation that correction of an energy deficit could not account for meal termination because the latter occurs before the former. Much of the research, as with that on meal initiation research, has focused on some form of conditioning. But other work, such as that cited previously, invokes higher order units of regulation, cognitively imposed. It is also likely that cultural constraints on meal composition and meal size, manifested both in serving sizes and by internalized standards, are the principal determinants of meal termination. These are, of course, also learned responses.

Social Factors

Most meals are eaten in the presence of others. Using eating diary data, de Castro and his colleagues have shown that nondieting males and females eat more when they are in the presence of others than when they are alone. Furthermore, there is an orderly increase in meal size as a function of the number of others present (de Castro & Brewer, 1992). This "social correlation" is evident for all meals of the day, as well as snacks; on weekdays and weekends; for meals eaten at home, in restaurants, and in other locations; and for meals ingested with and without alcohol (de Castro, 1991; de Castro, Brewer, Elmore, & Orozco, 1990). When subjects were instructed to eat alone for 5 days, intake decreased significantly, suggesting a causal role for the presence of others (Redd & de Castro, 1992). Feunekes et al. (1995) and de Castro (1990) present data indicating that the social enhancement of eating is mediated primarily by the fact that people eating together spend a longer time eating.

In his extensive program of research, de Castro has used eating diaries as a method for studying the effect of the presence of others on meal size. However, social facilitation has been also documented in observational studies (Klesges, Bartsch, Norwood, Kautzman & Haugrud, 1984; Krantz, 1979) and in the laboratory (Berry, Beatty, & Klesges, 1985; Clendenen, Herman, & Polivy, 1994; Edelman et al., 1986). In a field evaluation of the U.S. Army's T-ration, daily caloric consumption increased as a function of the number of meals soldiers reported eating socially (Salter et al., 1991). It is clear that under many circumstances people eat more when they are in the company of others than when they are alone.

However, the effects of the presence of others on meal size are much more complex than a simple social facilitation notion would suggest. A substantial body of research can be understood in the context of the notion that people see eating "lightly" as a means of making a good impression in a social situation (Chaiken & Pliner, 1987; Pliner & Chaiken, 1990). Aspects of the situation that increase the importance of impression management or the salience of a norm of minimal consumption decrease meal size. People eat smaller meals when they are with strangers or coworkers than with family or friends (Clendenen et al., 1994; de Castro, 1994). Soldiers eat smaller meals when they are eating with a noncommissioned officer who eats a small (vs. large) amount (Engell, Kramer, Luther, & Adams, 1990). Women and men eat less when they are with a member of the opposite sex (Conger, Conger, Costanzo, Wright, & Matter, 1980; Pliner & Chaiken, 1990), especially one of high social desirability (Mori, Chaiken, & Pliner, 1987). People eat less when they are implicitly or explicitly under observation (Herman, Polivy, & Silver, 1979; Polivy, Herman, Hackett, & Kuleshnyk, 1986) and when they are eating in the presence of a dieter (Polivy, Herman, Younger, & Erskine, 1979) or someone who eats minimally or not at all (Conger et al., 1980; Nisbett & Storms, 1974; Rosenthal & Marx, 1979). Indeed, social modeling of minimal consumption can be powerful enough to override 24-hour deprivation-induced hunger (Goldman, Herman, & Polivy, 1991). Those who are particularly sensitive about their weight (i.e., dieting or overweight) are particularly affected by the presence of others. Obese individuals eat less when eating with others (Krantz, 1979) than when alone, especially when their eating companions are of normal weight (De Luca & Spigelman, 1979). The inhibiting effect of a noneating companion is particularly strong for dieters (Herman et al., 1979).

Cultural Standards and Memory

A major determinant of meal size and meal content is the cultural definition of an appropriate meal. In the United States, a dinner with two entrees or two desserts is excessive. Fried eggs are appropriate for breakfast but not for dinner. Cultural standards surely influence the amount eaten. For example, after finishing a main course, it is inappropriate to eat another; after eating dessert, it is inappropriate to return to a main course item. Most generally, having just eaten what is defined as a culturally complete meal, one is by virtue of this memory discouraged from further ingestion. Little research has been done to verify these reasonable claims. We would predict that a person who ate a high-calorie meal without dessert would be more likely to continue to eat than a person who ate a meal of substantially lower calories but that was a complete meal with dessert. Pliner (1999) had participants observe a videotaped target person eating either while engaging in behaviors associated with meals (e.g., warming food, using dishes and utensils, sitting down to eat) or not associated with meals (e.g., eating food cold, eating out of containers without utensils, eating while standing); the amount of food eaten was identical in the two conditions. Participants' ratings of the target's likelihood of eating in the next 2 hours and the degree of hunger were lower in the meal than in the nonmeal condition. These findings suggest that, at least from an observer's perspective, if someone has eaten a meal, he or she is not expected to eat again or be hungry in the near future.

One recent study on densely amnesic people confirms the importance of memory as a determinant of meal continuation or initiation (Rozin, Dow, Moscovitch, & Rajaram, 1998). Amnesic subjects who do not distinctly remember that they have just eaten will eat a second and even third full lunch, if these are served in sequence with intervals of 10 to 30 minutes. What is absent from these people is the memory that they have just eaten. Apparently, this is a really important piece of information and, hence, determinant of eating. Apparently, physiological satiety signals after a first meal are insufficient to completely inhibit ingestion as well.

Location

Although location is not mentioned in most discussions of how much people eat, after a moment's thought, one realizes how important location is in determining both what and how much people eat. Because most human meal studies are in the laboratory, location has not emerged as a major variable. Meiselman and his colleagues have consistently argued for the importance of location and context in general. Many of the studies by Meiselman and his collaborators at Natick are done in natural eating settings, allowing for evaluation of location effects. A number of studies demonstrate important effects and interactions among palatability, appropriateness, and location. Bell & Meiselman (1995) reviewed the effects of location, and conclude by supporting its major importance. For example, arrangement of food on a serving line affects choice. The characteristics of the meal occasion significantly affect acceptability and choice (Bell & Meiselman, 1995), and the ethnic setting (type of restaurant, restaurant decor) influences selection (Bell, Meiselman, Pierson, & Reeve, 1994).

Coll, Meyer, & Stunkard (1979) observed Americans eating at eating establishments, from restaurants, to cafeterias, to ice cream parlors and fast food establishments. They reported that the particular location strongly influences both the amount and type of food consumed and suggested that for eating in public places, location may be the most powerful determinant of amount consumed. For example, intake in kilocalories was about twice as high in a restaurant as in a cafeteria. Similarly, intake patterns of airline meals are different when served in the air or on the ground (Green & Butts, 1945).

In the marketing literature and the food retail industry, it is well known that location of food in a food store influences food purchase and, ultimately, consumption. Locations including end of aisle and eye level enhance the likelihood of purchase (Kahn & McAlister, 1997). Because more than 50% of consumer food store purchases are unplanned, such local environmental factors have a large amount of potential variance to explain.

THE ANOMALIES OF BREAKFAST

Breakfast is literally the first meal of the day, that which breaks the overnight fast. For most individuals, it is eaten shortly after rising. It is known that (perceived) time of day, per se, can influence amount eaten. A classic study by Schachter and Gross (1968) manipulated subjects' perceptions of time of day by means of a clock "gimmicked" to run either at twice normal speed or at half normal speed. Subjects arriving at 5:00 PM, were given instructions for 5 minutes and then occupied for 30 minutes in the presence of the clock. At that point, those in the "fast-clock" condition were under the impression that it was 6:05, whereas those in the "slow-clock" condition believed it to be 5:20. All were then given the opportunity to snack on crackers. Obese subjects, behaving in conformity with Schachter's external hypothesis, ate more when they believed it was past dinner time (6:05) than when they believed it was before dinner time (5:20); normal weight subjects showed the reverse pattern, eating less in the fast-clock condition than in the slow-clock condition.

Several interesting anomalies exist about breakfast compared with other meals. It is the smallest meal, following the longest fast, and, in Western cultures, its culinary makeup is qualitatively different from other meals. We also suspect, although we know of no relevant data, that many people eat the same breakfast every day; breakfast items do not seem to show the decrease in palatability with repetition shown by other food items.

Breakfast as the Smallest Meal

In Western cultures meal size increases from breakfast to lunch to dinner (de Castro, 1987a). It is also the meal that is most likely to be skipped, at least among young people (Schachter, 1971; Singleton & Rhoads, 1982). As the first meal of the day after what is usually the longest period of deprivation (break fast), it should be the largest meal. The equivalent of breakfast for free-running rats (first meal of the dark cycle) *is* the largest meal. Yet, characteristically, for humans, across cultures, breakfast is the smallest meal. There are two perspectives on this paradox. One is a metabolic account. Although breakfast follows the longest fast, it also usually follows the largest meal. The actual state of energy depletion is presumably some function of time since the last meal, the size of the last meal, and the energy expenditure since the last meal (Booth & Mather, 1978). Evidence from rat research is abundant, dating from LeMagnen & Tallon (1966), that meal size is better predicted by time to next meal than to time from last meal. In this context, the larger size of supper makes sense. The second perspective on the smallness of breakfast is cultural. The surprising smallness of breakfast is a distinctive feature of *human* meal patterns and promotes a distinctively human account. Cuisine is also a distinctively human occurrence and is characterized by elaborate meal preparations. It may be that the metabolic and work-cycle pressures to eat promptly on awakening and get to work to maximize daylight hours conflict with the food preparation traditions and encourage a small breakfast. Of course, in modern society, with microwaves and the like, an elaborate breakfast can be created in minutes. Perhaps some sort of cultural conservatism discourages this modernization of breakfast.

Breakfast as Culinarily Unique

Breakfast is distinct in its contents as well as its size; it tends to be different from lunch and dinner. In many instances, different foods are considered suitable for breakfast compared with lunch and dinner. Indeed, tacit acknowledgment of this distinction can be seen in a recent campaign by the Canadian Egg Marketing Board to induce consumers to "eat breakfast for dinner."

Pliner (1999) showed subjects videotapes in which a "target person" was shown eating a meal containing soup, fried chicken, and an apple and asked them for their "best guess" as to whether she was eating breakfast, lunch, dinner, or a snack. Only 4% of participants guessed she was eating breakfast. Birch, Billman, and Richards (1984) predicted that the "acceptability" of a food would be related to cultural beliefs about its appropriateness for particular mealtimes. The strength of these cultural beliefs can be seen in the fact that all their adult subjects sorted eight foods identically: orange juice, scrambled eggs, and Cheerios were seen as appropriate for breakfast but not dinner; frozen green beans, cheese pizza, and macaroni and cheese were seen as appropriate for dinner but not for breakfast; bread and banana were seen as appropriate for either. When 3- to 5-year-old children sorted the foods, 70% of their categorizations conformed to those of the adults. Participants tasted the set of foods twice—once between 8 and 10 AM and once at 3:30 and 5:30 PM—and provided preference assessments on each occasion. For both children and adults, the "breakfast" foods were liked better in the morning than in the afternoon, whereas the reverse was true for "dinner" foods. However, different results were obtained by Kramer, Rock, and Engell (1992). Subjects ate four meals, two consisting of breakfast-type foods (e.g., bacon and eggs) and two consisting of lunch-type foods (e.g., turkey sandwich). One meal of each type was eaten at breakfast time (8:00 AM) and one at lunch time (noon). Hedonic ratings showed no hint of the predicted statistical interaction between food type and meal time; rather, the lunch foods were rated more positively at both mealtimes. Furthermore, amounts consumed showed no evidence of a food type–mealtime interaction. Instead, there were two main effects; participants ate more of the lunch foods, and they ate more at lunch time.

In terms of notions of cuisine, such as flavor principles (E. Rozin, 1973), breakfast is the anomalous meal. That is, if one attempted to assign someone to their culture on the basis of information about a particular meal they consumed, breakfast would be the least informative. The French, German, or American breakfasts differ less than their other meals. The evolution of Western breakfast foods is a major topic for research—research that has not been done. Suffice it to say that breakfast foods seem to involve minimal preparation and often center on a warm beverage, typically tea or coffee. The culinary anomaly of breakfast is largely a Western phenomenon. For most peoples of the world, in traditional cultures, breakfast is simply warmed over supper. There are not distinctive breakfast foods, and the warming over reduces the preparation time.

CONCLUSIONS: THE MEAL AS THE PRIVILEGED PSYCHOLOGICAL OR PHYSIOLOGICAL UNIT

We have reviewed abundant evidence, linguistic, culinary, practical, and memorial, that the meal is a special unit of eating at the psychological level. It is the basic or privileged unit, in that smaller units are described as subdivisions of meals, and there is no particular designation of higher units (e.g., daily or monthly intake). Both our day and our thinking are organized in terms of meals.

This said, the great preponderance of research on meals has to do with their physiological basis as opposed to their psychological reality. The extensive body of research on regulation of food intake, primarily in rats and humans, has used the meal as the unit. The focal question has been: what initiates a meal, what terminates a meal, and what determines the size of meals? The "what" in these questions almost always refers to physiological events, such as stomach fill, blood glucose levels, body temperature, flow of some nutrient metabolite from one body compartment or brain area to another. Although general schemes of food regulation, from Stellar (1954) on, have given appropriate attention to nonmetabolic factors, the focus has been metabolism. Research on rats and humans, under appropriately controlled conditions, has indicated roles for many physiological factors, from gastric fill to levels and flows of nutrients in liver, brain, and other tissues, in the determination of meal size, initiation, and termination. These findings constitute a major literature that is covered elsewhere in this volume. One of the most important findings in this area is that, all other things being equal (*ceteris parebis*), changes in the caloric density of a diet lead to compensatory changes in volume consumed. *Ceteris parebis*, meal size is controlled by various indices of nutritional status. But *ceteris parebis*, especially for humans, means extracting a meal from its natural context and holding constant what *we* take to be the principal determinants of meal size. This is most clearly illustrated by the multiple meal-taking of amnesics, who should be receiving normal physiological satiety signals, but whose failure to recall having recently eaten seems a sufficient stimulus for ingestion of a full meal (Rozin et al., 1998).

Let us consider, in a natural eating context, what are the principal determinants of whether someone will eat and how much he or she will eat. We believe that in natural conditions, the principal determinants are predominantly nonmetabolic. People will eat if it is a culturally appropriate mealtime, if they do not remember having just eaten, and if presented with food. The amount they eat will probably be primarily affected by the palatability and appropriateness of the food and the amount available. Secondary effects would include cultural constraints (e.g., how much it is appropriate to eat for this particular type of meal), degree of hunger, social setting, and location. The perceived energy density of the food may be a more important determinant of intake than the actual energy density (Wooley, 1972).

The remarkable phenomenon is that although intake in any given meal is predominantly controlled by nonmetabolic factors, people's weight varies very little from week to week, or even year to year. We take this remarkable fact to mean that the focus of action of metabolic factors may be over a "unit" that is much longer than a meal, perhaps a daily or even weekly intake. A recent study by Leann Birch and her colleagues (Birch, Johnson, Andresen, Petersen, & Schulte, 1991) on successive meal sizes of children over 6 days supports this idea. Birch concludes that although individual meals may vary greatly from day to day for the same child with the same choices, the 24-hour intake is relatively constant. The mean coefficient of variation for meal size is 33.6%, but for 24 hours it is only 10.4%. In most cases, Birch reports adjustment from meal to meal. Many laboratory studies indicate that rats (e.g., Carlisle & Stellar, 1969), human adults (Kissileff & Thornton, 1982b; Spiegel, 1973) and children (Birch & Deysher, 1985) regulate energy intake in meals in that they increase bulk intake when the caloric density of the diet is reduced. Some studies report effects on the first meal, but anything close to full compensation, if it occurs, takes place over a number of meals or even days. Louis-Sylvestre, Tournier, Chapelot, & Chabert (1994) found nearly complete energy adjustment in the 24 hours after the low-calorie dish was served and a return to the baseline level once the energy content of the dish was restored. Along the same lines, eating diary studies show small but significant negative correlations between the size of a meal and the size of the subsequent meal in both adults and children (Birch et al., 1991; de Castro & Kreitzman, 1985).

This is not to say that physiological factors do not exert their influences on meal size—there is little else for them to operate on! Rather, it means that physiological factors may often operate in a modulatory way, exerting their net effect over a sequence of meals. The metabolic compensation one pays for a highly palatable meal indulgence will be seen over the coming meals and days, not in the meal itself.

We conclude that it is reasonable to be very interested in meals, and if that *is* our interest, we need to pay much more attention to determinants of meal initiation, meal size, and meal termination in natural situations, in context.

REFERENCES

Anderson, N.H., & Norman, A. (1964). Order effects in impression formation in four classes of stimuli. *Journal of Abnormal and Social Psychology 69*, 467–471.

Appadurai, A. (1981). Gastro-politics in Hindu South Asia, *American Ethnologist, 8*, 494–511.

Ballard, B.D., Gipson, M.T., Guttenberg, W., & Ramsey, K. (1980). Palatability of food as a factor influencing obese and normal-weight children's eating habits. *Behavior Research and Therapy, 18*, 598–600.

Bell, R., & Meiselman, H. (1995). The role of eating environments in determining food choice. In D.W. Marshall (Ed.), *Food choice and the consumer* (pp. 292–310). London: Blackie Academic and Professional.

Bell, R., Meiselman, H.L., Pierson, B.J., & Reeve, W.G. (1994). Effects of adding an Italian theme to a restaurant on the perceived ethnicity, acceptability, and selection of foods. *Appetite, 22*, 11–24.

Bellisle, F. (1979). Human feeding behavior. *Neuroscience and Biobehavioral Reviews, 3,* 163–169.

Bellisle, F., & LeMagnen, J. (1980). The analysis of human feeding patterns: the Edogram. *Appetite, 1,* 141–150.

Bellisle, F., & LeMagnen, J. (1981). The structure of meals in humans: Eating and drinking patterns in lean and obese subjects. *Physiology & Behavior, 27,* 649–658.

Bellisle, F., Lucas, F., Amrani, R., & LeMagnen, J. (1984). Deprivation, palatability and the microstructure of meals in human subjects. *Appetite, 5,* 85–94.

Bernstein, I.L., Zimmerman, J.C., Czeisler, C.A., & Weitzman, E.D. (1981). Meal patterns in "free-running" humans. *Physiology & Behavior, 27,* 621–623.

Berry, S.L., Beatty, W.W., & Klesges, R.C. (1985). Sensory and social influences on ice cream consumption by males and females in a laboratory setting. *Appetite, 6,* 41–45.

Birch, L.L., Billman, J., & Richards, S.S. (1984). Time of day influences food acceptability. *Appetite, 5,* 109–116.

Birch, L.L., & Deysher, M. (1985). Conditioned and unconditioned caloric compensation: Evidence for self-regulation of food intake in young children. *Learning and Motivation, 16,* 341–355.

Birch, L.L., & Deysher, M. (1986). Conditioned compensation and sensory specific satiety: Evidence for self-regulation of food intake by young children. *Appetite, 7,* 223–231.

Birch, L.L. Fisher, J.O., & Grimm-Thomas, K. (1996). The development of children's eating habits. In H.L. Meiselman & H.J.H. MacFie (Eds.), *Food choice, acceptance and consumption* (pp. 161–206). London: Blackie Academic and Professional.

Birch, L.L., Johnson, S.L., Andresen, G., Petersen, J.C., & Schulte, M.C. (1991). The variability of young children's energy intake. *The New England Journal of Medicine, 324,* 232–235.

Birch, L.L., McPhee, L., Sullivan, S., & Johnson, S. (1989). Conditioned meal initiation in young children. *Appetite, 13,* 105–113.

Bobroff, E.M., & Kissileff, G.R. (1986). Effects of changes in palatability on food intake and the cumulative food intake curve in man. *Appetite, 7,* 85–96.

Booth, D.A. (1972). Conditioned satiety in the rat. *Journal of Comparative and Physiological Psychology, 81,* 457–471.

Booth, D.A. (1977). Appetite and satiety as metabolic expectancies. In Y. Katsuki, M. Sato, S.F. Takagi, & Y. Oomura (Eds.), *Food intake and chemical senses.* Baltimore: University Park Press.

Booth, D.A., & Davis, J.D. (1973). Gastrointestinal factors in the acquisition of oral sensory control of satiation. *Physiology and Behavior, 11,* 23–29.

Booth, D.A., Fuller, J., & Lewis, V. (1981). Human control of body weight: Cognitive or physiological? Some energy-related perceptions and misperceptions. In L.A. Cioffi, W.P.T. James, & T.B. Van Itallie (Eds.), *The body weight regulatory system: Normal and disturbed mechanisms* (pp. 305–314). New York: Raven Press.

Booth, D.A., Lee, M., & McAleavey, C. (1976). Acquired sensory control of satiation in man. *British Journal of Psychology, 67,* 137–147.

Booth, D. A., & Mather, P. (1978). Prototype model of human feeding, growth, and obesity. In D.A. Booth (Ed.), *Hunger models: Computable theory of feeding control* (pp. 279–322). New York: Academic Press.

Booth, D.A., Mather, P, & Fuller, J. (1982). Starch content of ordinary foods associatively conditions human appetite and satiation, indexed by intake and eating pleasantness of starch-paired flavours. *Appetite, 3,* 163–184.

Bruch, H. (1957). *The importance of overweight.* New York: W. W. Norton and Company.

Cabanac, M. (1971). The physiological role of pleasure. *Science, 173,* 1103–1107.

Cabanac, M. (1979). Sensory pleasure. *Quarterly Review of Biology, 54,* 1–29.

Campfield, L.A., & Smith, F.J. (1990). Systemic factors in the control of food intake: Evidence for patterns as signals. In: E. M. Sticker (Ed.), *Handbook of behavioral neurobiology: Vol. 10. Neurobiology of food and fluid intake* (pp. 183–206). New York: Plenum Publishing.

Carlisle, H.J., & Stellar, E. (1969). Caloric regulation and food preference in normal, hyperphagic, and aphagic rats. *Journal of Comparative and Physiological Psychology, 69,* 107–114.

Chaiken, S., & Pliner, P. (1987). Women, but not men, are what they eat: The effect of meal size and gender on perceived femininity and masculinity. *Personality and Social Psychology Bulletin, 13,* 166–176.

Clendenen, V.I., Herman, C.P., & Polivy, J. (1994). Social facilitation of eating among friends and strangers. *Appetite, 23,* 1–13.

Coll, M., Meyer, A., & Stunkard, A.J. (1979). Obesity and food choices in public places *Archives of General Psychiatry 36,* 795–797.

Conger, J.C., Conger, A.J., Costanzo, P.R., Wright, K.L., & Matter, J.A. (1980). Effects of social cues on the eating behavior of obese and normal subjects. *Journal of Personality, 48,* 258–271.

de Castro, J.M. (1987a). Circadian rhythms of the spontaneous meal pattern, macronutrient intake, and mood of humans. *Physiology and Behavior, 40,* 437–446.

de Castro, J. (1987b). Macronutrient relationships with meal patterns and mood in the spontaneous feeding behavior of humans. *Physiology and Behavior, 39,* 561–569.

de Castro, J. (1988). A microregulatory analysis of spontaneous fluid intake by humans: Evidence that the amount of liquid ingested and its timing is mainly governed by feeding. *Physiology & Behavior, 43,* 705–714.

de Castro, J.M. (1990). Social facilitation of duration and size but not rate of the spontaneous meal intake of humans. *Physiology and Behavior, 47,* 1129–1135.

de Castro, J.M. (1991). Social facilitation of the spontaneous meal size of humans occurs on both weekdays and weekends. *Physiology and Behavior, 49,* 1289–1291.

de Castro, J.M. (1994). Family and friends produce greater social facilitation of food intake than other companions. *Physiology and Behavior, 56,* 445–455.

de Castro, J.M., & Brewer, E.M. (1992). The amount eaten in meals by humans is a power function of the number of people present. *Physiology and Behavior, 51,* 121–125.

de Castro, J.M., Brewer, E.M., Elmore, D.K., & Orozco, S. (1990). Social facilitation of the spontaneous meal size of humans is independent of time, place, alcohol, or snacks. *Appetite, 15,* 89–101.

de Castro, J.M., & Kreitzman, S.M. (1985). A microregulatory analysis of spontaneous human feeding patterns. *Physiology and Behavior, 35,* 329–335.

Decke, E. (1971). Cited in S. Schachter, *Emotion, obesity, and crime* (p. 103). New York: Academic Press.

De Luca, R.V., & Spigelman, M.N. (1979). Effects of models on food intake of obese and nonobese female college students. *Canadian Journal of Behavioral Science, 11,* 124–129.

Edelman, B., Engell, D., Bronstein, P., & Hirsch, E. (1986). Environmental effects on the intake of overweight and normal-weight men. *Appetite, 7,* 71–83.

Engell, D., & Hirsch, E. (1991). Environmental and sensory modulation of fluid intake in humans. In D.J. Ramsey & D. Booth (Eds.), *Thirst: Physiological and psychological aspects* (pp. 382–390). London: Springer-Verlag.

Engell, D., Kramer, F.M., Luther, S., & Adams, S.O. (1990). The effect of social influences on food intake. Cited in E. Hirsch & F.M. Kramer (1993). Situational influences on intake. In B. Marriott, (Ed.), *Nutritional needs in hot environments* (pp. 215–240). Washington, DC: National Academy Press.

Feunekes, G.J.J., de Graaf, C., & van Staveren, W.A. (1995). Social facilitation of food intake is mediated by meal duration. *Physiology and Behavior, 58,* 551–558.

Fitzsimons, J.T., & LeMagnen, J. (1969). Eating as a regulatory control of drinking. *Journal of Comparative and Physiological Psychology, 67,* 273–283.

Goldman, S., Herman, C.P., & Polivy, J. (1991). Is the effect of social influence attenuated by hunger? *Appetite, 17,* 129–140.

Green, D.M., & Butts, J.S. (1945). Factors affecting acceptability of meals served in the air. *Journal of the American Dietetic Association, 21,* 415–419.

Green, J., Pollak, C.P., & Smith, G.P. (1986). Meal size and intermeal interval in human subjects in time isolation. *Physiology & Behavior, 41,* 141–147.

Guy-Grand, B., Lehnert, V., Doessans, M., & Bellisle, F. (1994). Type of test-meal affects palatability and eating style in humans. *Appetite, 22,* 125–134.

Hammer, L.D. (1992). The development of eating behavior in childhood. *Pediatric Clinics of North America, 39,* 379–394.

Heatherton, T., Herman, C.P., & Polivy, J. (1991). Effects of physical threat and ego threat on eating behavior. *Journal of Personality and Social Psychology, 60,* 138–143.

Hedderley, D.I., & Meiselman, H.L. (1995). Modelling meal acceptability in a free choice environment. *Food Quality and Preference, 6,* 15–26.

Herman, C.P., & Polivy, J. (1975). Anxiety, restraint, and eating behavior. *Journal of Abnormal Psychology, 43,* 647–660.

Herman, C.P., & Polivy, J. (1980). Restrained eating. In A.J. Stunkard (Ed.), *Obesity* (pp. 208–225). Philadelphia: W.B. Saunders Company.

Herman, C.P., Polivy, J., & Silver, R. (1979). Effects of an observer on eating behavior: The induction of "sensible" eating. *Journal of Personality and Social Psychology, 47,* 85–99.

Hetherington, M.M. (1996). Sensory-specific satiety and its importance in meal termination. *Neuroscience and Biobehavioral Reviews, 20*, 113–117.

Hill, S.W. (1974). Eating responses of humans during dinner meals. *Journal of Comparative and Physiological Psychology, 86*, 652–657.

Hill, S.W. & McCutcheon, N.B. (1975). Eating responses of obese and nonobese humans during dinner meals. *Psychosomatic Medicine, 37*, 395–401.

Hirsch, E.S. & Kramer, F.M. (1993). Situational influences on food intake. In B. Marriott (Ed.), *Nutritional needs in hot environments* (pp. 215–244). Washington, DC: National Academy Press.

Johnson, W.G. (1974). The effects of cue prominence and obesity on effort to obtain food. In S. Schachter & J. Rodin (Eds.), *Obese humans and rats* (pp. 53–60). Potomac, MD: Erlbaum Associates.

Kahn, B. E., & McAlister, L. (1997). *Grocery revolution. The new focus on the consumer.* Reading, MA: Addison-Wesley Publishing Co.

Kahneman, D., Wakker, P.P., & Sarin, R. (1997). Back to Bentham? Explorations of experienced utility. *Quarterly Journal of Economics, 112*, 375–405.

Kaplan, H.I., & Kaplan, H.S. (1957). The psychosomatic concept of obesity. *Journal of Nervous and Mental Disease, 125*, 181–201.

Kissileff, H. R., Guss, J. L., & Nolan, L. J. (1996) What animal research tells us about human eating. In H.L. Meiselman & H.J.H. MacFie (Eds.), *Food choice, acceptance and consumption* (pp. 105–160). London: Blackie Academic and Professional.

Kissileff, H.R., & Thornton, J. (1982a). A quadratic equation adequately describes the cumulative food intake curve in man. *Appetite, 3*, 255–272.

Kissileff, H.R., & Thornton, J. (1982b). Facilitation and inhibition in the cumulative food intake curve in man. In A.R. Morrison & P.L. Strick (Eds.), *Changing concepts of the nervous system.* (pp 585–607). New York: Academic Press.

Klesges, R.C., Bartsch, D., Norwood, J.D., Kautzman, D., & Haugrud, S. (1984). The effects of selected social and environmental variables on the eating behavior of adults in the natural environment. *International Journal of Eating Disorders, 3*, 35–41.

Kramer, F.M., Rock, K., & Engell, D. (1992). Effects of time of day and appropriateness of food intake and hedonic ratio at morning and midday. *Appetite, 18*, 1–13.

Krantz, D.S. (1979). A naturalistic study of social influences on meal size among moderately obese and nonobese subjects. *Psychosomatic Medicine, 41*, 19–27.

Lappalainen, R., & Epstein, L.W. (1990). A behavioral economics analysis of food choice in humans. *Appetite, 14*, 81–93.

LeMagnen, J. (1955). Sur le mecanisme d'etablissement des appetits calorique. *Comptes Rendus Acad Sci, 240*, 2436–2438.

LeMagnen, J. (1956). Hyperphagie provoquee chez la rat blanc par alteration du mechanism de satiete peripherique. *Comptes Rendus des Seances de la Societe de Biologie, 150*, 32–35.

LeMagnen, J. (1971). Advances in studies on the physiological control and regulation of food intake. In E. Stellar & J.M. Sprague (Eds.), *Progress in physiological psychology* (Vol. 4, pp. 203–261). New York: Academic Press.

LeMagnen, J., & Tallon, S. (1966). La periodicite spontanee de la prise d'aliments ad libitum du rat blanc. *J. Physiologie, Paris, 58*, 323–349.

Lester, L.S., & Kramer, F.M. (1991). The effects of heating on food acceptability and consumption. Cited in E.S. Hirsch & F.M. Kramer (1993). Situational influences on food intake. In B. Marriott (Ed.), *Nutritional needs in hot environments* (pp.215–244). Washington, DC: National Academy Press.

Lipsitt, L.P., & Behl, G. (1990). Taste-mediated differences in the sucking behavior of human newborns. In E.D. Capaldi & T.L. Powley (Eds.), *Taste, experience, and feeding* (pp. 75–93). Washington, DC: American Psychological Association.

Louis-Sylvestre, J., Tournier, A., Chapelot, D., & Chabert, M. (1994). Effect of a fat-reduced dish in a meal on 24-h energy and macronutrient intake. *Appetite, 22*, 165–172.

Lowe, M. (1982). The role of anticipated deprivation in overeating. *Addictive Behaviors, 7*, 103–112.

McArthur, L., & Burstein, B. (1975). Field dependent eating and perception as a function of weight and sex. *Journal of Personality and Social Psychology, 43*, 402–420.

McKenna, R.J. (1972). Some effects of anxiety level and food cues on the eating behavior of obese and normal subjects: A comparison of the Schachterian and psychosomatic concepts. *Journal of Personality and Social Psychology, 22*, 311–319.

Meigs, A. S. (1984). *Food, sex, and pollution: A New Guinea religion*. New Brunswick, NJ: Rutgers University Press.

Meiselman, H.L. (1996). The contextual basis for food acceptance, food choice and food intake: the food, the situation, and the individual. In H.L. Meiselman & H.J.H. MacFie (Eds.), *Food choice, acceptance and consumption* (pp. 239–263). London: Blackie Academic and Professional.

Meiselman, H.L., Hedderley, D., Staddon, S.L., Pierson, B.J. (1994). Effect of effort on meal selection and meal acceptability in a student cafeteria. *Appetite, 23,* 43–55.

Miller, L., Rozin, P., & Fiske, A. (1998). Food sharing and feeding another person suggest intimacy: Two studies of American college students. *European Journal of Social Psychology, 28,* 423–436.

Milstein, R.M. (1980). Responsiveness in newborn infants of overweight and normal weight parents. *Appetite, 1,* 65–74.

Mook, D.G., & Votaw, M.C. (1992). How important is hedonism? Reasons given by college students for ending a meal. *Appetite, 18,* 69–75.

Mori, D., Chaiken, S., & Pliner, P. (1987). "Eating lightly" and the self-presentation of femininity. *Journal of Personality and Social Psychology, 53,* 693–702.

Nicolaidis, S., Danguir, J., & Mather, P. (1979). A new approach of sleep and feeding behaviors in the laboratory rat. *Physiology & Behavior, 23,* 717–722.

Nisbett, R.E. (1968). Determinants of food intake in obesity. *Science, 159,* 1254–1255.

Nisbett, R.E., & Gurwitz, S. (1970). Weight, sex, and the eating behavior of human newborns. *Journal of Comparative and Physiological Psychology, 73,* 245–253.

Nisbett, R.E., & Storms, M. (1974). Cognitive and social determinants of food intake. In H. London & R.E. Nisbett (Eds.), *Thought and feeling: Cognitive alteration of feeling states*. Chicago: Aldine Publishing Company.

Panksepp, J. (1978). Analysis of feeding patterns: Data reduction and theoretical implications. In D.A. Booth (Ed.), *Hunger models. Computable theory of feeding control* (pp. 144–166). New York: Academic Press.

Pliner, P. (1999). *Food eaten in a "meal" context is perceived as more satiating*. Unpublished research, University of Toronto.

Pliner, P., & Chaiken, S. (1990). Eating, social motives, and self-presentation in women and men. *Journal of Experimental Social Psychology, 26,* 240–254.

Pliner, P., Herman, C.P., & Polivy, J. (1990). Palatability as a determinant of eating: Finickiness as a function of taste, hunger, and the prospect of good food. In E.D. Capaldi & T.L. Powley (Eds.), *Taste, experience, and feeding* (pp. 210–226). Washington, DC: American Psychological Association.

Pliner, P., Herman, C.P., Polivy, J., & Zakalusny, I. (1980). Short-term intake of overweight individuals and normal weight dieters and non-dieters with and without choice among a variety of foods. *Appetite, 1,* 203–213.

Polivy, J. (1976). Perception of calories and regulation of intake in restrained and unrestrained subjects. *Addictive Behaviors, 2.*

Polivy, J., & Herman, C.P. (1976). Clinical depression and weight change: A complex relation. *Journal of Abnormal Psychology, 85,* 338–340.

Polivy, J., Herman, C.P., Hackett, R., & Kuleshnyk, I. (1986). The effects of self-attention and public attention on eating in restrained and unrestrained subjects. *Journal of Personality and Social Psychology, 50,* 1253–1260.

Polivy, J., Herman, C.P., Younger, J.C., & Erskine, B. (1979). Effects of a model on eating behavior: The induction of a restrained eating style. *Journal of Personality, 47,* 100–117.

Rappoport, L., Peters, G.R., Downey, R., McCann, T., & Huff-Corzine, L. (1993). Gender and age differences in food cognition. *Appetite, 20,* 33–52.

Redd, E.M., & de Castro, J.M. (1992). Social facilitation of eating: Effects of instructions to eat alone or with others. *Physiology and Behavior, 52,* 749–754.

Richter, C.P. (1927). Animal behavior and internal drives. *Quart. Review of Biology, 2,* 307–343.

Rodin, J. (1975). Responsiveness of the obese to external cues. In G.A. Bray (Ed.), *Obesity in perspective* (Vol. 2, Part 2, pp. 61–72). Washington, DC: U.S. Government Printing Office.

Rogers. P.L., & Blundell, J.E. (1979). Effect of anorexic drug in food intake and the microstructure of eating in human subjects. *Psychopharmacology, 66,* 159–165.

Rogozenski, J.G. Jr., & Moskowitz, H.R. (1982). A system for the preference evaluation of cyclic menus. *Journal of Food Service Systems, 2,* 139–161.

Rolls, B.J. (1990). The role of sensory specific satiety in food intake and food selection. In E.D. Capaldi & T.L. Powley (Eds.), *Taste, experience, and feeding* (pp. 197–209). Washington, DC: American Psychological Association.

Rolls, B.J., Hetherington, M., Burley, V.J., & van Duijvenvoorde, P.M. (1986). In M.R. Kare & J.G. Brand (Eds.), *Interaction of the chemical senses with nutrition*. New York: Academic Press.

Rolls, B.J., Rowe, E.A., & Rolls, E.T. (1982a). How flavor and appearance affect human feeding. *Proceedings of the Nutrition Society, 41*, 109–117.

Rolls, B.J., Rowe, E.A., & Rolls, E.T. (1982b). How sensory properties of food affect human feeding behavior. *Physiology and Behavior, 27*, 137–142.

Rolls, B.J., Rowe, E.A., Rolls, E.T., Kingston, B., Megson, A., & Gunary, R. (1981). Variety in a meal enhances food intake in man. *Physiology and Behavior, 26*, 215–221.

Rolls, B.J., van Duijvenvoorde, P.M., & Rolls, E.T. (1984). Pleasantness changes and food intake in a varied four course meal. *Appetite, 5*, 337–348.

Rosenthal, B., & Marx, R.D. (1979). Modelling influences on the eating behavior of successful and unsuccessful dieters and treated normal weight individuals. *Addictive Behaviors, 4*, 215–221.

Ross, L.D. (1974). Effects of manipulating salience of food upon consumption by obese and normal eaters. In S. Schachter & J. Rodin (Eds.), *Obese humans and rats* (pp. 43–52). Potomac, MD: Erlbaum Associates.

Rozin, E. (1973). *Ethnic cuisine: The flavor principle cookbook*. Lexington, MA: The Stephen Greene Press.

Rozin, P., Dow, S., Moscovitch, M., & Rajaram, S. (1998). The role of memory for recent eating experiences in onset and cessation of meals. Evidence from the amnesic syndrome. *Psychological Science, 9*, 392–396.

Rozin, P., Kurzer, N.C., & Cohen, A. (1999). *The meanings of "food:" Free associations to the word "food" by Americans as a function of gender and generation*. Manuscript in preparation.

Rozin, P., Pliner, P., & Berman, L. (1999). *The meal as a psychological unit: Evidence from memory and linguistic categories across many languages*. Manuscript in preparation.

Rozin, P., Rode, E., & Ostovich, J. (1999). *Memory for meals: relations between experienced and remembered pleasure*. Unpublished research.

Rozin, P., & Tuorila, H. (1993). Simultaneous and temporal contextual influences on food choice. *Food Quality and Preference, 4*, 11–20.

Salter, C.A., Engell, D., Kramer, F.M., Lester, L.S., Kalick, J.J, Lester, L.L., Dewey, S.L., & Carette, D. (1991). Cited in E. Hirsch & F.M. Kramer. (1993). Situational influences on intake. In B. Marriott (Ed.), *Nutritional needs in hot environments* (pp. 215–240). Washington, DC: National Academy Press.

Schachter, S. (1971). *Emotion, obesity, and crime*. New York: Academic Press.

Schachter, S., & Friedman, L.N. (1974). In S. Schachter & J. Rodin (Eds.), *Obese humans and rats* (pp. 11–14). Potomac, MD: Erlbaum Associates.

Schachter, S., Goldman, R., & Gordon, A. (1968). The effects of fear, food deprivation, and obesity on eating. *Journal of Personality and Social Psychology, 10*, 91–97.

Schachter, S., & Gross, L. (1968). Manipulated time and eating behavior. *Journal of Personality and Social Psychology, 10*, 98–106.

Schutz, H.G. (1989). Beyond preference: Appropriateness as a measure of contextual acceptance of food. In D.M. H. Thomson (Ed.), *Food acceptability* (pp. 115–134). Essex, England: Elsevier Applied Science Publishers.

Singh, D., & Sikes, S. (1974). Role of past experience on food-motivated behavior of obese humans. *Journal of Comparative and Physiological Psychology, 86*, 503–508.

Singleton, N., & Rhoads, D.S. (1982). Meal and snacking patterns of students. *Journal of School Health, 52*, 529–534.

Slochower, J.A. (1976). Emotional labeling and overeating in obese and normal weight individuals. *Psychosomatic Medicine, 38*, 131–139.

Slochower, J.A. (1983). *Excessive eating: The role of emotions and environment*. New York: Human Sciences Press, Inc.

Smith, G.P. (1982). The physiology of the meal. In T. Silverstone (Ed.), *Drugs and appetite* (pp. 1–21). San Diego: Academic Press.

Spencer, J.A., & Fremouw, W.J. (1979). Binge eating as a function of restraint and weight classification. *Journal of Abnormal Psychology, 92*, 210–215.

Spiegel, T.A. (1973). Caloric regulation of food in man. *Journal of Comparative and Physiological Psychology, 84*, 24–37.

Spiegel, T.A., Shrager, E.E., & Stellar, E. (1989). Responses of lean and obese subjects to preloads, deprivation and palatability. *Appetite, 13*, 45–69.

Spiegel, T.A., & Stellar, E. (1990). Effects of variety on food intake of underweight, normal-weight, and over-weight women. *Appetite, 15*, 47–61.

Stellar, E. (1954). The physiology of motivation. *Psychological Review, 6*, 5–22.

Stellar, E., & Shrager, E.E. (1985). Chews and swallows and the microstructure of eating. *American Journal of Clinical Nutrition, 42*, 973–982.

Stunkard, A., Coll, M., Lundquist, S., & Meyers, A. (1980). Obesity and eating style. *Archives of General Psychiatry, 37*, 1127–1129.

Stunkard, A.J., & Levitz, L.S. (1975). The influence of caloric density and availability on the food selections of normal and obese subjects. Cited in Levitz, L.S. The susceptibility of human feeding behavior to external controls. In G.A. Bray (Ed.), *Obesity in perspective* (Vol. 2, Part 2, pp. 53–60). Washington, DC: U.S. Government Printing Office.

Tom, G., & Rucker, M. (1975). Fat, full, and happy: Effects of food deprivation, external cues, and obesity on preference ratings consumption, and buying intentions. *Journal of Personality and Social Psychology, 32*, 761–766.

Tomarken, A.J., & Kirschenbaum, D.S. (1984). Effects of plans for future meals on counterregulatory eating by restrained and unrestrained eaters. *Journal of Abnormal Psychology, 93*, 458–472.

Turner, M., & Collison, R. (1988): Consumer acceptance of meals and meal components. *Food Quality and Preference, 1*, 21–24

Wansink, B. (1996). Can package size accelerate usage volume. *Journal of Marketing, 60*, 1–14.

Weingarten, H. (1983). Conditioned cues elicit eating in sated rats: A role for learning in meal initiation. *Science, 220*, 431–433.

Weingarten, H. (1984). Meal initiation controlled by learned cues: Basic behavioral properties. *Appetite, 5*, 147–158.

Weingarten, H. (1985). Stimulus control of eating: Implications for a two-factor theory of hunger. *Appetite, 6*, 387–401.

Woods, S. C., & Strubbe, J. H. (1994). The psychobiology of meals. *Psychonomic Bulletin and Review, 1*, 141–155.

Woody, E.Z., Costanzo, P.R., Liefer, H., & Conger, J. (1981). The effects of taste and caloric perceptions on the eating behavior of restrained and unrestrained subjects. *Cognitive Therapy and Research, 5*, 381–390.

Wooley, O.W., Wooley, S.C., & Dunham, R.B. (1972). Can calories be perceived and do they affect hunger in obese and nonobese humans. *Journal of Comparative and Physiological Psychology, 80*, 250–258.

Wooley, S.C. (1972). Physiologic versus cognitive factors in short term food regulation in the obese and nonobese. *Psychosomatic Medicine, 31*, 62–68.

Wooley, S.C., & Wooley, O.W. (1973). Salivation to the sight and thought of food: A new measure of appetite. *Psychosomatic Medicine, 35*, 136–142.

Wurtman, J.J. (1987). Disorders of food intake: Excessive carbohydrate snack intake among a class of obese people. *Annals of the New York Academy of Sciences, 499*, 197–202.

Wurtman, R.J., & Wurtman, J.J. (1986). Carbohydrate craving, obesity, and brain serotonin. *Appetite, 7* (Suppl), 99–103.

Wurtman, J.J., Wurtman, J.R., Mark, S., Tsay, R., Gilbert, W., & Growdon, J. (1985). D-Fenfluramine selectively suppresses carbohydrate snacking by obese subjects. *International Journal of Eating Disorders, 4*, 89–99.

Yeomans, M.R. (1996). Palatability and the micro-structure of feeding in humans: The appetizer effect. *Appetite, 27*, 119–133.

Yeomans, M.R., Gray, R.W., Mitchell, C.J., & True, S. (1997). Independent effects of palatability and within-meal pauses on intake and appetite ratings in human volunteers. *Appetite, 29*, 61–76.

Zylan, K.D. (1996). Gender differences in the reasons given for meal termination. *Appetite, 26*, 37–44.

Chapter 3

Nutritional Definitions of the Meal

Cees de Graaf

In humans, food intake and, as a consequence, energy and nutrient intake, usually occurs at a limited number of discrete eating occasions. The number of eating occasions and the distribution of eating occasions across the day are influenced by a large number of environmental, psychological, and physiological factors. This book gives an excellent impression of all these different aspects, which are relevant for defining and describing meals. In search of nutritional data on this issue it appeared that there are data from the United States and Europe, but virtually no data are available from non-Western countries.

The definition of "meals," "snacks," and "other eating occasions" can be based on data on actual intake in terms of energy/nutrient intake and time constraints as has been done extensively by de Castro (1988), or it can be done in a more conventional way, as is done in most of the tables and figures in this chapter. The conventional meal pattern in most Western societies is characterized by three main meals across the day, breakfast in the morning, lunch at the beginning of the afternoon, and dinner at the beginning of the evening. In between these main meals there can be another number of eating moments, in which people eat snacks. In most tables of this chapter, *meals* refer to either breakfast, lunch, or dinner, and *snacks* refer to all other eating/drinking occasions. In this chapter, the definition and description of the meal from a nutritional perspective refers to the frequency, distribution, and variability of energy and nutrient intake across the day. It appears that although the frequency and distributions of intake vary to a great extent across countries, a number of constant factors exist in these patterns.

Little systematic research has been done on nutritional aspects of meals and meal patterns across different countries. This implies that most data across countries on the frequency and distribution of energy and nutrient intakes across the day are not comparable. Data obtained from the SENECA study, a European longitudinal study on nutrition and the elderly (de Groot, van Staveren, & Hautvast, 1991; de Groot, van Staveren, Dirren, & Hautvast, 1996) are a notable exception. Other data are also available, although they are scattered across the literature. It should be emphasized that the data presented in this chapter represent a snap-shot in time. Meal patterns and the nutritional composition of meals change under environmental pressures.

MEAL FREQUENCY IN DIFFERENT COUNTRIES

The frequency of eating moments depends to a large extent on socioeconomic and other factors. Some people suggest that a trend exists toward an increasing number of eating moments across the day (grazing) (de Wolf, 1995), although not many solid data support this view. In the SENECA study (Schlettwein-Gsell, Barclay, Osler, & Trichopoulou, 1991; Schlettwein-Gsell & Barclay, 1996), mean eating frequency of selected populations of elderly increased from 1988 to 1993 in the participating centers in The Netherlands, Italy, and Poland, but not in the center in Switzerland.

Figure 3–1 gives an overview of number of daily eating occasions from a sample of more than 3182 adults in the United States in 1987/88, excluding eating moments <70 kcal, from a recent paper of Longecker, Harper, & Kim (1997). From this figure it is clear that many citizens in the United States have three meals a day, and the average number of eating moments (including eating moments <70 kcal) was 3.42. In two other American studies in the beginning of the 1990s, average meal frequencies were around 3 (Edelstein, Barret-Connor, Wingard, & Cohn, 1992) and around 4.5 (Young & Wolf, 1990).

Table 3–1 gives an overview of frequency distributions of the number of eating or drinking moments per day for elderly in a number of different European countries in 1993 (Schlettwein-Gsell & Barclay, 1996). Eating/drinking moments in this table are defined from dietary records and have no lower limit with respect to energy intake. From this table it is clear that the number of eating moments per day varies roughly between 1 and 8, with an average well above 3. Meal frequencies are highest in Northwestern European countries, such as Denmark and The Netherlands, where people eat or drink about 6 times a day.

In populations of other age groups, an average number of about 6 eating occasions per day is also apparent in the United Kingdom (Gatenby, 1997), and 5 to 6

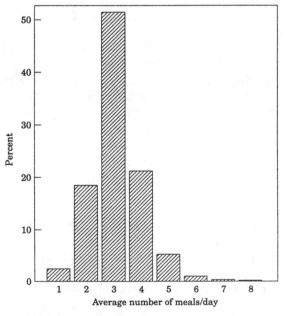

Figure 3–1 Distribution of eating frequencies from 3182 subjects from the Nationwide Food Consumption Survey in the United States 1987/1988. *Source:* Reprinted with permission from M.P. Longecker, J.M. Harper, and S. Kim, Eating Frequency in the Nationwide Food Consumption Survey (U.S.A.), 1987–1988, *Appetite*, Vol. 29, p. 57, © 1997, Academic Press.

Weight during the Day

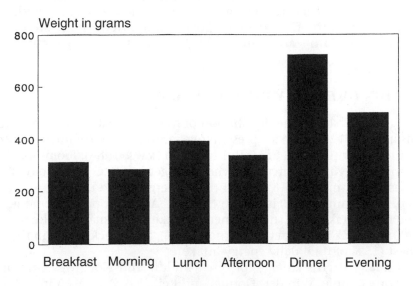

Eating Occasion

Figure 3–2 Distribution of energy intake across the day in a representative sample of Dutch people (*n* = 5898) in The Netherlands 1987/1988. *Source:* Reprinted with permission from Voedingscentrum, *Zo eet Nedrerland*, © 1998, Voedingscentrum, Den Haag, The Netherlands.

Table 3–1 Frequency Distributions of Number of Eating Moments across the Day in Elderly European Subjects in 1993 in %

Towns/ country	Roskilde/ Denmark	Haguenau/ France	Romans/ France	Padua/ Italy	Culemborg/ The Netherlands	Yverdon/ Switzerland	Marki/ Poland	Ballymoney, Limavady, Portstewart/ North Ireland
No. of subjects	110	106	136	138	118	144	119	62
No. eating occasions (including snacks)/day								
1	1	0	0	0	0	0	0	0
2	0	0	0	0	0	0	2	0
3	0	20	92	11	0	10	63	13
4	3	24	5	33	1	17	21	34
5	36	21	2	32	4	28	11	37
6	34	29	0	20	41	29	1	17
7	21	5	0	4	42	15	1	0
8	5	1	0	0	11	1	0	0
Mean frequency	5.8	4.7	3.1	4.7	6.5	5.3	3.5	4.6

Source: Reprinted with permission from D. Schlettwein-Gsell and D. Barclay, Dietary Habits and Attitudes, *European Journal of Clinical Nutrition*, Vol. 50, Supplement 2, pp. S56–S66, © 1996, Stockton Press.

eating occasions in Sweden (Anderson & Rossner, 1996). De Castro, Bellisle, Feunekes, Dalix, & deGraaf's (1997) cross-cultural comparison of meal patterns also showed greater eating frequencies in young adults in The Netherlands compared with France and the United States. Lower eating frequencies are found in Eastern and Southern European countries (e.g., Table 3–1 and Schlettwein-Gsell et al., 1991).

DISTRIBUTION OF ENERGY INTAKE ACROSS THE DAY

Figure 3–2 shows the distribution of food intake across the day in The Netherlands in 1987/1988 from a representative sample of approximately 6000 Dutch subjects older than 1 year of age (Ministerie van WVC, 1988). From this figure it is clear that the energy intake increases when it is getting later in the day. Breakfast is the smallest main meal (13% of total daily energy intake), lunch is the middle meal (23%), and dinner (34%) is the largest meal. A similar pattern emerges with respect to the food intake from snacks. Energy intake from snacks in the morning is lower (7%) than snack intake in the afternoon (9%), and snack intake in the evening is the highest (15% of total daily energy intake).

Similar patterns emerge for other Middle and Northwestern European countries. For example, Winkler, Doring, and Keil (1995) showed in a study in a south German town in 1984/1985 that breakfast, lunch, and dinner contributed about 17%, 29%, and 33% of total daily energy intake. The results of a recent cross-cultural study with a nonrepresentative sample of young adults suggest that a similar distribution of energy intake across the day applies for the United States (de Castro et al., 1997). Data from the southern United States (Louisiana) with children and adolescents are in line with this picture. Nicklas, O'Neil, and Berenson (1998) found that for children who consumed their breakfast at home, breakfast, lunch, and dinner contributed about 18%, 25%, and 32% of total daily energy intake.

From Table 3–2, which shows the distribution of energy intake across the day for various age groups in The Netherlands (Voedingscentrum, 1998), it is clear that the distribution of energy intake across the day in 1998 is similar to the distribution in 1987/1988 (Figure 3–2). This table also shows that this distribution is similar for the different age groups, although the contribution of snacks to total daily energy intake seems to decline with increasing age. Young children seem to have a rela-

Table 3–2 Contribution of Breakfast, Lunch, Dinner, and Snacks to Total Daily Energy Intake in The Netherlands, 1998, for Different Age Groups of Men and Women

	Breakfast	Lunch	Dinner	Snacks
Total group (n = 5985)	13	22	35	30
Men, 1–4 yr (n = 135)	18	20	29	32
Men, 13–16 yr (n = 137)	13	22	32	33
Men, 22–50 yr (n = 1252)	11	23	35	31
Men, >65 yr (n = 185)	14	23	35	27
Women, 1–4 yr (n = 119)	19	21	29	31
Women, 13–16 yr (n = 117)	14	23	32	31
Women, 22–50 yr (n = 1472)	12	21	37	30
Women, >65 yr (n = 236)	14	23	36	27

Source: Reprinted with permission from Voedingscentrum, Zo eet Nedrerland, © 1998, Voedingscentrum, Den Haag, The Netherlands

Table 3–3 Contributions of Different Eating Moments across the Day to Total Energy Intake in Elderly European Subjects in 1993 in %

Towns/ country	Roskilde/ Denmark	Haguenau/ France	Romans/ France	Padua/ Italy	Culemborg/ The Netherlands	Yverdon/ Switzerland	Marki/ Poland	Ballymoney, Limavady, Portstewart/ North Ireland
No. of subjects	110	106	136	138	118	144	119	62
Breakfast	19	20	18	13	15	19	30	21
During morning	3	2	1	2	9	2	4	3
Lunch	25	38	46	45	21	41	35	32
During afternoon	9	4	4	3	7	5	4	4
Dinner	35	30	30	36	33	30	26	29
During evening	8	5	1	1	14	3	1	12

Source: Reprinted with permission from D. Schlettwein-Gsell and D. Barclay, Dietary Habits and Attitudes, *European Journal of Clinical Nutrition,* Vol. 50, Supplement 2, pp. S56–S66, © 1996, Stockton Press.

tively higher breakfast intake, which may be because they usually do not eat and drink after dinner.

Table 3–3 shows that the distribution of energy intake across the day is similar for elderly subjects in Northwestern European countries in 1993, but different for elderly in Southern and Eastern European countries, where lunch is the main meal of the day. The data from de Castro et al. (1997) from France also suggest that lunch is the largest meal in France, whereas in the United States and The Netherlands it is dinner. A recent study from Finland suggests that in Finland, lunch is the largest meal (Roos & Prättälä, 1997).

The results in Table 3–3 show that the contribution of breakfast to total energy intake varies from 13% in Padua, Italy, to 30% in Marki, Poland. In all countries, breakfast is smaller than lunch. The main differences between the countries are in the contribution of lunch and dinner to total energy intake. The figures for the distribution of energy intakes across the day in the SENECA study in 1993 (Schlettwein-Gsell & Barclay, 1996) were similar to those obtained 5 years earlier (Schlettwein-Gsell, et al., 1991), suggesting that these meal patterns were stable during this follow-up period.

CONTRIBUTION OF MEALS AND SNACKS TO TOTAL DAILY ENERGY INTAKE

The data in Table 3–2 also show that on average in The Netherlands the main meals contribute about 70% to total energy intake, whereas 30% of energy intake comes from snacks. The contribution of snack intake to total energy intake was similar 5 years later in 1992 (Voorlichtingsbureau voor de Voeding, 1993) and also in 1998 (Voedingscentrum, 1998). The figure of 30% in The Netherlands is relatively high, although the results from the recent meal pattern study in Finnish adults suggested that approximately 45% of the energy intake was derived from foods besides the main meals (Roos & Prättälä, 1997) (Table 3–4). Data from Winkler et al. (1995) in a South German town in the mid 1980s suggested that snack intake contributed 21% to the total energy intake (Table 3–4).

Data for children and adolescents in the United Kingdom during the 1980s and 1990s suggest that snack intake contributed 23% to 31% of total energy intake, whereas snacks contributed only 17% of the total energy intake in the elderly (Gatenby, 1997; Summerbell, Moody, Shanks, Stock, & Geissler, 1995). Data from

Table 3–4 Contribution of Main Meals and Snacks to Total Energy Intake in Different Groups in Various Countries

Group (source)	Main Meals	Snacks
The Netherlands, 1998 (Voedingscentrum, 1998)		
Males, 16–19 yr (n = 142)	66	34
Females 16–19 yr (n = 139)	65	35
Males, 22–50 yr (n = 1252)	69	31
Females, 22–50 yr (n = 1472)	70	30
Males, > 65 yr (n = 185)	73	27
Females, > 65 yr (n = 236)	73	27
Finland, 1992 (Roos & Prättälä, 1997)		
Males, 25–64 yr (n = 870)	55	45[a]
Females, 25–64 yr (n = 991)	55	45[a]
England, 1970–1995 (Summerbell et al., 1995)		
Males, 13–14 yr, first half 1990s (n = 12)	71	29
Females, 13–14 yr, first half 90s (n = 21)	76	24
Males, 65–91 yr, 1970 (n = 24)	83	17
Females, 65–91 yr, 1970 (n = 64)	82	18
Scotland, first half 1990s (Ruxton Kirk & Belton, 1996)		
Children, 7–8 yr (n = 136)	74	26
Northern Ireland, second half 1980s (Robson et al, 1991)		
Adolescents, 12–15 yr (n = 1015)	69	31
Germany, 1984–1985 (Winkler et al, 1995)		
Males, 45–64 yr (n = 899)	79	21
United States (Bogalusa, LA) 1973–1988 (Nicklas et al., 1998)		
Children, adolescents, having breakfast at home (n = 193)	74	26
Children, adolescents, having breakfast at school (n = 200)	74	26
Children, adolescents, having no breakfast (n = 74)	69	31

[a] Including snacks and "other eating occasions" outside the main meals.

the 1992 food consumption survey in The Netherlands showed a similar age effect. The contribution of snack intake to total energy intake in teenage girls (13 to 19 years old) was approximately 35% to 37%, whereas in elderly Dutch women it was approximately 27% (Voorlichtingsbureau voor de Voeding, 1993). This age effect is exactly the same in the 1998 Dutch national food consumption survey (Table 3–5). It is unclear whether this age effect is linked to the suggested trend toward grazing.

The data from the SENECA study show large differences between European countries in snacks and meal intake. The contribution of snack intake to total energy intake in the elderly varied from 6% to 7% in towns in France and Italy to 30% in the participating center in The Netherlands (Schlettwein-Gsell & Barclay, 1996). The greater contribution of snack intake in the Northwestern European countries seems to be related to the greater frequency of eating moments in these countries.

MACRONUTRIENT COMPOSITION OF MEALS AND SNACKS

Not many data are available on the macronutrient composition of meals and snacks, although some recent papers throw more light on this issue (Gatenby, 1997; Navia et al., 1997; Nicklas et al., 1998; Roos & Prättälä, 1997; Summerbell, et al., 1995). Just as with all the data presented in this chapter, average values on nutrient composition of meals and snacks depend greatly on the food habits of particular groups. Snacks do not belong to a uniform group of foods with similar nutritional value, but snacks have a large variation in nutrient composition.

Table 3–5 Contribution to Macronutrients Energy Intake in % for Different Eating Occassions across the Day in The Netherlands, 1987/88.

	Fat[a]	Protein[a]	Carbohydrates[a]	Monosaccharides/disaccharides[b]	Alcohol
Breakfast	31	11	48	23	0
Snacks during morning	21	7	50	38	1
Lunch	40	14	41	14	0
Snacks during afternoon	20	5	57	43	4
Dinner	43	17	34	13	1
Snacks during evening	22	6	49	33	10

[a] Mean of individual values of contributions to energy intake.
[b] Calculated on the basis of mean intake in grams x 4 kcal divided by mean kcal intake.
Data were obtained from 3-day dietary records from a representative sample of 5898 subjects older than 1 yr.

Source: Reprinted with permission from Voedingscentrum, *Zo eet Nedrerland*, © 1998, Voedingscentrum, Den Haag, The Netherlands

Table 3–5 with data from The Netherlands in 1987/1988 shows that considerable differences exist in the macronutrient composition of the different eating occasions across the day (Ministerie van WVC, 1988). Breakfast is the main meal with a relatively high carbohydrate and low protein intake, whereas dinner is relatively high in fat, high in protein, and low in carbohydrates. Table 3–5 also shows that the snacks in general contain less fat, less protein, and more carbohydrates than the main meals. Monosaccharides and disaccharides have a high contribution to the energy intake of snacks, indicating that many snacks contain sugar and have a sweet taste. The profile of macronutrient composition of the various meals remained constant in the 1998 food consumption survey (Voedingscentrum, 1998).

Alcohol becomes more important as a source of energy later in the day, especially in the evening. From Table 3–5 it is also clear that few Dutch people drink alcoholic beverages with their dinner. Such figures will be different of course for southern European countries, where people habitually drink wine with either lunch or dinner (Hupkens, Knibbe, & Drop, 1993).

Recent data from the United Kingdom (Gatenby, 1997; Summerbell et al., 1995), Finland (Roos & Prättälä, 1997), Germany (Winkler et al., 1995), and the United States (Nicklas et al., 1998) suggest that in these countries, the distribution of macronutrients across the different meals and snacks is similar to The Netherlands. An analysis of Gatenby (1997) of some UK studies shows that on average the main meals contained more fat and protein, whereas the snacks contained more carbohydrates and less protein. Data from Nicklas et al. (1998) in the United States also suggest that breakfast and snacks in general contain relatively more carbohydrates and relatively less protein and fat than lunch and dinner. In line with this observation are data from Spain (Navia et al., 1997), which also showed that breakfast contains more carbohydrates but less protein and fat than either lunch or dinner.

MICRONUTRIENT COMPOSITION OF MEALS AND SNACKS

Sometimes it is thought that snacks have little nutritional value and only provide "empty" calories, without contributing much to the micronutrient intake. Although this may be true for some snacks, this is not the general rule, which emerges from actual data on the contribution of snacks to overall nutrient intake.

Data from the 1992 Dutch Food Consumption Survey show that whereas snacks contribute approximately 30% to energy intake, they contribute about 25%

to the calcium, iron, and phosphorus intake and approximately 15% to 30% to the intake of vitamin A, vitamin B_1, vitamin B_2, vitamin B_6, and vitamin C (Voorlicht-ingsbureau voor de Voeding, 1993).

The recent meal pattern study in more than 1700 Finish adults suggested that the vitamin C density/MJ of snacks was higher than the vitamin C density/MJ of main meals. The density of carotenoids in snacks was somewhat lower compared with the main meals. A comparison of nutrient intake of people with a conventional three-meal pattern and people with another meal pattern showed little differences in micronutrient intake between groups (Roos & Prättälä, 1997).

BREAKFAST

As is clear from the previous text and tables, breakfast is the meal with the relatively greatest carbohydrate content and the lowest protein and fat content. Although breakfast is the smallest main meal in most countries in the industrialized world, breakfast consumption has been identified as an important factor in nutritional well-being. Breakfast is also the meal that is sometimes skipped by people, and this skipping can have negative consequences for the overall nutritional quality of the diet. Skipping breakfast has been related to an impaired cognitive functioning in children (Pollitt & Mathews, 1998), an increased risk for obesity in children (Bellisle, Rolland-Cachera, Deheeger, & Guilloud-Bataille,1988), and an increased overall fat content in the diet (Macdiarmid, 1997).

Secular trends in the United States show that from the 1960s to the 1990s more children, especially adolescents, seem to skip breakfast (Siega-Riz, Popkin, & Carson, 1998). This trend is negative from a nutritional point of view. Providing breakfast at school might be an effective way of dealing with this issue (Nicklas et al., 1998).

VARIABILITY IN MEAL INTAKE

Energy and nutrient intake are variable between subjects and within subjects, across meals, and across days. A substantial amount of nutritional data is available on the variability of intake across 24 hours. Much less information is available on the variability of intake for each of the separate meals.

Between-subject variability represents the variability between the average daily energy/nutrient intakes of various persons of a particular group of subjects. It is clear that on average one subject ingests more energy and nutrients than the other. In many instances an average of 7 days of food intake is thought to represent a unbiased estimate for the true average daily energy and food intake (Tarasuk & Beaton, 1992a). From this perspective it is clear that the between-subject variability depends on the homogeneity of a study population. For example, the variability in intake within a group of small children and within a group of big young men will be smaller than when the variability is calculated across these two groups. However, a typical value for the between-subject coefficient of variation for the 24-hour energy intake of various groups of people is between 20% and 30 % (Nelson, Black, Morris, & Cole, 1989). Similar values are found for between-subject variability in 24-hour intake of proteins, fats, and carbohydrates, but values for between-subject variability of micronutrients can be considerably greater (Nelson et al., 1989).

It is clear that people do not eat the same every day, and considerable variation exists in energy and nutrient intake from one day to the next. One measure to characterize this variability is the within-subject variability for 24 hours, which represents the variability in energy/nutrient intake for a person from day to day, all other things being equal. Tarasuk and Beaton (1992a) showed that the mean of the within-subject coefficient of variation for 24-hour energy intake across 30 adult subjects from the famous 365-day Beltsville study was approximately 27%. With the data from the same study, Tarasuk and Beaton (1992b) showed that the day to day variability in energy intake is an individual characteristic. So, some people tend to eat the same every day whereas other people have a higher variability in 24-hour energy intake. An earlier study of de Boer et al. (1987) with 9-year-old boys from Finland, The Netherlands, Italy, the Philippines, and Ghana showed that mean within-subject coefficients of variation of 24-hour energy intake varied from 19% in The Netherlands to 24% in Italy. These figures are in line with data from Birch, Johnson, Andresen, Petersen, & Schulte (1991) and Pearcy and de Castro (1997) in young children. Mean within-subject coefficient of variations of 24-hour protein and carbohydrate intake are similar, but a greater variation exists in fat intake (de Boer et al., 1987).

Just as one can calculate the within-subject variability of energy/nutrient across days, one can also calculate the within-subject variability in energy/nutrient intake for the different meals. Is breakfast similar from day to day, or does a great variability exist in snack intake from day to day. Table 3–6 gives an impression of the average within-subject coefficient of variation for meals and snacks from 141 subjects who recorded their food intake for 6 to 9 separate days from four studies of Hulshof and colleagues (Hulshof, de Graaf, & Weststrate, 1995a,b; Hulshof, 1994).

Variability in energy/nutrient intake is much greater for snacks than for main meals, which is in line with the observation that main meals are consumed on a more regular basis than snacks. Variability across the meals does not change very much, although breakfast energy intake seems to be somewhat more stable than lunch and dinner intake.

The variability in carbohydrate, protein, and energy intake seems to be consistently lower than the variability in fat intake and much less than the variability in alcohol intake. For each eating occasion variability in fat intake is greater, showing that within persons fat intake is more variable from one day to the other than energy, protein, and carbohydrate intake. Variability in alcohol intake is high, showing that within this group of subjects, alcohol is not consumed on a strict daily basis.

From this table it is clear that the average variability is greater for separate meals than for total daily intake. This is mainly due to the greater mean of overall intake

Table 3–6 Mean of Within-Subject Coefficients of Variation in % (*n* = 141) for Energy and Macronutrient Intakes at Separate Eating Occasions During the Day. Subjects Recorded Their Intake During 6–9 Separate Days.

Energy/Nutrient	Eating Occasion						
	Breakfast	Morning	Lunch	Afternoon	Dinner	Evening	Total
Energy intake (MJ)	34	125	41	95	40	90	20
Carbohydrate	35	121	41	89	43	84	21
Protein	42	148	47	125	40	128	22
Fat	52	160	54	139	59	135	33
Alcohol	244	265	245	219	229	169	159

Source: Data from T. Hulshof, Fat and Non-Absorbable Fat and the Regulation of Food Intake, *PhD Thesis*, Wageningen Agricultural University, 1994.

compared with the mean intake of separate eating occasions and *not*, as is sometimes assumed (Birch et al., 1991; Pearcy & de Castro, 1997), due to compensatory responses in intake across the day.

HEALTH CONSEQUENCES OF DIFFERENT MEAL PATTERNS

All humans eat on a regular basis, and as a consequence, they will spend a considerable part of their life in a postprandial state. From that perspective it is obvious that meal patterns have a great effect on all kinds of nutritional parameters that are supposed to be related to health and chronic diseases. The recent increasing interest in the effect of meal patterns on human health is also apparent from a recent supplement to the *British Journal of Nutrition,* titled "Periodicity of Eating and Human Health"(Gibney, Wolever, Frayn, 1997). From this supplement it is clear that meal patterns have been suggested to be involved in various nutrition-related health issues, such as the development of obesity, dental caries, postprandial lipoprotein levels, diabetes, and physical and mental performance. The reader is referred to this supplement for an extensive treatment of these issues.

The relationship among meal patterns, energy balance, and obesity is not clear (Bellisle, McDevit, & Prentice, 1997). Cross-sectional correlational data suggested that a lower eating frequency was positively related to a higher prevalence of obesity (e.g., Fabry, Fodor, Hejl, Braun, & Zvolankova, 1964; Fabry et al., 1966). However, more recent data show that this inverse relationship is weak at most (Edelstein et al., 1992; Kant, 1995; Summerbell, Moody, Shanks, Stock, & Geissler, 1996). With respect to energy balance, well-controlled studies have shown that energy expenditure with a low eating frequency is similar to energy expenditure with a high eating frequency (e.g., Verboeket-van der Venne and Westerterp, 1991, 1993). So, the possible relationship between meal patterns and body weight must be mediated through possible effects on energy intake. However, until now no data have been available on the effect of meal frequency on spontaneous energy intake, so it is not possible to predict whether a change in meal frequency would be accompanied by a change in energy intake.

A high meal frequency has obvious advantages for the management of blood glucose levels as is necessary for subjects with type II diabetes. People with diabetes must take care not to have large fluctuations in blood glucose levels, and this is best achieved by having a high frequency of small meals (Jenkins, 1997). However, it not clear yet whether long-term adherence to a high-frequency meal pattern will ultimately result in better glucose tolerance (Jenkins, 1997). With respect to risk factors for cardiovascular diseases, such as increased total, and low-density lipoprotein cholesterol levels, the situation is similar as to the questions with respect to the prevalence of obesity and glucose tolerance. Some data suggest that a greater meal frequency is associated with beneficial effects on fasting lipoprotein levels, but the level of understanding and knowledge of effects precludes recommendations for the general public (Mann, 1997).

DISCUSSION

The data presented in this chapter show that meal patterns differ from country to country, confirming that the sociocultural environment and economic circumstances have drastic effects on these patterns. These different meal patterns result in differences in the distribution of nutrient intakes across the day. In Northwestern

European countries, and probably also in the United States, breakfast is the smallest main meal, and dinner is the largest main meal of the day. In Southern European countries, lunch is the most important meal of the day. In every country breakfast is smaller than lunch. Great differences also exist in average eating frequencies between countries, with the highest frequency in Northwestern European countries and lower eating frequencies in Southern European countries. The United States seems to take an intermediate position in this respect. The eating frequency is related to the relative contribution of snacks to total daily energy intake.

Not many data are available on this issue, but data from The Netherlands and other countries (Spain, United States) suggest that breakfast is the main meal, with a relatively high carbohydrate content and a relatively low fat and protein level, whereas the main (cooked) meal contains more fat and protein and less carbohydrates. Snacks do not form a uniform group with respect to nutrient content, but on average in The Netherlands, they have a relatively high carbohydrate content and a relative low fat content. Data from the United Kingdom, Germany, and the United States suggest a similar pattern. Snacks can also have a substantial contribution to the micronutrient intake of people.

The variability of total daily energy and macronutrient intake within individuals, as indicated by the average value of within-subject coefficient of variation, lies around 20% to 25% for energy, carbohydrates, and protein, and somewhat higher for fat. The mean within-subject coefficient of variation of energy and nutrient intake at separate meals (breakfast, lunch, dinner) is considerably greater, although breakfast energy intake seems somewhat more stable than lunch and dinner energy intake. The within-subject variability of energy intake from snacks is much greater than the variability in meal intake, confirming the regularity of the pattern of three main meals a day. The lower variability for total daily intake is a consequence of the higher mean for the total intake compared with the separate eating occasions.

Different meal patterns can have consequences on health, and several types of data suggest that a greater frequency of eating might have beneficial consequences for some nutrition-related chronic diseases. However, not enough reliable long-term data are available to allow for specific recommendations for the general public.

REFERENCES

Anderson, I., & Rossner, S. (1996). Meal patterns in obese and normal weight men: the 'Gustaf' study. *European Journal of Clinical Nutrition, 50,* 639–646.

Bellisle, F., McDevitt, R., & Prentice, A.M. (1997). Meal frequency and energy balance. *British Journal of Nutrition, 77* (Suppl. 1), S57–S70.

Bellisle. F., Rolland-Cachera, M.F., Deheeger, M., & Guilloud-Bataille, M. (1988). Obesity and food intake in children: evidence for a role of metabolic and/or behavioral daily rhythms. *Appetite, 11,* 111–118.

Birch, L.L., Johnson, S.L., Andresen, G., Petersen, J.C., & Schulte, M.C. (1991). The variability of young children's energy intake. *New England Journal of Medicine, 324,* 232–235.

de Boer, J.O., Knuiman, J.T., West, C.E., Burema, J., Räsänen, L., Scaccini, C., Villavieja, G.M., & Lokko, P. (1987). *Human Nutrition: Applied Nutrition, 41A,* 225–232.

de Castro, J.M. (1988). Physiological, environmental, and subjective determinants of food intake in humans: a meal pattern analysis. *Physiology and Behavior, 44,* 651–659.

de Castro, J.M., Bellisle, F., Feunekes, G.I.J., Dalix, A.M., & de Graaf, C. (1997). Culture and meal patterns: a comparison of the food intake of free-living American, Dutch, and French students. *Nutrition Research, 17,* 807–829.

de Groot, C.P.G.M., van Staveren, W.A., Dirren, H., & Hautvast, J.G.A.J. (Eds.) (1996). SENECA Nutrition and the elderly in Europe. *European Journal of Clinical Nutrition, 50,* (Suppl. 2), 1–126.

de Groot, C.P.G.M., van Staveren, W.A., & Hautvast, J.G.A.J. (Eds.). (1991). EURONUT-SENECA: Nutrition and the elderly in Europe. *European Journal of Clinical Nutrition, 45,* (Suppl. 3), 1–196.

de Wolf, E.M. (1995). Trend in nutrition: Grazing. *Netherlands Journal of Nutrition, 56,* 24–25.

Edelstein, S.L, Barret-Connor, E.L., Wingard, D.L., & Cohn, B.A. (1992). Increased meal frequency associated with decreased cholesterol concentrations; Rancho Bernardo, CA, 1984–1987. *American Journal of Clinical Nutrition, 55,* 664–669.

Fabry, P., Fodor, J., Hejl, Z., Braun, T., & Zvolankova, K. (1964). The frequency of meals: its relationship to overweight, hypercholesterolaemia, and glucose tolerance. *Lancet, 2,* 614–615.

Fabry, P., Hejda, S., Cerna, K., Osoncova, K., Pechor, J., & Zvolankova, K. (1966). Effects of meal frequency in schoolchildren: Changes in weight-height proportion and skinfold thickness. *American Journal of Clinical Nutrition, 18,* 358–361.

Gatenby, S.J. (1997). Eating frequency: Methodological and dietary aspect. *British Journal of Nutrition, 77 (Suppl. 1),* S7–S20.

Gibney, M.J., Wolever, T.M.S., Frayn, K.N. (Eds.). (1997). Periodicity of eating and human health. *British Journal of Nutrition, 77 (Suppl. 1),* S1–S129.

Hulshof, T. (1994). *Fat and non-absorbable fat and the regulation of food intake.* Unpublished doctoral dissertation, Wageningen Agricultural University, The Netherlands.

Hulshof, T., de Graaf, C., & Weststrate, J.A. (1995a). Short-term effects of high-fat and low-fat/high-SPE croissants on appetite and energy intake at three different deprivation periods. *Physiology and Behavior, 57,* 377–383.

Hulshof, T., de Graaf, C., & Weststrate, J.A. (1995b). Short-term satiating effect of the fat replacer sucrose polyester (spe) in man. *British Journal of Nutrition, 74,* 569–585.

Hupkens, C.L.H., Knibbe, R.A., & Drop, M.J. (1993). Alcohol consumption in countries of the European community: Uniformity and diversity in drinking patterns. *Addiction, 88,* 1391–1404.

Jenkins, D.J.A. (1997). Carbohydrate tolerance and food frequency. *British Journal of Nutrition, 77,* S71–S81.

Kant, A.K. (1995). Frequency of eating occasions and weight change in NHANES I Epidemiologic Follow-up Study. *International Journal of Obesity, 19,* 468–474.

Longecker, M.P., Harper J.M., & Kim, S. (1997). Eating frequency in the nationwide food consumption survey (U.S.A.), 1987–1988. *Appetite, 29,* 55–59.

Macdiarmid J.I. (1997). *Characterization and eating patterns of habitual high and low fat consumers.* Unpublished doctoral dissertation, Department of Psychology, University of Leeds, England.

Mann, J. (1977). Meal frequency and plasma lipids and lipoproteins. *British Journal of Clinical Nutrition, 77,* S83–S90.

Ministerie. van WVC. (1988). *Wat eet Nederland.* Den Haag, The Netherlands: Ministerie van WVC and Ministerie van Landbouw en Visserij.

Navia, B., Requejo, A.M., Ortega, R.M., Sobaler, A.M.P., et al. (1997). The relationship between breakfast and whole diet energy profiles in a group of preschool children. *Annals of Nutrition and Metabolism, 41,* 299–301.

Nelson, M., Black, A.E., Morris, J.A., & Cole, T.J. (1989). Between- and within-subject variation in nutrient intake from infancy to old age: Estimating the number of days required to rank dietary intakes with desired precision. *American Journal of Clinical Nutrition, 50,*155–167.

Nicklas, T.A., O'Neil, C.E., & Berenson, G.S. (1998). Nutrient contribution of breakfast, secular trends, and the role of ready-to-eat cereals: A review of data from the Bogalusa Heart Study. *American Journal of Clinical Nutrition, 67,* 757S–763S.

Pearcy, S.M., & de Castro, J.M. (1997). Food intake and meal patterns of one year old infants. *Appetite, 29,* 201–212.

Pollitt, E., & Mathews, R. (1998). Breakfast and cognition: An integrative summary. *American Journal of Clinical Nutrition, 67,* 804S–813S.

Robson, P.J., Strain, J.J., Cran, G.W., Savage, J.M., Primrose, E.D., & Boreham, C.A.P. (1991). Snacks energy and nutrient intakes of Northern Ireland adolescents. *Proceedings of the Nutrition Society, 50,* 180A.

Roos, E., & Prättälä, R. (1997). Meal pattern and nutrient intake among adult Finns. *Appetite, 29,* 11–24.

Ruxton, C.H.S., Kirk, T.R, & Belton, H.R. (1996). The contribution of specific dietary patterns to energy and nutrient intake in 7–8 year old Scottish schoolchildren III. Snacking habits. *Journal of Human Nutrition and Dietetics, 9,* 23–31.

Schlettwein-Gsell, D., & Barclay, D. (1996). Longitudinal changes in dietary habits and attitudes of elderly Europeans. *European Journal of Clinical Nutrition, 50 (Suppl. 2),* S56–S66.

Schlettwein-Gsell, D., Barclay, D., Osler, M., & Trichopoulou, A. (1991). Dietary habits and attitudes. *European Journal of Clinical Nutrition, 45 (Suppl. 3),* S83–S96.

Siega-Riz, A.M., Popkin, B.M., & Carson, T. (1998). Trends in breakfast consumption for children in the United States from 1965 to 1991. *American Journal of Clinical Nutrition, 67 (Suppl.),* 748S–756S.

Summerbell, C.D., Moody, R.C., Shanks, J., Stock, M.J., & Geissler, C. (1995). Sources of energy from meals versus snacks in 220 people in four age groups. *European Journal of Clinical Nutrition, 49,* 33–41.

Summerbell, C.D., Moody, R.C., Shanks, J., Stock, M.J., & Geissler, C. (1996). Relationship between feeding pattern and body mass index in 220 free-living people in four age groups. *European Journal of Clinical Nutrition, 50,* 513–519.

Tarasuk, V., & Beaton, G.H. (1992a) Statistical estimation of dietary parameters; implications of patterns in within-subject variation—a case study of sampling strategies. *American Journal of Clinical Nutrition 55,* 22–27.

Tasrasuk, V., & Beaton, G.H. (1992b). Day-to-day variation in energy and nutrient intake: Evidence of individuality in eating behaviour. *Appetite, 18,* 43–54.

Verboeket-van de Venne, W.P., & Westerterp, K.R. (1991). Influence of the feeding frequency on nutrient utilisation in man: Consequences for energy metabolism. *European Journal of Clinical Nutrition, 45,* 161–169.

Verboeket-van de Venne, W.P., & Westerterp, K.R. (1993). Frequency of feeding, weight reduction and energy metabolism. *International Journal of Obesity, 17,* 31–36.

Voedingscentrum. (1998). *Zo eet Nederland 1998.* Voedingscentrum, Den Haag: The Netherlands.

Voorlichtingsbureau voor de Voeding. (1993). *Zo eet Nederland in 1992.* Voorlichtingbureau voor de Voeding, Den Haag: The Netherlands.

Winkler, G., Doring, A., & Keil, U. (1995). Meal patterns in a Southern German population: Results from the WHO MONICA Augsburg Dietary Survey 1984/85. *Zeitschrift fur Ernährungswissenschaft, 34,* 2–9.

Young, T.B., & Wolf, D.A. (1990). Case-control study of proximal and distal colon cancer and diet in Wisconsin. *International Journal of Cancer, 46,* 832–838.

Part II

Biological Bases of the Meal

Chapter 4

Physiological Controls of Single Meals (Eating Episodes)

Harry R. Kissileff

OBJECTIVES

The main objective of this chapter is to describe how the nervous system controls eating behavior in humans. A subsidiary objective is to show how the basic biology underlying eating behavior can explain how social and cultural phenomena contribute to the control of eating in humans. Gaps in our knowledge and new directions for research are also described. Finally, this chapter has two goals for the food production industry and food providers. First, it will show that this information could be useful for guiding the development, marketing, and presentation of food products. Second, it will help food providers develop procedures for determining the optimal size for portions or servings, as well as enable those who give nutritional advice to optimize recommendations for types and amounts of foods to eat.

TERMINOLOGY AND BASIC PHYSIOLOGY

In this section terms that will be used throughout the chapter are defined.

Meals

Meals are occasions when a substantial amount of food is eaten in a short period (relative to a day). This is not a quantitative definition, and indeed attempts in animal studies (Clifton, 2000) to quantify what constitutes a meal rely on statistical procedures applied to the eating of single types of foods over several days. The time interval between successive acts of eating (i.e., licks, bites, chews, or swallows) that is used to separate meals is ultimately an arbitrary definition made by the experimenter, and consequently meals described by arbitrary criteria are not physiological units of eating. Other chapters in this book have dealt with definitions of the meal from their perspectives. From the physiological perspective "meal" is any "eating episode" as described in the next section.

Eating Episode

An eating episode is a more precise way to delineate the behavior whose underlying physiology we are trying to describe. An eating episode is any amount of eat-

The author was supported by the following grants while writing this paper: MH42206, DK26687, and DK53089. The author is grateful to Dr. Joseph R. Vasselli for critical and constructive comments on the manuscript.

ing in any time frame separated from other such episodes by unrelated behavior. It could range from a few bites to a banquet. Eating episodes can be quantified by amounts eaten per unit time, by punctuation from other activities associated with eating, and by periods of time when eating is occurring, contrasted with periods of time when it is not occurring. However, eating episodes are circumscribed, independent events, not a subset of a larger collections. From a physiological perspective, it is easier to study eating episodes when eating is the only event taking place and when only a single food is being consumed. Such episodes are exaggerations of events that take place whenever eating is done, but in typical situations where humans eat, the eating is often accompanied by other activities, such as conversing, reading, or entertainment by electronic media. It is not the case that eating episodes that take place in laboratories are unnatural. The laboratory is simply a convenient place to study what happens physiologically when eating takes place. Physiology must not only analyze larger eating episodes, such as banquets, into their component behavioral elements, such as chews and swallows, but must be able to generate explanations of what happens physiologically when the elements are assembled into a larger array, such as a banquet. *The reader should understand that the "meal" being analyzed in this chapter is the physiologist's "eating episode."*

Amount Consumed

The measurement of how much is eaten is the amount consumed. The amount consumed is ordinarily measured by weight or volume. Frequently, these weights and volumes are transformed into energy units by computation on the basis of manufacturers' specifications of nutritive content and energy density.

Food Attribute

A food attribute is a quality of a food that has the potential to have an independent effect on food intake, either in the course of a meal in which it is presented or in another course. Food attributes include sensed qualities (e.g., taste, temperature, texture); physical properties (whether the food is solid or liquid, viscosity); chemical components (e.g., additives either sensed or not, functional ingredients); or derived qualities, such as energy content, energy density, or macronutrient composition. Other chapters deal with sensory attributes (Chapter 5) and nutritional attributes (Chapter 3). Food attributes can also be social (as food value or food safety) and cultural. Other chapters deal with these attributes. This chapter deals with attributes that have an impact on physiology. A major problem in the investigation of controls of meals is the impact of food attributes on these controls.

Food Selection

Food selection refers to items chosen from an array.

Satiation/Satiety

Satiation and satiety have been used by students of ingestive behavior to describe a process of bringing a meal to a satisfying conclusion "satiation" and a state of inhibition "satiety." They have also been used to describe stages in the inhibition of a meal; the early stages are referred to as satiation, the later ones as satiety. They have also been used to separate mechanisms that terminate a meal from mechanisms that inhibit eating between meals (Blundell & Coscina, 1992). Consequently, the literature on this topic can be confusing. The use of these terms will be avoided here as much as possible because of their ambiguity and frequent lack of additional explanatory power. However, some statements about them cannot be completely ignored. For example, Smith has stated that "Unless we make a conscious choice to halt food intake, the physiological process that stops eating is unconscious. This process is called satiation"(Smith, 1998, p.3). Smith, therefore, believes that some automatic process brings meals to an end. However, although he uses the term "we" in his definition, it is clear from the context and from his other publications on the physiology of meals (see "History of the Meal as a Unit for the Physiological Study of Food Intake" and Smith, 1982) that he is talking about a meal of a single item in a rat. Meals in humans, as will be shown, are undoubtedly influenced by physiological processes analogous to those occurring in rats. However, these processes are more likely in humans to be part of the sum of signals that determine action at any given time than to be stimuli for reflexive inhibition of eating. Because humans typically eat meals composed of more than one item and often switch from one to another in the course of a meal (Hadigan, Kissileff, & Walsh, 1989), the decision to stop eating is probably a conscious one, but one that is more likely to be made because of the physiological effects of the food the longer one has been eating.

Hunger

Hunger is an ambiguous word, sometimes referring to the presumed physiological state that predisposes an individual to eat and sometimes referring to the report of a visceral feeling. Presumably the report has an underlying physiological basis, which is the presumed state.

Appetite

Although some authors distinguish appetite from hunger, in this chapter they are used interchangeably.

Palatability

Palatability is another ambiguous term. Sometimes it refers to a property of a food, whereas at other times it refers to a subject's report of how much the food is liked. Other chapters in this book deal with this concept more than this chapter. The author has previously proposed (Kissileff, 1990) that palatability is best understood in the framework of ingestive behavioral research in which an eater or drinker, a food or fluid, a measure of response (action by the person, measuring device), and an observer/interpreter, all contribute to the outcome. Palatability should therefore be qualified by whether the change in behavior is a result of a change in the food or fluid, or in the eater or drinker. For example, statements that "one food is more palatable than another" imply that a change in the food has been made and the eater's response to that change is a measure of the effect of the change on palatability. Thus palatability is partly intrinsic to the food. Likewise, the statement that "illness experienced after a food with a certain flavor is eaten makes that food less palatable" implies that palatability of the food can have a learned component.

Basic Physiology Needed To Understand Intake Control

Nerves Muscles and Synapses

The basic anatomical units from which behavioral function is built are nerves, muscles, and synapses. Skeletal muscles move the arms, hands, and fingers, by which means the individual places food in the mouth. Chewing also takes place by means of muscles in the cheeks that move the jaws up and down or side to side. Muscles consist of contractile tissue attached at two points to bones. When they contract, the bones move. Nerves are conductive tissues that send signals from one part of the body to another. They have two main parts, a cell body and long projections called axons. The signals are sent in the form of electrical impulses that travel either toward (sensory) or away from (motor) the central nervous system in an analogous way that signals travel in a telephone cable back and forth from a central switching station. Synapses are gaps between nerves and muscles or between two nerve cells into which a chemical substance, called a neurotransmitter, is secreted when the nerve impulse reaches the synapse. The transmitter is released in response to a nerve impulse. Afterwards, it is taken back into the presynaptic element for reuse. The neurotransmitter is an essential link in the signaling system because without it the neural signal would stop when it reached the end of the nerve cell.

Neurotransmitters and the Brain

The central nervous system of humans is enclosed by bony tissue (the cranium and spinal column). The brain receives impulses from the rest of the body (termed periphery) mainly through the spinal cord for parts of the body below the shoulders and through cranial nerves for the head. The cranial nerves are particularly important for meals because they carry signals from the tongue and olfactory bulb to the muscles of the jaw that control mouth movements. Neurotransmitters in the central nervous system are divided into numerous classes on the basis of their chemical structure. These transmitters can be modulated by pharmacological agents, and much of the pharmaceutical industry is built around the development of transmitter mimics and blockers of the activity of neurotransmitters involved in the control of eating behavior. Examples of drugs and transmitter systems that they affect are amphetamine, which affects the noradrenaline transmitter system, naltrexone, which blocks the endogenous opioid system, and fenfluramine, which prevents reuptake of serotonin back into the synapse from which it was released.

Reflexes and Conditioning

Reflexes are the basic elements of behavior. A reflex act is a fixed built-in connection between a stimulus and a response. A stimulus is an agent, act, or influence that produces a functional reaction in a receptor or irritable tissue. A receptor is a sensory nerve terminal that responds to stimuli. The term also refers to subcellular units that combine with chemical agents. An example of a reflex is the salivation that occurs when food stimuli contact the tongue. Another is the flexion of a limb in response to a noxious stimulus, such as a pinprick or hot object. Most reflexes that concern eating are visceral and are not under conscious control. However, the paradigm of the reflex has often been invoked to explain eating behavioral phenomena such as the starting and stopping of a meal (see "Initiation and Termination of Meals," later in this chapter). Even the amount consumed in a bout of drinking has been conceptualized as a reflex in explanations of changes in intake in response to taste stimuli that have been followed by pleasant or unpleasant consequences (Sclafani, 1997).

Conditioning (see also "Integration of Learned and Unlearned Cues") takes place when a neutral stimulus such as the sound of a tone is presented shortly before a stimulus (called unconditioned) is presented that elicits the reflexive response. The classic example is the bell (conditioned stimulus) presented before meat powder (unconditioned stimulus) that caused a dog to salivate in Pavlov's laboratory (Pavlov, 1927). After several pairings, salivation was elicited only when the bell was sounded without the meat powder. In studies of eating behavior, the amount consumed in a meal or in a fixed period of time (e.g., 30 minutes) is treated as the outcome measure of an unconditioned reflex response that terminates eating in response to (a) the postingestive effects of foods or fluids consumed and/or (b) specific manipulations such as infusion or injections of hormones or other chemical agents. In studies of eating employing conditioning paradigms, tastes and odors, which are paired with unconditioned stimuli that follow eating or drinking, are considered conditioned stimuli. A classic example is the so-called conditioned aversion paradigm (Nachman & Ashe, 1973; but see Grigson, Lyuboslavsky, Tanase, & Wheeler et al., 1999, for current status) in which a lithium chloride injection (unconditioned stimulus) is injected into the experimental subject after it has consumed a food or fluid with a distinctive taste (conditioned stimulus) or odor. The next time the animal encounters that odor in a food or fluid, its intake of that food or fluid is reduced (conditioned response), even without the lithium injection. Often another stimulus is presented without the postingestive stimuli in which case the stimuli are termed discriminative, the positive one (CS+) being associated with postingestive nutritive or aversive stimuli and the negative one (CS-) being associated with no postingestive effects or no nutritive postingestive effects (see Sclafani, Fanizza, & Azzara, 1999 for further discussion and examples). The use of the terminology of conditioning to explain these complex behavioral responses is heuristic. It should not be inferred that complex sequences of behavior that guide choices and control intakes are combined into a simple conditioned reflex or even a series of conditioned reflexes. The determinants of amounts consumed in these situations, and in human eating in general, remain to be explained.

Hormones

The body possesses another communication system besides the nervous system. A system of chemical messengers, called hormones, provides a mechanism for events that occur in one part of the body to be communicated to another. The system works in the following manner. When a stimulus for a particular hormone is received in the gland or organ where the receptor is located, by means of a complex sequence of chemical (enzymatically controlled) reactions, the hormone is secreted into the bloodstream, or in the case of some hormones, such as cholecystokinin (CCK), which is secreted by cells in the small intestine, both into the blood stream and into tissues surrounding it. When the hormone arrives at its target, receptors there take it up, and by another series of specific chemical reactions, translate the message into action. For example, the hormone epinephrine, secreted by the adrenal gland in response to stress, circulates in the bloodstream. When it arrives at the liver, it begins a chemical chain of reactions that results in the breakdown of glycogen (a stored form of carbohydrate) into simple sugar (glucose), which then circulates to other body tissues (e.g., muscles), where its breakdown (oxidation) provides energy for these tissues to perform their functions. Specific examples of hormones that contribute to the control of eating in meals will be provided later in the discussion of the liver and circulation.

Units of Anatomical Organization

The body is made up of cells, the smallest body components that can reproduce themselves; tissues, which are aggregates of cells performing the same function (e.g., muscle, bone, skin); and organs, complete functional units (e.g., stomach, pancreas, liver).

ORGANIZATION AND APPROACHES

After a brief overview of the history of the meal concept in physiology and the methods by which meals have been studied in humans, four approaches to the study of the control of meals will be described. These approaches are not unique to this problem, but they provide a convenient way of presenting the data. The approaches are ordered from basic to applied. The first approach is to carefully describe or catalog the behavior that is to be studied. *Description* usually takes the form of records of motor acts; goal-directed activities, such as searching for or preparing food; or measurement of the rate at which the amount consumed occurs or changes over the course of the meal. *Construct-driven studies* are based on description. They are attempts to explain the records of behavior by resorting to theoretical constructs, often believed to be processes taking place within the mind (as opposed to brain) of the individual. Hunger, satiety, and palatability are constructs based on observed behavior that are used to explain the occurrence or intensity of the behavior. Constructs usually lead to the next level of study, mechanisms. *Mechanisms and mechanistic studies* attempt to explain behavior by means of anatomically definable structures operating in predictable sequences. The issue of what variables should be measured (Kissileff, 1992) in response to manipulations is important and not completely resolved. Booth and Westrate have recommended that causal processes can only be measured by varying a particular source of the influence on that process, independently of other potential influences, and observing some effect on an aspect of ingestion or on some symbolic reference to the eater's actions toward foods (Booth & Weststrate, 1994). These actions usually take the form of verbal ratings. However, it is not clear whether the various aspects of ingestion will all respond the same way to any given source of influence. The final approach to the control of meals is the *applied*. Studies that use this approach are designed to measure food intake or verbal reports when appetite-controlling drugs are administered, to compare populations with potential disturbances in eating control mechanisms, or to evaluate a functional property of the food, such as reduction in nutrient density (Bell, Castellanos, Pelkman, Thorwart, & Rolls, 1998; Rolls, Pirraglia, Jones, & Peters, 1992).

History of the Meal as a Unit for the Physiological Study of Food Intake

Curt Richter was the first to note that rats ate their daily ration in "feeding periods," which have come to be known as meals. Brobeck (1955) introduced the field to the notion that daily intake (still in the rat) was the product of the number of meals and their size. In this seminal paper Brobeck introduced the notions that meal size and meal frequency (i.e., number of meals) are the outputs of systems that respond to appetite and satiety, specifically that larger meals and shorter intermeal intervals were the result of increased appetite, whereas smaller meals and longer intermeal intervals indicated decreased appetite and increased satiety. Meal frequency and intermeal interval are inversely related: the longer the intermeal interval, the lower the frequency. This notion led to much animal work in which the control of single meals was studied to identify the reflex pathways through which eating was facilitated and inhibited (Smith, 1999).

Work on eating in humans in which intakes were actually measured in the laboratory, in an attempt to determine the physiology underlying the control of intake, began with the development of the "feeding machine" in which human subjects ate a liquid diet by pressing a button to receive a 7-ml portion (Hashim & Van Itallie, 1964; 1965). Records of the eating pattern, which turned out to be meals, were kept but never fully analyzed. Studies of outpatients were soon conducted in several other laboratories and attempts were made to determine how individual meals were controlled (Jordan, Wieland, Zebley, Stellar, & Stunkard, 1966; Meyer & Pudel, 1972). Solid food–monitoring procedures were introduced to make the food in these meals more similar to what people were used to eating (Kissileff, Klingsberg, & Van Itallie, 1980; Stellar & Schrager, 1985).

Smith (1982) further elaborated on the meal as the functional unit of feeding and provided a rigorous working framework for the physiological study of the meal. His framework was built around mechanisms initiating and terminating the meal and stages of satiety. He asked some critical questions about how food comes to inhibit eating: Where do food stimuli act? Which food stimuli are adequate? What neural and endocrine mechanisms are activated, and what central mechanisms respond to and integrate the peripheral mechanisms? The meal as a unit of analysis has also received attention in the clinical literature, and Blundell and Coscina point out that biological regulations of appetite rest on the ability of the individual to monitor eating behavior or its consequence, body weight. Disorders of body weight regulation either of overweight and underweight are implicitly disturbances in eating. Consequently, "Behavioral regulation ultimately entails control of meal size" (Blundell & Coscina, 1992, p. 96). Nevertheless, Brobeck's rule that both meal size and meal number can be involved in changes of food intake still holds, and it is not necessarily true that meal size is the only variable that counts in the overall weight regulation picture.

Problems with Approaches

All approaches to the control of meals have certain problems in common. First, the choice of a food to measure immediately limits the ability of the investigator to generalize findings to all foods. This problem is particularly acute when only a single item is used for a meal. On the other hand, mechanisms must be developed out of simple elements, and then complexity should be added. After a mechanism is understood at its most basic level, the next step is to determine how that mechanism responds when additional components are added.

Changing One Food Attribute Often Changes Others

Second, when attributes of foods are varied in an attempt to determine the impact of these variations on food intake, it is impossible to conclude with certainty that the variation in that attribute is responsible for the effect. Changing nutrient density or composition will change other aspects of the food whose impact may be unknown or impossible to measure. For example, simply adding water changes not only the nutrient density but also the energy density and organoleptic properties (taste, smell, and aroma) and may change the rate of gastric emptying, absorption, and possibly metabolism. The addition of thickeners, flavor enhancers, and so on all have the same problem. Furthermore, when changes in the attribute change the amount consumed, the amount consumed, itself, results in a change in the stimulation provided by the food. Consequently, comparisons of intakes of foods with different attributes, at a single episode, or even repeatedly over the same 24-hour period, or some longer period, amount to little more than description. These descriptions, however, can be extremely valuable in providing observations on which to base mechanistic hypotheses that could lead to the development of a more general theory of intake control.

Composition of the Diet Does Not Determine Intake

A third major problem with the various approaches to the controls of meals is the erroneous perception that the composition of foods, per se, is a determinant of the amount of food consumed. Although I proposed in 1984 (Kissileff, 1984) that foods vary along a small number of (physiologically relevant) dimensions each of which makes a contribution to the satiating process, I did not mean to imply that the foods themselves or their components determined how much was eaten. The satiating efficiency of a food was never intended to be interpreted as a fixed property of the food. Rather it was a measure of the food's ability to reduce subsequent food intake under controlled conditions. I attempted to clarify this problem in 1985 (Kissileff, 1985) by pointing out that foods satisfy and reduce subsequent food intake by the effects they have on the sensory and motor apparatus that controls the acts of eating. Foods also influence subsequent food intake by processes involving learned controls (see, for example, Louis-Sylvestre et al., 1989), which have as yet received insufficient attention. It is, therefore, possible that the same amount of the same food can have different effects on different occasions, depending on the prior presentation of that food and the postingestive effects it has had (see "Integration of Learned and Unlearned Cues"). The evaluation of the satiating effectiveness of new foods must, therefore, use not only single exposures but repeated applications under different experimental conditions (i.e., with different test meals, different contexts, or different response measures). This problem can be solved by designing studies in which the postulated organs controlling intake are directly stimulated in the same manner because it is believed that foods stimulate them. Very few studies have stimulated the organs independently of the food, and those that have did not attempt to mimic food stimuli, per se. This research direction is therefore ripe for exploitation.

The Confounding Problem

The fourth major problem with approaches to the controls of meals is in the interpretation of mechanisms of effects when intake is the variable being used to evaluate the effect of a food component or manipulation of a compartment (such as the oral cavity, stomach, or intestine) on intake. The intake being used to measure the effect also contributes to the stimulation of the compartment. As a result, the manipulation and the measurement are confounded. I refer to this as the "confounding problem." It is impossible to prevent food stimuli during an intake test from affecting the very organs one is trying to manipulate, although mixing of the stimulation from the measurement and manipulation can be reduced by using control conditions (e.g., measuring an effect on food intake with and without a given taste, distention, or intestinal stimulation or with different gradations of the manipulation, such as different amounts of gastric distention, or concentrations of tastants). This is the analogous problem that plagued studies of metabolism before the use of radioactive tracers. For example, without a tracer it would be impossible to determine whether the rise in glucose that occurred when food was eaten was attributable to absorption of glucose from digested food or from its release from the liver as a result of hormonal stimulation. This problem could not be solved simply by varying the amount of starch in the food because the resulting hormonal effects could covary with amount of starch and with the rate of digestion and absorption. In animal studies the purported effect of an intestinal peptide released during eating on subsequent eating was evaluated using an animal in which the food eaten drains from the stomach, thereby never reaching the intestine, the site of stimulation of the hormone. It was believed that the effects of the intestinal peptide were being measured without additional confounding from the peptide's release during the meal (Gibbs, Young, & Smith, 1973). However, as described later ("Intestinal Stimulation"), this animal preparation may have eliminated an important part of the normal controls. One way to eliminate this problem completely in humans is to study the effects of manipulations by asking subjects to make verbal ratings of how much they would eat or how they feel during the manipulation without letting them eat or to let them simply taste food to simulate eating. This method has been applied with success (Melton, Kissileff, & Pi-Sunyer, 1992) (see "The Liver and Circulation" for more details).

METHODS OF STUDY

A variety of methods have been used to study the control of eating in humans. Previous reviews (Hetherington & Rolls, 1987; Kissileff, Guss, & Nolan, 1996) have detailed these methods, so the present review will merely provide a brief sketch to give the reader some idea how data using the various approaches has been obtained.

Measurement of Intake at Single Meals by Varying the Food

The most obvious way to study the amount consumed in a meal is simply to measure how much of a presented food or array of foods is eaten. Measuring the amount consumed is typically done by weighing the item or items to be eaten before and after they are presented to the subject. Additional fine grain measurements can be made with a weighing or volume recording device that monitors rate of consumption while it is occurring (Kissileff et al., 1980; Silverstone, Fincham, & Brydon, 1980; Jordan et al., 1966). Further details about measurements of intake in meals are given later ("Descriptive Studies"). Typical examples of this approach are the work of Rolls' (Rolls et al., 1992) and Blundell's (Burley & Blundell, 1991; Burley, Cotton, Weststrate, & Blundell, 1994) groups on the fat substitute olestra. We have referred to this method of evaluating food attributes as the concurrent evaluation procedure (Kissileff, 1985).This procedure has been used to describe the effects of (1) changes in food attributes such as nutrient or energy density, (2) changes in subject liking of the food (i.e., palatability), or (3) the addition or deletion of specific components such as fiber, fat, or carbohydrate. As mentioned earlier, simple measurement of food intake, in which the presumed independent variable is a food attribute, confounds the variable measured with the variable being manipulated (see section on "The Confounding Problem"), results in loss of control over the manipulation, and prevents the experimenter from getting at the mechanism by which the attribute affects intake. One positive aspect of this approach is that it is simple and is a reasonable place to start when asking whether a new food component may have any functional effect at all. But it is important to recognize that finding such an effect is only the beginning.

The Preloading Paradigm

One solution to the problem of evaluating the effects of food components on food intake in a meal is to present the item to be tested separately from the food or foods whose intake is measured. This can be done by designing a two-stage meal in which the test food item is presented first in a fixed quantity or against a control given on another eating occasion, and the food whose intake serves as the measured effect is presented second. The item presented first is called a "preload" (Kissileff, 1984) because previous animal studies using this paradigm physically administered the item to be tested by placing it into the experimental subject rather than having the individual consume it (see Kissileff et al., 1996 for historical details), and the second item or items is called the "test meal." This procedure has three advantages over concurrent evaluation for evaluating satiating effects of foods. First, a constant food (i.e., test meal) is used as the assay measure to evaluate the satiating effects of different types of test items (preloads). Therefore, the confounding problem that the item being tested is also the item being measured is eliminated. For example, if food 1 had sucrose added to it and food 2 was a control (nothing added), the subject might eat the same amount of each food in a concurrent evaluation test because they have a similar taste (Rolls, Hetherington, & Burley, 1988), but when the two foods were presented as preloads before an identical test meal, the subject might eat less after sucrose than after aspartame because the sucrose has metabolic effects that the aspartame does not. Conversely, if food 1 had fiber added to it and food 2 was a control (nothing added), the subject might eat less after food with the fiber in a concurrent evaluation test because it has an aversive texture, but when the two foods are presented as preloads before an identical test meal, the subject might eat the same amounts of each because the fiber fails to influence metabolic processes that control intake (see "The Liver and Circulation"). It cannot be concluded unequivocally from these types of studies that intake is controlled by the metabolic effects (see "The Liver and Circulation") rather than sensory effects of foods (see Chapter 5).

The second advantage is that the amount of preload being tested is controlled and therefore the same amount of stimulation of the intake-controlling receptors is provided for each preload. Thus, in the case of the fiber described previously, the fact that when left to eat what they want, subjects may choose to eat different amounts of two foods being tested results in differential stimulation and consequently a biased test. Furthermore, it is easier with the preloading manipulation to present different components in controlled amounts to evaluate their satiating effectiveness per unit preload given. The third advantage is that people typically eat more than a single food item in a meal. Therefore, a preload followed by a second course, even if the second course test meal is only one item, is a better approach than simply modifying a single course and measuring its intake.

An extension of the preloading principle is the administration of food or food components directly to a body compartment before or during a meal. This procedure has been used to probe mechanisms, mainly by placing food or food components in the stomach, intestine, and bloodstream (intravenous). In animal studies, foods or food components (such as simple sugars or fatty acids or amino acids) have also been placed in the brain and the portal vein, but these compartments are difficult, if not impossible, to reach in humans. As a result, our ability to carry out mechanistic studies in people is limited (see Kissileff et al., 1996 for further discussion of the administration of nutrients to different compartments, but also see "Mechanistic Studies and Mechanisms for Control of Eating in Meals"). A recent study using this approach (Cecil, Francis, & Read, 1998) has demonstrated that allowing the subject to eat a food (preload) normally actually provides better satiation, as measured by reduction in the intake of a subsequent meal, than either gastric or intestinal stimulation with the same amount of the same preload. Presumably, activation of all the normal sensory pathways in their proper sequence is needed for optimal satiating effects.

DESCRIPTIVE STUDIES

Descriptive studies form the basis for subsequent construct-driven and mechanistic studies. In this section two types of studies will be described: (1) studies of the rates of eating during the meal, and (2) studies in which the eating episode has been decomposed into microstructural entities, such as chews, and bites, or other motor movements. Jordan et al. (1966) were the first to describe the rate of eating. They found that subjects eating a liquid diet ate faster at the beginning of the meal and gradually slowed toward the end. These findings were replicated by Meyer and Pudel (1972), who showed in addition that obese subjects did not display this gradually slowing. They suggested that the slowing indicated a disturbance in satiety. Kissileff and colleagues followed up this hypothesis but developed a universal eating monitor that could be used with both solids and liquids (Kissileff et al., 1980). Their findings (summarized in Kissileff & Thornton, 1982) were that the slowing was not as apparent with solid foods as with liquids, that not all obese persons exhibited the failure to slow down, and that men were more likely to eat faster at the beginning of meals and slow down than women. Similar results (rapid eating at the beginning then slowing down) were also found when individual chews and bites were studied across the meal (Bellisle & Le Magnen, 1981; Stellar & Schrager, 1985). These microstructural variables were closely linked to both the physical and hedonic qualities of the food (Bellisle, Lucas, Amrani, & Le Magnen, 1984; Bobroff & Kissileff, 1986), as well as to deprivation. Further descriptive studies have used periodic interruptions in a meal in which subjects are asked to make ratings (Hill, Magson, & Blundell, 1984; Wentzlaff, Guss, & Kissileff, 1995). These ratings appear to depend on both amounts consumed and the quality of the food (Yeomans, 1998). The ratings also appear to be useful in discriminating mechanisms of drug action (Yeomans & Gray, 1997) and in identifying possible underlying mechanisms for overeating in patients with bulimia nervosa (Kissileff et al., 1996).

CONSTRUCT-DRIVEN STUDIES AND THEIR PROBLEMS

Construct-driven studies are those that were designed to explain some of the phenomena discovered from purely descriptive studies, although they were mainly based on existing findings in animals. Four construct-driven problems that are central to the physiological control of meals will be described: (1) the role of energy density and content, (2) signals that control the initiation and termination of eating, (3) the relationship of meals to metabolic controls of eating, (4) integration of learned and unlearned cues controlling eating. The reason that these are construct driven, as opposed to mechanistic, is that they do not need to specify organs and tissues, or if they do, the specification is not precise because foods provide the initial source of stimulation but the locus of the receptors is unknown. Examples of these sorts of problems will emerge as the constructs unfold. The analysis of the mechanisms underlying learned cues places this topic on the borderline between construct and mechanism.

Do Energy Content and Density Determine Amount Consumed?

The Concept of Homeostasis and Its Role in the Meal

Claude Bernard, a nineteenth-century physiologist, proposed that constancy of the internal environment (by which he meant the cellular environment in living tissues) is the essential condition for life. Walter Cannon termed this condition "homeostasis," the maintenance of relative constancy in the internal environment. To preserve this constancy energy intake must be equivalent to energy output, among other inputs and outputs. This physiological rule has generated the hypothesis that intake at each meal, which is the foundation of energy homeostasis, must also be equivalent to the energy used since the last meal or anticipated energy used until the next meal (Le Magnen, 1992). Booth (1994) points out that "this concept [homeostasis] is only scientifically useful so long as it is used to seek out the mechanisms by which regulation of the internal environment is achieved. Instead, a lot of research has been, and continues to be, directed at demonstrating or refuting the occurrence of good regulation" (Booth, 1994, p. 51). Perhaps part of the reason that this research continues without providing mechanistic information is that investigators think they are demonstrating effects of foods on food intake, and industrial concerns are eager to know whether their products could be useful in energy intake reduction strategies. Furthermore, it is easy to measure reductions in food intake when foods of a given energy content and density are eaten or given as a preload but difficult to know where in the body, if any place, the information on energy content is registered. However, as Booth has forcefully stated: "Satiety tests that do not elucidate the mechanism of action are useless because conclusions are limited to the conditions of testing which necessarily differ from those of clinical use" (Booth, 1994, p.58). It is therefore critical that investigators who serve industry and regulators and consumers who use this information be aware of the limitations of purely construct-driven as opposed to mechanistically designed experiments. Conclusions from an experiment that demonstrates that a particular food or food component affects subsequent or concurrent food intake through a mechanism (see "Mechanistic Studies and Mechanisms for Control of Eating Meals," later in this chapter) can be generalized to any situation in which that mechanism is presumed to be operating, whereas conclusions from a study that simply generates another phenomenon will be limited to the conditions under which the study was conducted.

Another problem with this approach is that changing the energy content or density of foods always involves changes in some other attribute or attributes, such as water content, organoleptic properties, or macronutrient content. Separating the attribute of energy from these other variables is virtually impossible, although attempts have been made to demonstrate that the effects of these variables is negligible when sensory tests are made. Nevertheless, a number of attempts have been made to test the hypothesis that the size of a meal (in energy units) is conserved (i.e., maintained constant under constant conditions) by means of feedback signals that control eating in relation to energy being consumed. Probably an equal number (although it would be difficult to track down all the studies) indicate that individuals simply consume a constant weight or volume of food regardless of energy density. Examples of each will be given, but the conclusion from all this work must ultimately be that either the theory is totally useless, or the conditions under which the predictions come true have not been adequately specified. It is likely that the conditions have been inadequately specified because on theoretical grounds it is impossible for the energy in food to have a significant impact on further consumption because the meal typically ends within 5 to 30 minutes, long before the food eaten is even absorbed, let alone metabolized. No theory of the energy homeostasis has addressed the mechanism by which energy in the food consumed activates the controls of eating, except the conditioning theory of Booth (1972, 1977, 1994), which has not been fully tested (see "Integration of Learned and Unlearned Cues").

Evidence For and Against Homeostatic Control of Meal Size

Booth and Jarman were the first to provide convincing evidence that the amount consumed in a meal was related to metabolic processes that result in maintaining energy intake constant (under constant conditions). Their basic procedure was to place 0.5 to 1.5 g of starch or glucose (both are carbohydrates) solutions into rats' stomachs. They found that, after the carbohydrates were completely absorbed from the intestine, food intake was depressed in proportion to the energy provided. Because the stomach was empty, the effect could not be attributable to gastric distention or any osmotic effect of the solutes (the movement of water, in this case from fluid bathing stomach tissues into the stomach, across a membrane from a compartment where the solute concentration is lower to the one where it is higher). Furthermore, 3-O-methyl glucose, a nonmetabolized carbohydrate, was ineffective in reducing intake compared with saline. Infusions into the duodenum and hepatic portal vein likewise reduced intake, further demonstrating that the intake reductions were not attributable to gastric distention. Although the stimulus for reduction in intake was presumably metabolic, neither the site nor the metabolic product responsible for the effect was identified. Similar studies in which humans consumed starch or glucose have yielded similar intake-reducing results, thereby demonstrating putative metabolic control over meal size (Booth, 1981).

Further evidence for the maintenance of a constancy of energy intake in meals comes from studies in which the energy density of single-item or multiple-item meals has been altered. The first of these types of studies used volunteers living on a metabolic ward who obtained all their food by pressing a button for each 7-ml portion. All subjects maintained weight and increased their intakes (Campbell, Hashim, & Van Itallie, 1971) when the diet was diluted. These results were confirmed and extended by Spiegel (1973) in free living subjects. She found that when subjects' liquid formula diet was diluted by 50% from 1 kcal/ml to 0.5 kcal/ml, half the subjects maintained their energy intake on the dilute diet, whereas the other half reduced energy intake by the amount of the dilution. That is, the regulators ate on average 87% of what they were eating before dilution, whereas the nonregulators ate only 50%. However, this response required several days to occur and was a result of a combination of increase in meal size and meal number. This work has been extended to multiple items in a paradigm in which subjects' meals were scheduled (Porikos, Hesser, & Van Itallie, 1982) or when food was freely available (Foltin, Fischman, Emurian, & Rachlinski, 1988). Subjects in the Porikos studies ate about 85% of baseline when diluted foods and drinks were substituted for normal. In this case the dilution was only about 25%. In the studies of the Foltin group, subjects compensated perfectly, eating 100% of the baseline when the foods were diluted. Here the average dilution was not reported. However, in all these studies with multiple-item meals, it is really impossible to specify the amount of dilution because that depends on the energy densities of items eaten and these could be different during periods when diluted foods were offered than when full energy was offered.

On the other hand, extensive recent work of Rolls and colleagues (e.g., Bell et al., 1998) showed that individuals consumed a constant weight of food regardless of energy content. Thus, when people were given meals containing several items in which the overall energy density ranged from .8 to 1.3 kcal/g, there was no corresponding change in energy intake. Mean daily intakes ranged from 1350 to 1334 g/day, with a corresponding increase in energy. The amounts consumed in individual meals (three plus a snack) were approximately equal and about 25% of the total daily intake. One possible reason for these discrepant findings is that the response to energy dilution may require multiple exposures in order to change (Louis-Sylvestre et al., 1989).

Initiation and Termination of Meals

Many studies have been driven by the idea that a particular type of physiological manipulation affects the initiation or termination of a meal. Such studies are examples of studies driven by constructs (the start and termination of meals, which are extrapolated from descriptive information regarding the organization of eating over time). The idea that food intake is controlled by signals that initiate and terminate meals is pervasive but not axiomatic. Many studies have been designed with the hypothesis that they are studying signals that initiate and terminate meals. However, no evidence exists that any particular signal or set of signals initiates and terminates meals. Typically, what is measured is not when a meal stops but how much is eaten, and it is then inferred from the amount eaten that a manipulation has facilitated or inhibited a signal for meal termination. This is not direct evidence. One would have to observe the influence of a signal on stopping of eating to demonstrate that it affected such a mechanism. One study in which this was done, with an intestinal infusion, provided a demonstration that the start of the infusion acted as a switch to terminate eating, and stopping the infusion acted to release it from inhibition (Liebling, Eisner, Gibbs, & Smith, 1975). In another study, procaine (an anesthetic) was injected into the brains of rats, and they stopped eating temporarily (Epstein, 1960), but such direct studies of the mechanisms that start and stop eating are the exception. Electrical brain stimulation has likewise been shown to trigger eating as long as the stimulation lasts (Miller, 1957; see also Grossman, 1967, pp 366–68, for other examples and references).

Measures of meal size, in relation to manipulations that increase or decrease it, do not in themselves provide evidence that these manipulations tap into a system controlling the termination of meals any more than signals that bring an individual to begin eating tap into a system controlling the onset of meals. The most parsimonious framework for these types of studies is the hypothesis that there are facilitatory and inhibitory influences for actions of all sorts, including eating. When the sum of the facilitatory influences exceeds the sum of the inhibitory influences, eating will take place, and it will stop when this inhibition matches facilitation. Measurements of meal sizes and intermeal intervals in response to various stimuli simply tell how strong these influences are on the facilitation and inhibition. They also yield practical information that could be useful in constructing foods or diet plans. However, from the standpoint of mechanism, physiological stimuli from internal sources, such as changes in fuel oxidation rate or gastric distention, are integrated with other signals, such as concern about one's next appointment or whether one is at a party, to determine a final common decision to eat or to stop eating. Rats may not have to deal with such problems, but lack of concern about the animal's natural eating has led to a situation in which the physiological controls that have been studied have not been integrated with the rest of the animal's behavioral repertoire as they should be. No evidence exists to reject the hypothesis that effects that appear to be specific to meal size or meal frequency are not simply examples of different dose-effect responses on facilitatory and inhibitory systems. Thus, a manipulation that increases meal size at some dose may at a higher or lower dose affect meal frequency as well.

Attempts To Link Meals to Long-Term Control of Energy Balance

Because a meal is often completed long before the nutrients are metabolized and available as a source of energy yet the body maintains a relative constancy of energy intake and output, some investigations have proposed that the link between energy input and output is some signal or combination of signals generated during eating that inhibits eating and ultimately brings the meal to an end. Others have ignored that link and focused mainly on the mechanisms that inhibit eating. A third approach is from a purely empirical standpoint, in which the question is asked whether particular nutrients when served as part of a meal influence intake of the meal, thus completely sidestepping the issue of mechanism as has been described earlier (Rolls et al., 1992). A fourth group of investigators, and probably most authors of this book, assumes that meals stop for totally nonphysiological reasons, such as disappearance of food from the plate, the need to engage in other activities, social regulation and conformity, or just being "tired of eating." In these cases, regulation of energy balance is presumably maintained either by timing of the next eating event or by changes in metabolic efficiency. Collier, for example, believes that the system controlling intake is driven by economic considerations related to a kind of cost-benefit strategy, whereas metabolism adapts to whatever is consumed (Collier, 1986). Pliner and Rozin have argued similarly that meals are simply opportunities to eat the most in the shortest period of time that food is available.

Integration of Learned and Unlearned Cues

An important question for understanding the physiology of meals is the role of learned behavior. Learning may be limited in animals, especially in laboratories where the stimuli are limited. However, in the real animal world and in humans, the amount consumed in a meal may be strongly conditioned by prior experiences with the same foods.

Conditioned Satiety

Booth (see, for example, Booth, 1994; Booth, Gibson, Toase, & Freeman, 1994) has developed a scheme by which the amount consumed in a meal is determined by a complex stimulus consisting of the combination of visceral state and sensory cues. He has shown in rats (Booth, 1972) and to a limited extent in humans (Booth, Lee, & McAleavy, 1976) that the amount consumed is changed by exposing the individual to a specific odor in a concentrated starch solution an animal drinks on one day, whereas another odor is given in a dilute solution on a separate day. After several such trials with each odor and concentration, an intermediate concentration is given and the animals (and people) consume more after the flavor that had been placed in the dilute starch than in the concentrated. It is not clear how this effect takes place, but Sclafani (1990) has extended this work and shown that the stimulus is beyond the stomach, not in the mouth. His rats received gastric infusions of water after they consumed one flavor and gastric infusions of starch after they consumed another. After several trials, they were permitted to drink from separate tubes containing each flavor without a gastric infusion. Rats drank less of the flavor that had been paired with water. When the experiment was repeated, but the starch infusion was given along with an agent that prevented the absorption of starch, the difference in choice disappeared. These experiments provide a rudimentary framework for the way in which food choices and ultimately intakes in meals could be explained. Foods that provide a certain postingestive stimulation, whether it be a metabolic, humoral, or neural stimulation, influence, by means of conditioning, the response of the individual to those foods on a subsequent occasion.

Linking Cultural and Physiological Controls

This work is still at the conceptual stage because we do not know how the stimuli and responses interact with the neural systems that facilitate and inhibit eating despite the fact that activation of brain areas can be measured in response to these sorts of conditioned effects (Houpt, Philopena, Joh, & Smith, 1996). A third type of study suggests that internal stimuli associated with the consequences of ingestion can serve as conditioned stimuli. Davidson (1998) calls these occasion setters and proposes that they modulate signals that facilitate and inhibit eating. This framework has great promise for providing a mechanistic basis, linking both internal and external stimuli with controls of eating behavior. It is conceivable that the context of a banquet, for example, allows gastral distension or its signal to go beyond levels that would terminate an ordinary meal.. It is possible that individuals whose eating is normally restrained may have different physiological responses in the presence of disinhibitors than individuals whose eating is not normally restrained. Thus, the idea that cognitive and associated sensory, as well as internal, stimuli could serve as occasion setters to modulate facilitation and inhibition of eating could explain how conditioning links cultural controls to physiological ones.

MECHANISTIC STUDIES AND MECHANISMS FOR CONTROL OF EATING MEALS

Sequence of Anatomical Sites Affecting Eating and Their Relation to the Mechanism of Eating

Initiation of Eating and Stimulation of the Mouth

The act of eating in humans probably begins when visceral signals (indicating a reduction in the availability of metabolic fuels or specific nutrients and/or memories of prior eating occasions) are coupled with a relatively empty gastrointestinal tract and visual stimuli indicating the presence of consumable food. Evidence is not available from either brain scans or electroencephalography indicating the sequence of neural events with which a meal commences, and animal studies must be relied on to surmise what may be taking place in the human brain. The higher centers of the cortex probably make the decision to initiate the reflexes involved in placing food into the mouth. In people with a normal operating system (i.e., neural and muscular elements intact), the commonplace act of eating is usually taken for granted until something goes wrong with the operating system, such as a stroke or accident, which prevents the normal reflex pathways from being facilitated. The sequence of motor acts by which food is placed in the mouth, whether directly by hand or by use of utensils, has not been studied in relation to the visceral stimuli that evoke feelings, expressed as verbalizations, such as "I'm hungry" or "I'd like something to eat." How visceral and cognitive stimuli interact with the development of these systems in humans would be a tremendous help in further understanding the transition from reflex to voluntary control of eating. Studies of animals with brain damage have shown that normal eating patterns can be severely disrupted by lesions to areas that are believed responsible for both motivational and motor control (Teitelbaum & Epstein, 1962) The fact that animals can use alternatives when the normal pathways are blocked suggests the control of eating even in the rat, unless the cortex is severed from the hindbrain (Grill & Norgren, 1978), is not simply an automatic reflexive act but is controlled by higher brain centers operating by stimulating and inhibiting centers that coordinate the reflexes involved in swallowing once food is in the mouth.

From Mouth to Bloodstream

Chewing, salivation, and swallowing, which take place in the mouth, pass the bolus of food through the esophagus to the stomach. As food leaves the esophagus and enters the stomach, a reflex, termed "receptive relaxation" (Cannon & Lieb, 1911), occurs that allows the stomach to distend. The presence of food in the stomach also facilitates further relaxation, thereby preventing the buildup of pressure in the stomach. Food also is broken down by chemical processes in the stomach, although this processing is limited to what can be accomplished by acid hydrolysis. This digested food is then known as chyme, which is emptied at approximately 2 kcal/min into the duodenum (Hunt & Knox, 1968). The stomach contains receptors that detect its own stretch, but no solid evidence exists that it can detect different nutrients (Philips & Powley, 1996). Apparently, previous investigators (Deutsch, 1983) believed that the stomach contained chemoreceptors because they ligated the duodenum so far from the stomach that the nutrients were detected by receptors in the duodenum, and not by receptors in the stomach. Philips and Powley eliminated that problem by use of a more anteriorly placed ligation. When the stomach can no longer relax to accommodate additional food, gastric pressure begins to rise. This pressure rise could serve as a signal to inhibit further eating, but no conclusive evidence exists that it does.

From Duodenum to Brain

After passing through the duodenum, the chyme reaches the small intestine where it is digested by enzymes secreted from the pancreas, and the fats in the chyme are emulsified by secretions from the liver. Although the chyme is in the intestine, some hormones that have been shown to affect food intake are released (see "Intestinal Stimulation," "Gastric-Intestinal Interaction," and "The Liver and Circulation"). These include the gut peptides CCK, glucagon, gastrin releasing peptide (GRP), glucagon-like peptide-1 (GLP-1), and the pancreatic hormones, insulin and amylin. Digested chyme is absorbed into the portal vein (amino acids and carbohydrates) or into the lymphatic system (emulsified fatty acids). Nutrients in the portal vein reach the liver, whence they travel to the vena cava, heart, general circulation, and eventually reach the brain. In the brain, nutrients can result in changes in the concentrations of neurotransmitters (Wurtman & Wurtman, 1995). The time required for all these processes is variable, but the major delay is in the delivery of chyme to the small intestine. Consequently, the participation of nutrients to the control of ingestion is relatively limited during a meal that may last between 5 minutes and an hour or two, depending on the circumstances. Finally, nutrients reaching the bloodstream and stimuli from visceral organs provide further stimulation to the brain, causing levels of neurotransmitters to change. Some of these neurotransmitter changes have also been linked to inhibition of eating (see "The Brain").

Metabolism

The final result of digestion and absorption of chyme is delivery of the digestion products to the tissues, which need them for growth and activity. In the tissues, a process called metabolism converts these products into energy that can then be used to power the muscles that move limbs and are responsible for all bodily movements. Metabolism is a process that takes place within cells, and the conversion of glucose and fatty acids into energy takes place in a specific cellular structure called the mitochondrion. The energy generated from the breakdown of glucose and fatty acids is captured by the mitochondrion in a cycle involving chemical compounds called adenosine phosphates, often referred to as the body's energy currency. When energy is used to power a cellular process such as the contraction of a muscle, the triphosphate form (ATP) is converted to diphosphate (ADP), and when energy is generated from the breakdown of glucose and fatty acids, the diphosphate is converted back to triphosphate. From the standpoint of food intake control, the simple forms of the major macronutrients (i.e., products of digestion = fats, proteins, and carbohydrates) have all been shown to have effects on food intake when administered independently (see Kissileff & Van Itallie, 1982 and Kissileff et al., 1996). The cellular mechanisms responsible for these effects must ultimately involve the chemical processes that produce energy. Consequently, these energy precursors are called "metabolic fuels" in analogy with a machine that uses fuel to power it. Sufficient evidence exists that these processes, which take place in the liver where they might be sensed (Friedman, 1990), as well as in other tissues, can have rapid effects on food intake in humans, and this may provide the link between the body's need for energy and the control of food intake (Kahler, Zimmermann, & Langhans, 1999) (see "The Liver and Circulation"). The final step in the pathway of the food from mouth to bodily tissues is the exhaustion of the fuel supply. The mechanism for this important step in the initiation of another meal is treated in the section "The Liver and Circulation" as well.

Framework for Presentation of Mechanisms

Unfortunately, none of the construct-driven or mechanistic approaches has satisfactorily explained, at the level of neuromuscular controls, what stimuli in what magnitude actually guide eating to a particular goal intake. Instead, we can only cite evidence that measured application of stimulation to a given organ can reduce or enhance intake (in relation to some standardized condition) at meals in humans. Therefore experiments will be described that show that stimuli from five sources can modulate intake: the oropharynx, the stomach, the intestine, the liver and the circulation (by means of metabolic controls), and the brain. Where the evidence allows it, how some of these sources are integrated will be shown.

Oral Stimulation

Changes in the tastes and flavors of foods have long been known to change intake (Drewnowski, 1997), but little information is available on the relationship between specific aspects of these changes and changes in intake, and even less information is available on the mechanism by which these effects occur. One approach to mechanisms of control has been to compare intake of the tested food with verbal ratings of liking in a brief exposure taste test before the food was eaten in the meal (Bobroff & Kissileff, 1986). Only a poor relationship existed between ratings of liking and intakes across subjects ($r^2 = .34$), but a strong relationship was found between intake and ratings of liking within subjects ($r^2 = .68$). Within subjects, intake changed by 100 g/unit change in liking on a 9-point scale. A unique feature of this study is that the food (a yogurt shake) was adulterated with a minute amount of cumin (an inappropriate spice for yogurt among Americans). Similar types of studies have been done with sugars and salt with similar results (poor relationship between intake and hedonic ratings between subjects) (Hellemann & Tuorila, 1991; Lucas & Bellisle, 1987; Perez, Dalix, Guy-Grand, & Bellisle, 1994). However, with sugars and salt, osmotic and energy differences, as well as taste differences, could affect the intakes. Thus, although we know that a pleasurable meal will be consumed in greater quantity than an unpleasant one (and indeed this circularity often defines whether a meal is pleasant or not), the relationship between liking and amount consumed has not been quantified sufficiently for a mechanism to be tested. Two possible mechanisms have been proposed, but they have not been satisfactorily tested. One mechanism is that increased palatability leads to greater insulin release, which in turn stimulates more eating (see following), and the other is that increased palatability leads to increased oxidation, which also increases intake.

Gastric Stimulation

The work of Geliebter and associates (Geliebter, Westreich, & Gage, 1988) is the only investigation of the role of pure gastric stimulation, unaccompanied by nutrients, on food intake. Although only eight subjects (four obese and four not obese, with two of each sex in each group) were studied, the overall results suggested that there is a threshold at 400 ml of distention by the water-filled balloon; between 400 and 600 ml intake was reduced by approximately .5 ml for each milliliter of distention (50% efficiency on a volume basis, in non-obese subjects). Above 600 ml of distention no additional reduction in intake was found. These results are consistent with gastric distention contributing to inhibition of eating, but they do not support the hypothesis that a critical level of gastric distention provides a signal for satiety as proposed or implied by previous investigators (Deutsch, Young, & Kalogeris, 1978; McHugh & Moran, 1985). Studies in which the pylorus was reversibly blocked by placing a noose around it, in rats, are consistent with the results of filling the stomach with a balloon, but these studies are less quantitative because intakes were used as the measured variable and served as the manipulation (i.e., the confounding problem again) (Davis, Smith, & Sayler, 1997, 1998; Philips & Powley, 1996; Rauhofer, Smith, & Gibbs, 1993; Seeley, Kaplan, & Grill, 1995). These types of studies have the advantage that the stomach is filled naturally by consumption because the blockade prevents the stomach from emptying and thereby creates distention. This type of study has not been conducted in humans. Studies in which intake was not directly monitored, but feelings of hunger and fullness were reported, are consistent with the findings of Geliebter in that only partial satiety was felt with gastric stimulation alone, but meal-like satiety was experienced when nutrients were infused into the intestine while the stomach was distended (Feinle, Grundy, & Read, 1997; Read, French, & Cunningham, 1994). According to Read and colleagues, distention alone induces an uncomfortable feeling of gastric pressure, not the pleasant feeling of satisfaction normally experienced after a meal (Read et al., 1994; Khan & Reed, 1992). However, it should be noted that the reports of the Read group are based on air-filled balloons, whereas those of Geliebter and colleagues, in which discomfort is not reported except at extreme levels, are based on water-filled balloons.

Intestinal Stimulation

The pioneering work of N. W. Read has shown that infusion of nutrients into the duodenum can reduce food intake in humans. Both carbohydrates (Lavin & Read, 1995) and fats (Welch, Saunders, & Read, 1985; Welch, Sepple, & Read, 1988), as well as a mixed food (soup) (Cecil et al., 1998), are effective. Corresponding infusions of fats and carbohydrates into the circulation are relatively ineffective on single meals, hence the stimulus for the effect originates in the intestine. The next step in the process of uncovering the mechanism is to determine whether the intestinal infusion affects intake by a neural pathway from the intestine to the brain, by the release of a hormone, or both. Work with animal preparations has demonstrated convincingly that intestinal chemoreceptors working by both neural pathways and hormones have powerful inhibitory effects on food intake (see Greenberg, 1998 for an extensive review). Work with humans has been somewhat more limited, but it is clear that at least two peptides, CCK (Kissileff, Pi-Sunyer, Thornton, & Smith, 1981), glucagon (Geary, Kissileff, Pi-Sunyer, & Hinton, 1992), and possibly a third, GRP as shown by bombesin infusion (Muurahainen, Kissileff, & Pi-Sunyer, 1993), are capable of reducing food intake in humans when infused intravenously. The effects of CCK are reversible when a CCK-a (a receptor subtype of CCK found in the gut) antagonist is also infused. It is likely that some of the effects of nutrient infusion are mediated by CCK-a receptors because food intake reduction is attenuated (Lieverse, Jansen, Masclee, Rovati, & Lamers, 1994), and the feeling of fullness is blunted during fat infusions when a CCK-a antagonist is injected (Feinle, D'Amato, & Read, 1996).

Gastric-Intestinal Interaction

Several laboratories have now obtained evidence that normal satiation involves a combination of gastric and intestinal stimuli. These experiments indicate that the intestinal stimulus is primarily humoral. In humans (Muurahainen, Kissileff, Lachaussee, & Pi-Sunyer, 1991), monkeys (Moran & McHugh, 1982), and rats (Moran & McHugh, 1988) the intestinal peptide, CCK, reduces food intake in part by facilitating signals of gastric distention. In the studies with monkeys and humans, at doses where CCK and gastric distention with a dilute food (soup = .4 kcal/g) or saline solution (monkeys) did not reduce food intake by itself, the combination resulted in a significant reduction. It has been suggested, on the basis of recordings from sensory neurons in the stomach (gastric vagal afferents), that signals of gastric distention are amplified in the presence of elevated levels of CCK (Schwartz, McHugh, & Moran, 1991). Consequently, the effects of a soup preload that fills the stomach would produce a greater feeling of fullness when CCK was infused after the preload (Muurahainen et al., 1991). A critical experiment tested this hypothesis: Quantitative verbal reports of fullness increased at a more rapid rate when CCK rather than saline was infused during increases in gastric distention by means of a water-filled balloon placed in the stomach (Melton et al., 1992).

The Liver and Circulation—Metabolic Controls

Most of our understanding of the role of metabolism in the control of food intake still comes from animals, but a recent human study establishes the relevance of this mechanism to humans. The identification of the liver as an organ in which metabolic control affects food intake began with Russek's (Russek, 1963) hypothesis that the liver mediated the decreased food intake after norepinephrine injection, which depleted liver glycogen. Russek modified his view in 1981, and it is instructive to review what he said: "Hunger would normally appear when intestinal absorption and liver glycogen and liver pyruvate [an intermediary glucose breakdown product that generates ATP] decrease to a certain critical level. Absorption of glucose and/or amino acids from the intestine will immediately increase liver pyruvate, hyperpolarize the hepatocyte's membrane, reduce hunger discharges [leading to a signal to the brain that fuel is available] and induce satiety" (Russek, 1981, p. 138). Thus increases in food intake would be expected when metabolic fuel production (i.e., breakdown of glycogen and/or glucose or fatty acid oxidation) was blocked, and food intake decrease would be expected when metabolic fuel oxidation was increased. Friedman and Tordoff (Friedman & Tordoff, 1986) have shown that food intake can be increased by preventing metabolic fuel oxidation and that the effects of blocking two fuel sources appear to be supra-additive. That is, if a certain amount of intake increase occurs from blocking the oxidation of fat and the oxidation of carbohydrate, by injecting chemical inhibitors of the metabolic processes for each fuel, blocking both of them equivalently more than doubles the effect. The idea here is that even if one metabolic fuel oxidation pathway is blocked, another can take over, but when both are blocked simultaneously, the common metabolic pathway is totally blocked. Therefore, the signal must be a common path farther down the chain of reactions. Conversely, infusion of sugars that are directly used by the liver into the portal vein of rats reduces their food intake (Tordoff & Friedman, 1988), presumably by stimulating the receptors postulated by Russek. Further evidence for the cellular basis of food intake control by metabolic stimuli has been summarized (Langhans, Damaske, & Scharrer, 1985). A recent study has shown that a blockade of fatty oxidation in the liver can increase food intake in humans (Kahler et al, 1999). When the metabolic inhibitor etomoxir was injected 2.5 hours before a high-fat breakfast in subjects adapted to a high-fat diet, breakfast intake was increased by 22%.

Another approach to metabolic controls of meals has examined the metabolic stimuli that trigger the start of a meal. Although many animal studies relate the start of eating and the increase in food intake to a reduction in glucose utilization (Mayer, 1955; Novin, 1994), only recently has evidence been published that reduction in peripheral glucose levels (which is believed to mirror glucose availability for cellular processes) links this reduction to the start of a meal (Campfield, Smith, Rosenbaum, & Hirsch, 1996) in humans. These findings have been confirmed and extended by Melanson and colleagues (1999), who showed that when a high-fat diet was given, the delay in the request for a meal was doubled from 65 to 126 minutes, but the drop in blood glucose still preceded the meal request. This delay in meal onset corresponded with a delay in decline of blood glucose, but in both cases the meal was preceded by the same transient decline. Nevertheless, the investigators could not rule out the possibility that the longer delay after the fat meal was attributable to a more prolonged presence of nutrients in the intestine (Read et al., 1994), which would also delay meal onset.

No account of metabolic controls of food intake would be complete without consideration of insulin. After a meal, and even at the outset of the meal, insulin, the major endocrine secretion of the pancreas, rises in response to stimulation of the head receptors (cephalic phase) and in response to the rising blood glucose level as carbohydrate is digested and absorbed. Insulin, in turn, promotes the uptake of glucose by muscles and adipose tissue, thereby restoring the glucose level to normal. Many animal studies have been done on the role of glucose and insulin (see Woods, Seeley, Porte, & Schwartz, 1998) in the control of food intake, but few studies have been done in humans. When glucose availability is reduced either by insulin or by a glucose analog that cannot be metabolized (2-deoxy d-glucose or 2-DG), animals begin to eat or eat more at a meal. Humans, like animals, have increased hunger in response to glucose decline stimulated by 2-DG (Thompson & Campbell, 1970). However, humans have not been shown to change food intake in response to a fourfold elevation of insulin levels without a drop in blood glucose induced by intravenous infusion of insulin and glucose (Woo, Kissileff, & Pi-Sunyer, 1984). Vanderweele (1994) has suggested that insulin may only be effective as a satiety agent if it reaches the liver by way of the portal vein. Portal vein infusions are not normally done in healthy human subjects, so we must rely on data from animals to follow up this hypothesis.

The Brain

It almost goes without saying that the ultimate control of eating in a meal resides in the brain. However, untangling the various interconnections, sequences of activation, and neurotransmitters involved in the human brain without prohibitive invasive procedures is virtually impossible. We must therefore rely on evidence from animals, indirect evidence from observations of brain activities that coincide with eating, and human pharmacological manipulations whose site of action can be inferred from animal studies. Basically, neurochemical circuits in the brain facilitate and inhibit eating, as shown by their manipulations increasing or decreasing food intake during individual meals. Different sets of transmitters and sites appear to be involved in the facilitation than in the inhibition. The major inhibitor that has been studied in humans is serotonin (Blundell & Lawton, 1995; Lawton, Wales, Hill, & Blundell, 1995), which appears from animal data to inhibit food intake when injected into the brain along a pathway that extends from the midbrain to the hypothalamus (Leibowitz & Alexander, 1998; Simansky, 1996). Neuropeptide Y appears to be the most potent stimulant of meal size increase (Kalra & Kalra, 1990; Leibowitz, 1994). A complete neuroendocrine circuit underlying the metabolic controls of food intake has been proposed by Seeley & Schwartz (1999) and should be consulted for a more complete up-to-date review of the important developments in this area. Among these developments is the identification of leptin, the protein product of the *ob* gene produced in adipose tissue, which may be involved in controlling meal size by means of a CCK-sensitive mechanism in relation to the body's energy stores (Emond, Schwartz, Ladenheim, & Moran, 1999).

APPLICATIONS

Eating Pathology

Having described the physiological controls of normal eating, we may reasonably ask whether disturbances in amounts consumed in binge-eating disorder, anorexia nervosa, or bulimia nervosa could have a physiological basis. The first step in this investigation has been taken. Excessive intakes of patients who report that they overeat have been measured in the laboratory where they are now amenable to further study (Goldfein, Walsh, Lachaussee, Kissileff, & Devlin, 1993; Guss, Kissileff, Walsh, & Devlin, 1994; Kissileff, Walsh, Kral, & Cassidy, 1986; Walsh, Kissileff, Cassidy, & Dantzic, 1989). These studies have revealed that patients with bulimia nervosa have both deficient CCK release and a slowing of gastric emptying (Devlin et al., 1997), and that they needed to eat more food than controls to reach the same level of satiety (Hadigan, Walsh, Kissileff, LaChaussée, & Devlin, 1992 and Kissileff et al., 1996). Further mechanistic studies are being conducted.

Construction of Satiating Foods and Physiologically Based Portion Sizes

The average rate of energy delivery of nutrients to the intestine from the stomach is 2 kcal/min, and the average rate of energy consumption is about 2000 kcal/ day. The number of meals eaten will, therefore, determine the individual's energy requirement per meal. Comfortable distention of the stomach before discomfort sets in is about 1000 ml (Geliebter, 1988). The more distinct flavor/taste cues (i.e., items) in a meal, the more the individual will eat (Rolls et al., 1981). On the basis of these facts physiologically optimal parameters exist for determining serving sizes. Of course, choices such as the number of portions and the number of meals exist as well. Because of these variables, it is not possible to determine optimal portion size nor is it possible to determine portion size without knowing the individual's daily resting and active metabolic rates. If we assume that the individual has three meals and a snack each day, each of which average about 500 kcal and the interval between meals is about 4 hours, energy balance will take place at every meal, and the individual should never feel "hungry." If the portion size of any given item is divided equally across the total energy of the meal, the size of portion is equal to the total daily energy expenditure, divided by the number of meals (eating events), divided by the number of items. A meal of 500 kcal will be completely emptied at no longer than 250 min. Longer intermeal intervals should be preceded by larger meals to accommodate Le Magnen's rule that meal size predicts the size of the next intermeal interval (Bernstein, Zimmerman, Czeisler, & Weitzman, 1981; Le Magnen, 1981). Consequently, smaller meals would need to be followed by smaller intervals. If there were four courses in a 500-kcal meal, the optimal size of each would be 125 kcal, and the optimal average nutrient density to fill the stomach adequately would be .5 kcal/ml. What we still do not know is how the number of items and their sensory properties impact the amount that will be eaten. Probably no fixed answer to this question exists because the body is constantly re-evaluating and relearning its optimal portion size on the basis of the effects of prior experiences. How far back these are remembered and how great an impact they have at any given meal is largely empirical. Furthermore, the actual energy need is based on total energy expenditure, which varies with activity. Thus, in theory, it is possible to predict optimum portion size and meal size for a given set of circumstances, but unless these circumstances are imposed and prescribed, portion size cannot be set optimally.

Integration of Cognitive and Physiological Controls

It is important to recognize that physiological controls are always operating, whether we think about them or not. We cannot breathe, walk, eat, or do anything without the brain and the inputs it receives from the body and from the environment. Much of the contents of this book ignores physiology because the concepts being considered in food design, presentation, context of eating, and so on can all be discussed without considering physiology directly. Furthermore, the intake control system is plastic, not rigid (i.e., reflex driven as a stimulus-response input/output system as some animal physiologists see it). Consequently, when social and cognitive factors enter into the intake-controlling equation, it is important for physiologists and sociologists to recognize that the integration of these types of controls is not a negation of the importance of the systems they are studying but rather a new problem to be solved. This is particularly important in the realm of modification of eating behavior for therapeutic purposes, to treat obesity, diabetes, or bulimia. People eat in a social milieu most (but not all) of the time. How these cognitive and social factors have an impact on the physiological controls of intake is a major and fundamental unsolved problem for the next millennium. Whether conditioning theory in which external and internal cues can become occasion setters for changes in thresholds of physiological responses or whether physiological responses, such as sensitivity to gastric distention, changes with the eating occasion, remains to be seen.

Implications of Physiological Controls for Food and Drug Manufacturers

Underlying the development of functional foods for energy-intake reduction and appetite-suppressing drugs are the assumptions that somehow these items are capable of overriding the strong cognitive and social factors that control intake and that food intake is ultimately based on physiological control, which these products can influence. Even if these assumptions are not voiced, they must ultimately be heeded because, in the end, physiology does prevail. What is not currently clear is the relative strengths of the various stimuli to which the consumer responds. Appetite-suppressing agents may work in the laboratory, but it is not known whether or how much they interact with controls of the physiological system that come from the cognitive and/or social stimuli. For example, how will the pleasantness of an appealing food at a party interact with increased sensitivity to gastric distention. Are they simply additive, or at some level will one overwhelm the other? Experiments to test such hypotheses are needed *before* specific foods or drugs are designed. It cannot be reiterated too strongly that satiety tests without mechanisms have no general applicability.

REFERENCES

Bell, E.A., Castellanos, V.H., Pelkman, C.L., Thorwart, M.L., & Rolls, B.J. (1998). Energy density of foods affects energy intake in normal-weight women. *American Journal of Clinical Nutrition, 67,* 412–420.

Bellisle, F., & Le Magnen, J. (1981). The structure of meals in humans: Eating and drinking patterns in lean and obese subjects. *Physiology and Behavior, 27,* 649–658.

Bellisle, F., Lucas, F., Amrani, R., & Le Magnen, J. (1984). Deprivation, palatability and the micro-structure of meals in human subjects. *Appetite, 5,* 85–94.

Bernstein, I.L., Zimmerman, J.C., Czeisler, C.A., & Weitzman, E.D. (1981). Meal patterns in "free-running" humans. *Physiology and Behavior, 27,* 621–623.

Blundell, J.E., & Coscina, D.V. (1992). Commentary: Part I Integration of signals responsible for eating patterns. In G.H. Anderson & S.H. Kennedy (Eds.), *The biology of feast and famine: Relevance to eating disorders* (pp. 93–101). London: Academic Press, Inc.

Blundell, J.E., & Lawton, C.L. (1995). Serotonin and dietary fat intake: effects of dexfenfluramine. *Metabolism, 44,* 33–37.

Bobroff, E.P., & Kissileff, H.R. (1986). Effects of change in palatability on food intake and the cumulative food intake curve in man. *Appetite, 6,* 85–96.

Booth, D.A. (1972). Conditioned satiety in the rat. *Journal of Comparative and Physiological Psychology, 81,* 457–471.

Booth, D.A. (1977). Appetite and satiety as metabolic expectancies. In Y. Katsuki, M. Sato, S.F. Takagi, & Y. Oomura (Eds.), *Food intake and chemical senses* (pp. 317–330). Tokyo: University of Tokyo Press.

Booth, D.A. (1981). The physiology of appetite. *British Medical Bulletin, 37,* 135–140.

Booth, D.A. (1994). *The psychology of nutrition.* London: Taylor & Francis Publishers.

Booth, D.A., Gibson, E.L., Toase, A., & Freeman, R.P.J. (1994). Small objects of desire: The recognition of appropriate foods and drinks and its neural mechanisms. In C.R. Legg & D.A. Booth (Eds.), *Appetite: Neural and behavioural bases* (pp. 98–126). London: Oxford University Press.

Booth, D.A., Lee, M., & McAleavy, C. (1976). Acquired sensory control of satiation in man. *British Journal of Psychology, 2,* 137–147.

Booth, D.A., & Weststrate, J.A. (1994). Concepts and methods in the psychobiology of ingestion. In M.S. Westerterp-Plantenga, E.W.H.M. Fredrix, & A.B. Steffens (Eds.), *Food intake and energy expenditure* (pp. 31–46). Boca Raton, FL: CRC Press.

Brobeck, J.R. (1955). Neural regulation of food intake. *Annals of the New York Academy of Sciences, 63 art. 1,* 44–55.

Burley, V.J., & Blundell, J.E. (1991). Evaluation of the action of a non-absorbable fat on appetite and energy intake in lean, healthy males. *International Journal of Obesity, 15,* 1.

Burley, V.J., Cotton, J.R., Weststrate, J.A., & Blundell, J.E. (1994). Effect on appetite of replacing natural fat with sucrose polyester in meals or snacks across one whole day. In H. Ditschuneit, F.A. Gries, H. Hauner, V. Schusdziaria, & J.G. Wechsler (Eds.), *Obesity in Europe 1993* (pp. 227–233). London: John Libbey & Company, Ltd.

Campbell, R.G., Hashim, S., & Van Itallie, T.B. (1971). Studies of food-intake regulation in man, responses to variations in nutritive density in lean and obese subjects. *New England Journal of Medicine, 285,* 1402–1407.

Campfield, L.A., Smith, F.J., Rosenbaum, M., & Hirsch, J. (1996). Human eating: Evidence for a physiological basis using a modified paradigm. *Neuroscience and Biobehavioral Reviews, 20,* 133–137.

Cannon, W.B., & Lieb, C.M. (1911). The receptive relaxation of the stomach. *American Journal of Physiology, 29,* 270–273.

Cecil, J.E., Francis, J., & Read, N.W. (1998). Relative contributions of intestinal, gastric, oro-sensory influences and information to changes in appetite induced by the same liquid meal. *Appetite, 31,* 377–390.

Clifton, P.G. (2000). Meal patterning in rodents: Psychopharmacological and neuroanatomical studies. *Neuroscience and Biobehavioral Reviews* (in press).

Collier, G. (1986). The dialogue between the house economist and the resident physiologist. *Nutrition and Behavior, 3,* 9–26.

Davidson, T.L. (1998). Hunger cues as modulatory stimuli. In N.A. Schmajuk & P.C. Holland (Eds.), *Occasion setting: Associative learning and cognition in animals* (pp. 223–248). Washington, DC: American Psychological Association.

Davis, J.D., Smith, G.P., & Sayler, J.L. (1997). Reduction of intake in the rat due to gastric filling. *American Journal of Physiology, 272,* R1599–R1605.

Davis, J.D., Smith, G.P., & Sayler, J.L. (1998). Closing the pylorus decreases the size of large meals in the rat. *Physiology and Behavior, 63,* 191–196.

Deutsch, J.A. (1983). Dietary control and the stomach. *Progress in Neurobiology, 20,* 313–332.

Deutsch, J.A., Young, W.G., & Kalogeris, T.J. (1978). The stomach signals satiety. *Science, 201,* 165–167.

Devlin, M.J., Walsh, B.T., Guss, J.L., Kissileff, H.R., Liddle, R.A., & Petkova, E. (1997). Postprandial cholecystokinin release and gastric emptying in patients with bulimia nervosa. *American Journal of Clinical Nutrition, 65,* 114–120.

Drewnowski, A. (1997). Taste preferences and food intake. *Annual Review of Nutrition, 17,* 237–253.

Emond, M., Schwartz, G.J., Ladenheim, E.E., & Moran, T.H. (1999). Central leptin modulates behavioral and neural responsivity to CCK. *American Journal of Physiology, 276,* R1545–R1549.

Epstein, A.N. (1960). Reciprocal changes in feeding behavior produced by intra-hypothalamic chemical injections. *American Journal of Physiology, 199,* 969–974.

Feinle, C., D'Amato, M., & Read, N.W. (1996). Cholecystokinin-A receptors modulate gastric sensory and motor responses to gastric distension and duodenal lipid. *Gastroenterology, 110,* 1379–1385.

Feinle, C., Grundy, D., & Read, N.W. (1997). Effects of duodenal nutrients on sensory and motor responses of the human stomach to distension. *American Journal of Physiology, 273,* G721-G726.

Foltin, R.W., Fischman, M.W., Emurian, C.S., & Rachlinski, J.J. (1988). Compensation for caloric dilution in humans given unrestricted access to food in a residential laboratory. *Appetite, 10,* 13–24.

Friedman, M.I. (1990). Making sense out of calories. In E.M. Stricker (Ed.), *Handbook of Behavioral Neurolobiology Volume 10 Neurobiology of Food and Fluid Intake.* (pp. 513–529). New York: Plenum Publishing Corp.

Friedman, M., & Tordoff, M. (1986). Fatty acid oxidation and glucose utilization interact to control food intake in rats. *American Journal of Physiology, 251,* R840–R845.

Geary, N., Kissileff, H.R., Pi-Sunyer, F.X., & Hinton, V. (1992). Individual, but not simultaneous, glucagon and cholecystokinin infusions inhibit feeding in men. *American Journal of Physiology, 262,* R975–R980.

Geliebter, A. (1988). Gastric distension and gastric capacity in relation to food intake in humans. *Physiology and Behavior, 44,* 665–668.

Geliebter, A., Westreich, S., & Gage, D. (1988). Gastric distension by balloon and test-meal intake in obese and lean subjects. *American Journal of Clinical Nutrition, 48,* 592–594.

Gibbs, J., Young, R.C., & Smith, G.P. (1973). Cholecystokinin elicits satiety in rats with open gastric fistulas. *Nature, 245,* 323–325.

Goldfein, J.A., Walsh, B.T., Lachaussee, J.L., Kissileff, H.R., & Devlin, M.J. (1993). Eating behavior in binge eating disorder. *International Journal of Eating Disorders, 14,* 427–431.

Greenberg, D. (1998). Intestinal satiety. In G.P. Smith (Ed.), *Satiation: From gut to brain* (pp. 40–70). New York: Oxford University Press.

Grigson, P.S., Lyuboslavsky, P.N., Tanase, D., & Wheeler, R.A. (1999). Water-deprivation prevents morphine-, but not LiCl-induced, suppression of sucrose intake. *Physiology and Behavior, 67,* 277–286.

Grill, H.J., & Norgren, R. (1978). Chronically decerebrate rats demonstrate satiation but not bait shyness. *Science, 201,* 267–269.

Grossman, S.P. (1967). *Physiological psychology.* New York: John Wiley & Sons.

Guss, J.L., Kissileff, H.R., Walsh, B.T., & Devlin, M.J. (1994). Binge eating behavior in patients with eating disorders. *Obesity Research, 2,* 335–363.

Hadigan, C.M., Kissileff, H.R., & Walsh, B.T. (1989). Patterns of food selection during meals in women with bulimia. *American Journal of Clinical Nutrition, 50,* 759–766.

Hadigan, C.M., Walsh, B.T., Kissileff, H.R., LaChaussée, J.L., & Devlin, M.J. (1992). Behavioral assessment of satiety in bulimia nervosa. *Appetite, 18,* 233–242.

Hashim, S.A., & Van Itallie, T.B. (1964). An automatically monitored food dispensing apparatus for the study of food intake in man. *Federation Proceedings, 23(1),* 82–84.

Hashim, S.A., & Van Itallie, T.B. (1965). Studies in normal and obese subjects with a monitored food-dispensing device. *Annals of the New York Academy of Sciences, 131,* 654–661.

Hellemann, U., & Tuorila, H. (1991). Pleasantness ratings and consumption of open sandwiches with varying NaCl and acid contents. *Appetite, 17,* 229–238.

Hetherington, M., & Rolls, B.J. (1987). Methods of investigating human eating behavior. In N.E. Rowland & F.M. Toates (Eds.), *Techniques in the behavioral and neural sciences, Volume 1: Feeding and drinking* (pp. 77–109). New York: Elsevier Science.

Hill, A.J., Magson, L., & Blundell, J.E. (1984). Hunger and palatability: Tracking subjective experience before, during and after consumption of preferred and less preferred food. *Appetite, 5,* 361–367.

Houpt, T.A., Philopena, J.M., Joh, T.H., & Smith, G.P. (1996). c-Fos induction in the rat nucleus of the solitary tract by intraoral quinine infusion depends on prior contingent pairing of quinine and lithium chloride. *Physiology and Behavior, 60,* 1535–1541.

Hunt, J.N., & Knox, M.T. (1968). Regulation of gastric emptying. In C.F. Code (Ed.), *Handbook of physiology, Section 6, Volume IV* (pp. 1917–1935). Washington, D.C.: American Physiological Society.

Jordan, H.A., Wieland, W.F., Zebley, S.P., Stellar, E., & Stunkard, A.J. (1966). Direct measurement of food intake in man: A method for the objective study of eating behavior. *Psychosomatic Medicine, 28,* 836–842.

Kahler, A., Zimmermann, M., & Langhans, W. (1999). Suppression of hepatic fatty acid oxidation and food intake in man. *Nutrition, 15,* 819–828.

Kalra, S.P., & Kalra, P.S. (1990). Neuropeptide Y: A novel peptidergic signal for the control of feeding behavior. *Current Topics in Neuroendocrinology, 10,* 192–217.

Khan, M., & Reed, D. (1992). The effect of duodenal lipid infusions on gastric pressure and sensory responses to balloon distention [abstract]. *Gastroenterology, 102,* 467.

Kissileff, H.R. (1984). Satiating efficiency and a strategy for conducting food loading experiments. *Neuroscience Biobehavior Review, 8,* 129–135.

Kissileff, H.R. (1985). Effects of physical state (liquid-solid) of foods on food intake: Procedural and substantive contributions. *American Journal of Clinical Nutrition, 42,* 956–965.

Kissileff, H.R. (1990). Some suggestions on dealing with palatability: Response to Ramirez. *Appetite, 14,* 162–166.

Kissileff, H.R. (1992). Where should human eating be studied and what should be measured? *Appetite, 19,* 61–68.

Kissileff, H.R., Guss, J.L., & Nolan, L. (1996). What animal research tells us about human eating. In H. Meiselman & H. MacFie (Eds.), *Food choice, acceptance and consumption* (pp. 104–160). New York: Blackie Academic (Division of Chapman & Hall).

Kissileff, H.R., Klingsberg, G., & Van Itallie, T.B. (1980). Universal eating monitor for continuous recording of solid or liquid consumption in man. *American Journal of Physiology, 238,* R14–R22.

Kissileff, H.R., Pi-Sunyer, F., Thornton, J., & Smith, G. (1981). C-terminal octapeptide of cholecystokinin decreases food intake in man. *American Journal of Clinical Nutrition, 34,* 154–160.

Kissileff, H.R., & Thornton, J. (1982). Facilitation and inhibition in the cumulative food intake curve in man. In A.J. Morrison & P. Strick (Eds.), *Changing concepts of the nervous system* (pp. 585–607). New York: Academic Press.

Kissileff, H.R., & Van Itallie, T.B. (1982). Physiology of the control of food intake. *Annual Review of Nutrition, 2,* 371–418.

Kissileff, H.R., Walsh, T., Kral, J., & Cassidy, S. (1986). Laboratory studies of eating behavior in women with bulimia. *Physiology and Behavior, 38,* 563–570.

Kissileff, H.R., Wentzlaff, T.H., Guss, J.L., Walsh, B.T., Devlin, M.J., & Thornton, J.C. (1996). A direct measure of satiety disturbance in patients with bulimia nervosa. *Physiology and Behavior, 60,* 1077–1085.

Langhans, W., Damaske, U., & Scharrer, E. (1985). Different metabolites might reduce food intake by the mitochondrial generation of reducing equivalents. *Appetite, 6,* 143–152.

Lavin, J.H., & Read, N.W. (1995). The effect on hunger and satiety of slowing the absorption of glucose: relationship with gastric emptying and postprandial blood glucose and insulin responses. *Appetite, 25,* 89–96.

Lawton, C.L., Wales, J.K., Hill, A.J., & Blundell, J.E. (1995). Serotoninergic manipulation, meal-induced satiety and eating pattern: Effect of fluoxetine in obese female subjects. *Obesity Research, 3,* 345–356.

Le Magnen, J. (1981). The metabolic basis of the dual periodicity of feeding in rats. *Behavioral Brain Science, 4,* 561–607.

Le Magnen, J. (1992). *Neurobiology of feeding and nutrition.* Orlando, FL: Academic Press.

Leibowitz, S.F. (1994). Specificity of hypothalamic peptides in the control of behavioral and physiological processes. *Annals of the New York Academy of Sciences, 739,* 12–35.

Leibowitz, S.F., & Alexander, J.T. (1998). Hypothalamic serotonin in control of eating behavior, meal size, and body weight. *Biology and Psychiatry, 44,* 851–864.

Liebling, D.S., Eisner, J.D., Gibbs, J., & Smith, G.P. (1975). Intestinal satiety in rats. *Journal of Comparative and Physiological Psychology, 89,* 955–965.

Lieverse, R.J., Jansen, J.B.M.J., Masclee, A.M., Rovati, L.C., & Lamers, C.B.H.W. (1994). Effect of a low dose of intraduodenal fat on satiety in humans: studies using the type A cholecystokinin receptor antagonist loxiglumide. *Gut, 35,* 501–505.

Louis-Sylvestre, J., Tournier, A., Verger, P., Chabert, M., Delorme, B., & Hossenlopp, J. (1989). Learned caloric adjustment of human intake. *Appetite, 12(2),* 95–104.

Lucas, F., & Bellisle, F. (1987). The measurement of food preferences in humans: Do taste-and-spit tests predict consumption? *Physiology and Behavior, 39,* 739–743.

Mayer, J. (1955). Regulation of energy intake and body weight. *Annals of the New York Academy of Sciences, 63 art. 1,* 15–43.

McHugh, P.R., & Moran, T.H. (1985). The stomach: A conception of its dynamic role in satiety. *Progress in Psychobiology & Physiological Psychology, 2,* 197–232.

Melanson, K.J., Westerterp-Plantenga, M.S., Saris, W.H., Smith, F.J., & Campfield, L.A. (1999). Blood glucose patterns and appetite in time-blinded humans: Carbohydrate versus fat. *American Journal of Physiology, 277,* R337–R345.

Melton, P.A., Kissileff, H.R., & Pi-Sunyer, F.X. (1992). Cholecystokinin (CCK-8) affects gastric pressure and ratings of hunger and fullness in women. *American Journal of Physiology, 263,* R452–R456.

Meyer, J.-E., & Pudel, V. (1972). Experimental studies on food intake in obese and normal weight subjects. *Journal of Psychosomatic Research, 16,* 305–308.

Miller N.E. (1957). Experiments in motivation. *Science, 126,* 1271–1278.

Moran, T.H., & McHugh, P.R. (1982). Cholecystokinin suppresses food intake by inhibiting gastric emptying. *American Journal of Physiology, 242,* R491–R497.

Moran, T.H., & McHugh, P.R. (1988). Gastric and nongastric mechanisms for satiety action of cholecystokinin. *American Journal of Physiology, 254,* R628–R632.

Muurahainen, N.E., Kissileff, H.R., Lachaussee, J., & Pi-Sunyer, F.X. (1991). Effect of a soup preload on reduction of food intake by cholecystokinin in humans. *American Journal of Physiology, 260,* R272–R280.

Muurahainen, N.E., Kissileff, H.R., & Pi-Sunyer, F.X. (1993). Intravenous bombesin reduces food intake in man. *American Journal of Physiology, 264,* R350–R354.

Nachman, M., & Ashe, J.H. (1973). Learned taste aversions in rats as a function of dosage, concentration, and route of administration of LiCl. *Physiology and Behavior, 10,* 73–78.

Novin, D. (1994). Regulatory control of food and water intake and metabolism by the liver. In D.A. Booth (Ed.), *Neurophysiology of ingestion* (pp. 19–32). New York: Pergamon Press.

Pavlov, I.P. (1927). *Conditioned reflexes.* New York: Dover Press.

Perez, C., Dalix, A.M., Guy-Grand, B., & Bellisle, F. (1994). Human responses to five concentrations of sucrose in a dairy product: immediate and delayed palatability effects. *Appetite, 23,* 165–178.

Philips, R.J., & Powley, T.L. (1996). Gastric volume rather than nutrient content inhibits food intake. *American Journal of Physiology, 271,* R766–R779.

Porikos, K.P., Hesser, M.F., & Van Itallie, T.B. (1982). Caloric regulation in normal weight men maintained on a palatable diet of conventional foods. *Physiology and Behavior, 29,* 293–300.

Rauhofer, E.A., Smith, G.P., & Gibbs, J. (1993). Acute blockade of gastric emptying and meal size in rats. *Physiology and Behavior, 54,* 881–884.

Read, N., French, S., & Cunningham, K. (1994). The role of the gut in regulating food intake in man. *Nutrition Reviews, 52, 1,* 1–10.

Rolls, B.J., Hetherington, M., & Burley, V.J. (1988). The specificity of satiety: The influences of foods of different macronutrient content on the development of satiety. *Physiology and Behavior, 43,* 145–153.

Rolls, B.J., Pirraglia, P.A., Jones, M.B., & Peters, J.C. (1992). Effects of olestra, a noncaloric fat substitute, on daily energy and fat intakes in lean men. *American Journal of Clinical Nutrition, 56,* 84–92.

Rolls, B.J., Rowe, A., Rolls, E.T., Kingston, B., Megson, A., & Gunary, R. (1981). Variety in a meal enhances food intake in man. *Physiology and Behavior, 26,* 215–221.

Russek, M. (1963). An hypothesis on the participation of hepatic glucoreceptors in the control of food intake. *Nature (London)*, *197*, 79–80.

Russek, M. (1981). Current status of the hepatostatic theory of food intake control and body weight regulation. *Appetite*, *2*, 137–143.

Schwartz, G.J., McHugh, P.R., & Moran, T.H. (1991). Integration of vagal afferent responses to gastric loads and cholecystokinin in rats. *American Journal of Physiology*, *261*(1 pt. 2), R64–R69.

Sclafani, A. (1990). Nutritionally based learned flavor preferences in rats. In E.D. Capaldi & T.L. Powley (Eds.), *Taste, experience, and feeding* (pp. 139–156). Washington, D.C.: American Psychological Association.

Sclafani, A. (1997). Learned controls of ingestive behaviour. *Appetite*, *29*, 153–158.

Sclafani, A., Fanizza, L.J., & Azzara, A.V. (1999). Conditioned flavor avoidance, preference, and indifference produced by intragastric infusions of galactose, glucose, and fructose in rats. *Physiology and Behavior*, *67*, 227–234.

Seeley, R.J., Kaplan, J.M., & Grill, H.J. (1995). Effect of occluding the pylorus on intraoral intake: A test of the gastric hypothesis of meal termination. *Physiology and Behavior*, *58*, 245–249.

Seeley, R.J., & Schwartz, M.W. (1999). Neuroendocrine regulation of food intake. *Acta Paediatrica Supplement*, *88*, 58–61.

Silverstone, T., Fincham, J., & Brydon, J. (1980). A new technique for the continuous measurement of food intake in man. *American Journal of Clinical Nutrition*, *33*, 1852–1855.

Simansky, K.J. (1996). Serotonergic control of the organization of feeding and satiety. *Behavior and Brain Research*, *73*, 37–42.

Smith, G.P. (1982). The physiology of the meal. In T. Silverstone (Ed.), *Drugs and appetite* (pp. 1–21). London: Academic Press.

Smith, G.P. (1998). Introduction. In G.P. Smith (Ed.), *Satiation from gut to brain* (pp. 3–9). New York: Oxford University Press.

Smith, G.P. (1997). Control of food intake. In A.C. Ross (Ed.), Modern nutrition in health and disease (pp. 631–643). Baltimore: Williams & Wilkins.

Spiegel, T.A. (1973). Caloric regulation of food intake in man. *Journal of Comparative and Physiological Psychology*, *84*, 24–37.

Stellar, E., & Schrager, E.E. (1985). Chews and swallows and the microstructure of eating. *American Journal of Clinical Nutrition*, *42*, 973–982.

Teitelbaum, P., & Epstein, A.N. (1962). The lateral hypothalmic syndrome: Recovery of feeding and drinking after lateral hypothalmic lesions. *Psychological Reviews*, *69*, 74–90.

Thompson, D.A. & Campbell, R.G. (1997). Hunger in humans induced by 2-Deoxy-D-glucose: Glucoprivic control of taste perference and food intake. *Science*, *198*, 1065–1067

Tordoff, M.G., & Friedman, M.I. (1988). Hepatic control of feeding: Effect of glucose, fructose, and mannitol infusion. *American Journal of Physiology*, *254*, R969–R976.

Vanderweele, D.A. (1994). Insulin is a prandial satiety hormone. *Physiology and Behavior*, *56*(3), 619–622.

Walsh, B.T., Kissileff, H.R., Cassidy, S.M., & Dantzic, S. (1989). Eating behavior of women with bulimia. *Archives of General Psychiatry*, *46*, 54–58.

Welch, I., Saunders, K., & Read, N.W. (1985). Effect of ileal and intravenous infusions of fat emulsions on feeding and satiety in human volunteers. *Gastroenterology*, *89*, 1293–1297.

Welch, I., Sepple, C.P., & Read, N.W. (1988). Comparisons of the effects on satiety and eating behaviour of infusion of lipid into the different regions of the small intestine. *Gut*, *29*, 306–311.

Wentzlaff, T., Guss, J.L., & Kissileff, H.R. (1995). Subjective ratings as a function of amount consumed: A preliminary report. *Physiology and Behavior*, *57*(6), 1209–1214.

Woo, R., Kissileff, H.R., & Pi-Sunyer, F.X. (1984). Elevated postprandial insulin levels do not induce satiety in normal weight humans. *American Journal of Physiology*, *247*, R745–R749.

Woods, S.C., Seeley, R.J., Porte, D.J., & Schwartz, M.W. (1998). Signals that regulate food intake and energy homeostasis. *Science*, *280*, 1378–1383.

Wurtman, R.J., & Wurtman, J.J. (1995). Brain serotonin, carbohydrate-craving, obesity and depression. *Obesity Research*, *3* (Suppl. 4), 477S–480S.

Yeomans, M.R. (1998). Taste, palatability and the control of appetite. *Proceedings of the Nutritionist Society*, *57*, 609–615.

Yeomans, M.R., & Gray, R.W. (1997). Effects of naltrexone on food intake and changes in subjective appetite during eating: Evidence for opioid involvement in the appetizer effect. *Physiology and Behavior*, *62*, 15–21.

Chapter 5

Sensory Combinations in the Meal

Harry T. Lawless

When most people think of the first time they heard music by Antonio Vivaldi, they will most likely recall the famous violin concerti, *The Four Seasons*. These well-known pieces are actually part of a larger set of 12 concerti that Vivaldi named "*Il Cimento Dell'armonia e Dell'inventione*," roughly translated as "the struggle between harmony and invention." This was a musical remark on the need for both predictable, familiar structure in music and the certain degree of inventiveness that makes good music interesting to the listener. When I think of human food consumption, I think of a few similar "struggles" that capture the nature of human eating behavior.

First, we have the paradox that, as omnivores, we need to seek novel food sources and combinations to keep us interested and well nourished. When food is abundant, we seek new and interesting combinations of items. On the other hand, like other foragers, we are often wary of a potential food that has no track record of safety. Indeed, making an animal sick after consuming a novel food or drink will render that item aversive to that individual for all time, a phenomenon known as "bait shyness" or learned aversion. We learn what is safe and nutritious. Witherly (1987) stated that our very existence "depends upon the ability to recognize, remember and select an adequate diet" (p. 403). So we struggle with the balance between neophilia (interest in what is new) and the need for diverse sensory experiences against our caution and neophobia lest we consume something dangerous to health. This is the paradox of the omnivore.

One solution to this problem is to provide some common themes with variations that diverge from the basic building blocks or provide additional flavors on top of the common themes. Elisabeth Rozin (1983) used the term "flavor principles" to denote the common sensory building blocks that recur in the culinary traditions of a given culture. Thus soy, sesame, chili, and garlic are common in Korean cuisine. Witherly (1987) proposed that the uses of sauces in many cuisines was one way to achieve this sense of comfortable sameness while allowing for variations on the theme. He noted that many sauces based on emulsified fat (butter or a similar roux) provide aids to mastication by having a lubricating property in addition to their basic flavor delivery. He goes on to say that "By selectively flavoring sauce bases, a chef promotes neophilia by adding variety to a foodstuff. Hence a sauce maximizes oral pleasure while contributing to assimilation and satiety" (p. 410).

The meal can be thought of as a combination of diverse sensory experiences where we have sensory "struggles." The most important is perhaps the struggle between the harmony of the foods to be consumed together (or in short and successive time intervals) and the desire for sensory contrast. Sensory contrast versus sensory balance is a real challenge for the culinary artist, to simultaneously maintain a degree of complexity while harmonizing the component parts. It is not surprising that great food is likened to great works of art or great music, at least among those of us who are attentive and in tune to our sensory experiences with food.

The goal of achieving simultaneous harmony with contrast arises from our need to have some relief from sameness. I recall one particular appetizer that came with an obligatory modicum of a very strong distilled spirit, which was listed as a component of the appetizer on the menu and was fully intended by the chef to be consumed with the food. The food was a paper-thin serving of gravad lox with capers and the spirit was a generous ounce or so of aquavit. This nice Scandinavian blend of fish and alcohol provided a kind of foil. The fish was moderately strong in flavor and oily. The neat spirits, sipped after a few bites of fish, cut through the oil, cleansing the palate and setting the gustatory meters back to their wandering zero. One can think of many examples of foods or foods and beverages that act this way. The alternation of hot curries with sweet, cool chutneys comes to mind as another example of including a foil in a meal.

We will return to the theme of harmony with contrast in this chapter, but first let us look at several physiological underpinnings for how the contrast effects come to work in the sensory experience of people while eating a meal.

SENSE MODALITIES

Probably no other common experience assaults all of the senses as does the consumption of foods. We may appreciate the visual impression of fine art or the sound of great music. A visit to the Grand Canyon may reveal great vistas, smells of pine, the coolness of the canyon in shadow, and the sound of hawks overhead. The smells of a marketplace on an overseas trip combine with exotic sights and sounds. However, these complex sensations are the exception to everyday life rather than the rule. In the simple act of eating we manage to stimulate all the senses at roughly the same time. We appreciate the appearance or presentation of the food, its aroma, flavor, texture, temperature, and even how crisp it sounds when we chew. We do not normally think of foods as having auditory properties but they do (Vickers, 1991). I often think I can tell something about the quality of a fine champagne by listening to the fizz. Many small bubbles give off a higher pitched fizz than the gross clumpy fat bubbles of the sort you get from club soda. Combining foods in a meal adds levels of complexity by providing harmony and contrast of many sensory attributes.

Appearance is often the first sense modality to arouse our interest in a food. Color can often predominate our expectations about flavor, and alterations in shape or color can relieve the sense of sameness (Witherly, 1987). Color can be a predictor of quality and arouses our expectations about other sensory attributes, for example, the association of deep red color with the ripeness of fruit (Lavin & Lawless, 1998). Unexpected colors may evoke initial caution until the food is determined to be safe, palatable, and nutritious. Blue corn chips are somewhat common now in restaurants serving southwestern American cuisine but were once considered novel and somewhat evil looking among diners who had not encountered them before.

After the visual sense, olfaction is the next "distance sense" to be aroused when we encounter food. P. Rozin (1982) points out that the sense of smell has a dual purpose—to tell us about odors in the environment at a distance and to inform us about flavors in the oral cavity. Smell is powerfully associated with bodily functions (hunger, satiation, sex) and strong emotion (Witherly, 1987). Because of the enormous range of different smell qualities we are able to experience, the variations that are possible in human food sensations are greatly expanded by the use of this modality. This may be one reason why people who lose their sense of smell lose their appreciation of the finer qualities of food sensations.

The sense of smell is often taken for granted in the experience of eating. This occurs because much of what we think we taste in the mouth is actually an olfactory experience. Flavors in the mouth become airborne when the food is mixed around, warmed, diluted by saliva, and so on. The vapors travel around in the oral cavity, perhaps down into the lungs, and reach the nasal passages from the rear as we swallow (a kind of pumping action) or exhale. The old French gastronome, Brillat-Savarin perhaps captured this best:

> For myself, I am not only convinced that there is no full act of tasting without the participation of the sense of smell, but I am also tempted to believe that smell and taste form a single sense, of which the mouth is the laboratory and the nose is the chimney; or, to speak more exactly, of which one serves for the tasting of actual bodies and the other for the savoring of their gases (1825/1972, p. 39).

Much of what we expect to be taste, then, is really smell but seems to be a taste because it resides first in the mouth. A lemon does not taste like a lemon. It tastes sour, somewhat bitter, and a little sweet. The true lemon flavor arises from the volatile, aromatic compounds that are sensed by the nose.

MIXTURE INTERACTIONS

In the psychology laboratory, researchers have attempted to understand the complex interactions occurring in food flavor by breaking the problem down into different sense modalities (taste vs. smell) and interactions of simple unitary components like sweetness and saltiness. Later they attempt to get back to the complexity of real life by examining more complex situations after the simpler more isolated effects are understood. Two main classes of interactions have been studied: simultaneous interactions when components are mixed together, and sequential interactions, when one item follows another. Both of these kinds of interactions are present when eating a meal, and like the effects of taste on smell, they further modify one another.

Mixture effects can occur simply because foods are not isolated pure chemicals but complex combinations of thousands of compounds with potential impact on the senses of taste, smell, and texture. Mixtures can also occur in the mouth during a meal. We may take a bite of two items at roughly the same time or in close succession, so that they are present in the oral cavity at the same time. Perhaps the residual matter from the previous bite is still present even when we switch to a new item or sip a beverage after a bite of solid food. Of course, we often create mixtures on purpose by adding seasonings, putting gravy on potatoes, sugar in coffee, and so on.

Controlled studies of tastes in mixtures have yielded one recurring and consistent effect: Most taste qualities partially mask or suppress one another (Bartoshuk, 1975; Lawless, 1977). If I take a sugar solution that is 3% sucrose by weight, it will taste moderately sweet. If a solution contains one-half of 1% citric acid, it will be moderately sour. If, however, I make up a solution that contains both 3% sugar and 0.5% citric acid, it will be less sweet than our first solution containing only 3% sucrose and less sour than our solution containing only 0.5% citric acid. Notice that I made this mixture in such a way as to keep the concentration of each component equal to the level in the unmixed or pure solutions for comparison. I did not make up 3% sucrose and 0.5% citric acid and then pour them together. Doing so would naturally reduce the tastes by simple dilution. The effect we see here is one of partial masking, commonly called mixture suppression, in which tastes are less intense when mixed compared with equal levels tasted alone (Lawless, 1986).

In most cases the classical four basic tastes will suppress one another when present in a mixture. The saltiness of your pretzel will be a good companion and foil to the bitterness of a stout or dark porter beer. A spoonful of sugar helps the medicine go down, not only because the sugar is hedonically pleasing to taste but also because the sweetness partially masks the bitterness of the medicine. These kinds of masking events are with us all the time in complex stimuli. They occur for smells and tastes, as shown in careful studies with smell mixtures (Cain, 1975). This suppression between individual sensory characteristics can be a good thing. Putting gravy on your brussel sprouts to cover the bitter taste and sulfury cabbage smell is a trick that my children learned to master quite early.

There are three major exceptions to the suppression rule, in which true synergies or interactions occur when the flavor impact is greater than expected. Small amounts of salt appear to enhance sweet tastes. This is likely due to an intrinsic sweet taste of salt (Bartoshuk, Murphy, & Cleveland, 1978). We are usually not aware of this because we use salt in sufficient quantities to mask the sweet taste by the more potent salty effects of sodium chloride. It is also known that some combinations of sweeteners will give a slightly greater sensory intensity than one would predict from a knowledge of the response to the individual items when tasted alone (Ayya & Lawless, 1992).

However, a more pronounced effect is seen in the mixtures of umami tastants. Umami is a Japanese term roughly translated as "delicious taste" associated with the glutamate salts and the salts of certain ribonucleotides. Witherly (1987) suggests that this is the oral signal for the presence of protein. This is not unreasonable because amino acids are the breakdown products of protein in foods, and various aging and fermentative processes are used to enhance the flavors of foods by breaking up proteins to more flavorful constituents in this way. Cheesemaking and soy sauce brewing are two examples. The origin of the umami concept begins with a Japanese scientist, Ikeda, who at the turn of the twentieth century noted that some traditional broths had great flavor impact. These broths contained monosodium glutamate (MSG) from kombu, the sea tangle, disodium inosinate from dried bonito flakes (katsuobushi), or disodium guanylate from shiitaake mushrooms. When MSG is added to either of the ribosides, a whopping synergistic effect is seen because of an enhanced binding effect at the taste receptors (Cagan, 1981; Yamaguchi, 1967). This has important implications for food flavors because glutamate is a naturally occurring amino acid with relatively high concentrations in some foods (tomatoes, cheeses, soy sauce).

In the sense of smell and with volatile flavors (aromas, flavor by mouth), the mixtures effects are not as straightforward as with taste because smells tend to blend or fuse in some cases, whereas taste qualities like salty and bitter often remain quite separable or distinguishable in a food. When odors are perceived as separate "notes" within a mixture, they show masking or suppression like the basic taste qualities (Cain, 1975; Cain & Drexler, 1974). Well-blended flavors are not so clear. Study of the chemical components of food flavors has yielded extensive lists of chemical compounds that can be isolated from a natural product such as coffee. However, this analysis has given almost no understanding of how the characteristic coffee aroma comes about or what the important and contributing compounds are in a flavor. To understand the important players, we need to have sensory analysis combined with chemical analysis. The human has to evaluate the separated compounds to see which are, in fact, impactful, smelly, and above the threshold of detection at the concentration in which they are found in the food (Piggott, 1990). Various methods combining chemical analysis and human sniffing have been developed to do just this (Acree, Barnard, & Cunningham, 1984). These methods identify the "dramatis personae" in the play of flavor interactions, but they do not tell us the script or how they work together. This is still an area of much ongoing research by flavor scientists. Henick-Kling and Acree (1994) have suggested that the flavor compounds can be classified in three groups. Nominal compounds are those that smell like the whole product and may contribute to the fundamental characteristic aroma (citral is a nominal for lemongrass). Congeners are compounds that modify the fundamental aroma to produce a complex variation on the theme. For example, one might say, "this compound adds a floral, green note to the apple aroma." Diversifiers contribute in ways that are not so easily named, although the odor has clearly changed. "Now the apple aroma is fresh."

The entire enterprise of food preparation can be seen as an attempt to make harmonious mixtures. Consider the humble tossed salad. If we look at just the vegetable ingredients, this is a recipe for blandness at least as such salads are constructed in the United States. Nonetheless, we have one important sensory ingredient to start—good crisp texture and crispness contrasted with other textures (the soft expressible juiciness of a cherry tomato, for example). But how do we fix the flavor problem? We dress the salad. Classically, dressings take a highly impactful taste stimulus such as vinegar and mix it with various other components—the coating, unctuous protective effect of oil, or various other components, not the least of which is sugar if you read the label of many prepared commercial salad dressings. So immediately we are faced with finding an antidote to the overly forward impact of a strong acid. Eventually this enterprise leads to dressings that overtake the entire subtle flavor of the salad. The vegetables are completely masked, and the dish is now defined by the buttermilk notes of our creamy ranch dressing or even the intense cheesy flavor of grated parmigiana.

Interactions between Taste and Smell

Do taste and smell interact? Of course they do; every cook knows that! But how and under what conditions? An apparent discrepancy between the food literature and sensory science is the issue of how taste and smell interact. It is a common belief among food scientists, as well as consumers, that taste and smell are somehow related or are closely linked in the perception of flavor. Some of this assumed relationship derives from the generic use of the word "taste" to mean all aspects of food flavor. However, if we restrict the use of the word taste to mean sensations from nonvolatile substances perceived in the oral cavity, we can ask whether these taste sensations interact with aromas and volatile flavors that are sensed by the olfactory apparatus. This taste/smell distinction is thus made on anatomical grounds. A number of experiments have addressed this question.

When you isolate taste and smell in a simple mixture, it is hard to find evidence of much interaction or suppression effects. Intensity ratings show about 90% additivity (Murphy & Cain, 1980). When framed as a simple question about the ways in which tastes and smells combine to produce overall impressions of flavor strength, little evidence exists for interactions between the two modalities. They just add up. The one notable exception to this rule is that some smells will contribute to judgments of "taste" magnitude. This is a common illusion with aroma by mouth. When a flavorous solution is placed into the mouth, people have a hard time distinguishing the volatile sensations as odor and misattribute them to taste. The effect is eliminated by pinching the nostrils shut during tasting, which prohibits the passage of volatile materials up into the nose and effectively cuts off these volatile flavor impressions. Aside from this mislabeling, the scientific evidence points to more independence of taste and smell than interaction, in contrast to popular belief.

However, different results emerge with real products rather than simple model solutions of flavor chemicals. Von Sydow, Moskowitz, Jacobs, & Meiselman (1974) and Perng and McDaniel (1989) examined ratings for taste and odor attributes in fruit juices that varied in added sucrose. Ratings for pleasant odor attributes increased and those for unpleasant odor attributes decreased as sucrose concentration increased. Sucrose also suppressed "harsh" tastes such as bitterness, sourness, and astringency. Such unpleasant tastes may have drawn attention away from volatile characteristics in juices of low sweetness and harsher character. An unpleasantly strong acid taste tends to capture your attention. When the juices were sweeter and more "in balance," panelists' attention may not have been so distracted by the harsh tastes. Then the chance of recognizing the flavors that are present is better. The interactions also may depend on the particular flavorants and tastants that are combined. Wiseman and McDaniel (1989) reported some enhancement of fruitiness of orange and strawberry solutions by aspartame compared with little or no effect for sucrose and a somewhat greater enhancement for orange than strawberry. Frank and Byram (1988) found sweetness to be enhanced by strawberry odor but not by peanut butter odor. Later studies with greater numbers of tastants showed general suppression of sodium chloride saltiness by volatile flavors, but more complex interactions with other tastants (Frank, Wessel, & Shaffer, 1990). Further research is needed. The degree of cultural experience panelists have with particular combinations may play a role (Stevenson, Prescott, & Boakes, 1995). When we frequently experience two flavors together, they become associated and tend to support each other's perception in subsequent dishes that we eat. Some of the enhancement of sweet tastes by sweet smells (those associated with heated sugar) may depend on our learned pairings of certain smells and tastes.

SENSORY ADAPTATION

Everyone has had the common experience of putting his or her foot into a hot bath and finding it nearly intolerable at first, but that hot sensation declines after a few moments. This is an example of sensory adaptation. The senses tend to adjust to an ambient level of constant stimulation so that we do not have to pay attention to things that are not changing. Another common experience is when we go from broad daylight into a dark movie theater. It takes a minute for the eyes to adjust to the lower level of light, but after a time we can maneuver to our seat without difficulty. Most people can recall the experience of coming into another person's home and smelling the characteristic odor of that family's environment. What styles of food you cook, whether there are pets, babies in diapers, smokers, and so on can all affect this odor. However, after about 10 minutes or so, we are usually unaware of the smell. This is an example of olfactory adaptation. The flavor senses show adaptation but only under certain conditions.

Adaptation occurs in the taste sense as well. You are generally unaware of the salt in your saliva, but if the mouth is rinsed with distilled water, that concentration of salt would usually cause a discernible taste. Taste adaptation is most clearly demonstrated under controlled conditions of constant stimulation, as when a stream of water is flowed over the tongue surface or a sample is immobilized on the tongue (Gent, 1979; McBurney, 1966). However, when the stimulus is more intermittent in nature, the sense of taste is not so easily attenuated, and taste intensity may be maintained. For example, Meiselman and Halpern (1973) showed that when solutions were pulsed over the tongue, they would tend to increase or enhance in intensity rather than show the falloff characteristic of adaptation. This may be one purpose or effect of having multiple different foods and a beverage available during a meal. Alternating bites of different foods and sips of a beverage may help to reset the taste sense and avoid dulling the palate as might occur if we were to eat an entire portion of one food at a time.

RELEASE FROM SUPPRESSION

We can combine the two principles of sensory adaptation and suppression in mixtures to illustrate how food flavors can change merely as a function of the sequence in which we eat them. This makes the meal an interesting experiment in tasting. Having your sweet iced doughnut and then drinking your orange juice is not so nice an experience as when the potently acidic juice is consumed first, and then the sweet item. When we fatigue or adapt the tongue to one specific taste (such as sweetness), we not only affect the response to subsequent sweet items but to the balance of other items that contain both sweetness and some other taste qualities. If the sweetness is adapted in a complex mixture, the other tastes are no longer suppressed. In a sour-sweet mixture like juice, taking away the sweet impact means the sourness is no longer inhibited. It is released from its formerly suppressed state. So it leaps forward and becomes more intense and much more apparent, perhaps even unpleasantly so.

Figure 5–1 shows how this effect can be seen in a controlled experiment (Lawless, 1982). When the tongue is adapted under controlled conditions, a reduction in sweet perception or any other taste quality can sometimes be achieved. The first bar shows the response to the sweetness of sucrose in water at a concentration level of about 3% by weight. The second bar shows how the sweetness is reduced when that same level of sucrose is present in a mixture with the bitter substance quinine. A reduction of about one-third in the sweet taste intensity occurs. Next, a solution of quinine is flowed over the tongue to adapt away the bitter response. What does the mixture taste like now? The bitterness of the mixture declines, and the sweet taste returns to the level it would be perceived at if just presented alone. That is, the bitterness is now "invisible" (or nearly so) to the tongue so there is no mixture as far

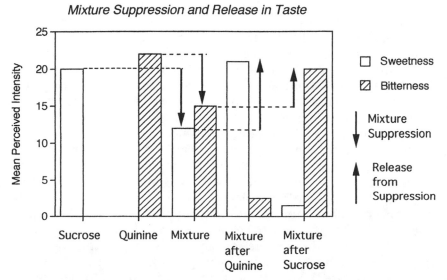

Figure 5–1 Sweetness of .32 M sucrose, .0001 M quinine, and a mixture containing .32 M sucrose and .0001 M quinine (equimolar to their single comparison solutions). The bitterness and sweetness are reduced in the mixture, a case of mixture suppression. After adaptation to quinine, the bitterness is reduced and the sweetness is increased, a release from the suppression effect. After adaptation of the tongue to sucrose, the sweetness is reduced and bitterness is increased. *Source:* Reprinted with permission from H.T. Lawless, Paradoxical Adaptation to Taste Mixtures, *Physiology and Behavior*, Vol. 29, pp. 149–152, © 1982, Elsevier Science.

as the brain is concerned and no suppression. The sweetness bounces back up to its unmixed level.

The hatched bars show the same effect for bitterness. When presented alone, a given level of bitter intensity is achieved. This is done with a much lower concentration than we had of sucrose because the quinine is a potent taste stimulus. A little goes a long way. When this concentration is presented with sucrose, we see the mixture suppression effect. If the tongue is adapted to sucrose, the sweetness is gone and so is its masking effect on the bitterness of quinine. The bitterness now rebounds to its unmixed level.

Because most foods are mixtures, these kinds of contrasts probably occur to one degree or another in any meal. Perhaps this is why most people tend to alternate bites of different items rather than eating an entire portion of one thing at a time. The effect can be demonstrated with a lemon and any beverage that has acid and sugar like a fruit drink or a semi-dry wine. Cut the lemon into wedges. If you are doing this with wine or another alcoholic beverage, put the knife away. Next try the drink to get a baseline idea of its sweet/sour balance. Now take a good healthy bite of a lemon wedge and suck out the juice and swallow. Quickly, try the drink again. You will probably notice that it has become distinctly sweeter and less acid, that the balance has changed. For some wines, this may be a poor adjustment. Previously tasted acidity makes a wine seem flabby. For other drinks like cheap tequila, the lemon may improve matters considerably.

Does this release effect happen for smell? An old perfumer's trick is sometimes used when trying to analyze a competitor's fragrance and the blend is just too complicated to see all of the notes clearly. The trick is to pick a note that you are sure about and then fatigue the nose to that particular component. Then the other smells become more readily apparent. Perhaps this is due to a release from the suppression effect. Figure 5–2 shows an odor test much like the sugar-quinine experiment discussed previously (Lawless, 1987). When vanilla and cinnamon were mixed, they suppressed one another. When the nose was partially adapted to vanilla, the cinnamon note returned to a higher level. Conversely, when the nose was partially adapted to cinnamon, the vanilla note rose back.

So what can this teach us about the sequence of items we consume in a meal? Part of the adventure of creating a good meal is to arrange possible sequences that may create some contrasting effects but not unpleasantly. Diners should be aware that these effects can occur. Probably we have some experiential knowledge or gut level appreciation of them. Chocolate chip cookies and milk are a lot more harmonious than chocolate chip cookies and orange juice. The problem of many a sommelier is to bring the wine to be tasted when the diners are in the middle of the salad course. When the tongue is adapted to the sharpness and acidity of a vinaigrette, even a finely structured wine may seem to be a little too sweet and flabby. Better to rinse the mouth or chew a piece of bread to clear the palate before sampling the wine if a decision is to be made to keep or return the bottle. The release-from-suppression effect makes a wine after vinegar a bad sequence. Conversely, the wine that seemed fine with your pasta or entree may seem overly harsh if paired with a sweet dessert. Better to finish the table wine before the dessert

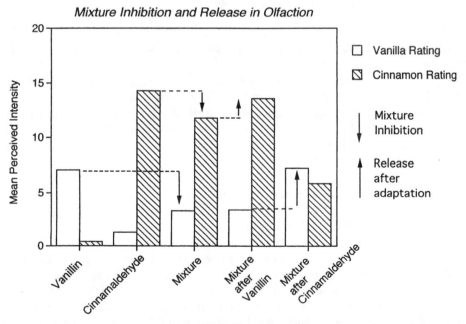

Figure 5–2 Perceived intensity of cinnamaldehyde and vanillin expressed as ratings of cinnamon odor strength and vanilla odor strength, respectively. When mixed in the vapor phase, the individual odor notes are suppressed. After presenting a vanillin sample to adapt the nose, the cinnamon intensity of the mixture increases. After presenting a cinnamaldehyde sample to adapt that component, the vanilla intensity increases. *Source:* Reprinted with permission from H.T. Lawless, An Olfactory Analogy to Release form Mixture Suppression in Taste, *Bulletin of the Psychonomic Society*, Vol. 25, pp. 266–268, © 1987, Psychonomic Society.

Exhibit 5–1 Sensory Interactions—Definitions and Descriptions

> *Mixture Suppression:* Individual taste and smell characteristics are less intense in a blend than if you tasted the same ingredient alone.
>
> *Adaptation:* The senses become less responsive when subjected to constant stimulation.
>
> *Release from Suppression:* After adaptation to one component of a mixture, the other components are less suppressed, and so they increase in intensity or become more apparent.

course or order some port. Exhibit 5–1 gives a brief summary and description of the effects of mixtures, adaptation, and release.

THE IMPORTANCE OF TOUCH, TEXTURE, TEMPERATURE, AND PAIN

Most of the previous discussion has centered on the sensations in a meal that arise from chemical stimulation of the senses of taste and smell. However, a great deal of the experience of different foods and beverages is tactile or thermal in nature. Consider an entire meal in which every food item and beverage are consumed at room temperature (cold pizza and warm, flat soda for example). This situation would be even worse if the food were homogenized in a blender to make a pureé. Humans like contrast, and texture contrast is critical to food enjoyment. The combination of crisp texture with juiciness (ability to express moisture when chewed) is not always easy to achieve in foods, but when it can be done it is often extremely successful—witness the "chicken nugget" item popular in many American fast food restaurants. The genius behind many successful confections lies in the contrasting textures provided. One example would be the brittle candy shell surrounding a chocolate phase that covers a peanut. This little delight provides three contrasting texture experiences as one bites through—the hard but brittle shell followed by the softer chocolate and finally the resistive peanut center. This notion of texture contrast is developed in a theory by Hyde and Witherly (1993) used to explain the appeal of foods like ice cream. However, the same principles of texture contrast can be extended to varying items in a meal. We like combinations that alternate soft with chewy (potatoes and meat) or soft and crunchy (pasta and garlic bread).

In addition to textural and thermal characteristics, one must also consider the tactile experiences in foods that border on the painful. Large proportions of the world's population consume foods high in spices that stimulate the senses of touch and pain in the mouth and nose (P. Rozin, 1990). These spices, like black pepper, chili pepper, mustard, horseradish, cinnamon, cumin, and coriander, have aromatic properties to be sure. However, they also stimulate a parallel sensory system in the nose and mouth mediated not by the taste and smell nerves but by the nerves responsible for touch, temperature, and pain. These are mainly the trigeminal nerves that terminate in the tissues of the eyes, nose, and mouth (see Green & Lawless, 1991 for a review). The lacrimation or tearing you experience when peeling onions is one example of this kind of stimulation. Psychologists for decades have been puzzled as to why an initially aversive experience such as chili pepper burn becomes appetitive after some experience with the sensations (P. Rozin, 1990).

Very hot or pungent spices provide a kind of thrill and a layer of added sensory experience on top of the tastes and smells in a meal. Although some evidence exists for a masking effect on tastes (Karrer and Bartoshuk, 1995; Lawless and Stevens, 1984), most of the flavorous experiences remain intact for those people accustomed to consuming very spicy food. Of course, to the person who has little or no experience with hot and spicy cuisine, the painful sensations may be quite distracting and detract from their appreciation of the other flavors that are present. In psychological terms, there is certainly a commanding attentional component to the experience of a hot chili burn the first few times you experience it. However, after a time, we learn that these sensations are safe and that they are just an added dimension to our enjoyment of food. Tolerance to the heat develops so that higher levels become desirable and some have speculated that the endorphin system is aroused, which might in part explain the addictive quality or perhaps "high degree of loyalty" shown by individuals who develop a taste for such aggressive flavor experiences. Some have suggested that an experience akin to the "runner's high" occurs with eating hot spices (Weil, 1975).

A good chili burn does more than stimulate the senses. In addition to the potential endorphin arousal, pepper heat also evokes a variety of defensive bodily reflexes, including sweating, salivating, and even tearing if the burn is hot enough. I have even heard anecdotes about a really hot curry making your ears ring. So we need to consider not only the primary sensations that are derived from food flavor and texture but the secondary experiences that arise as a part of the body's reflex arousal. Wine paired with food is another good example of a sensory combination that provides multiple sensory and physiological consequences. Witherly (1987) points out that wine is a potent stimulator of saliva, digestive tract activity, and appetite, all of which contribute to our appreciation of and desire for food. Salivary flow is in itself an important modulator of our sensory experience in a meal. It plays a "pivotal role" in our appreciation of various tastes by solubilizing them, serves a lubricating effect that makes mastication more efficient, and acts to enhance the various texture changes like melting that provide contrast during consumption of foods like chocolate and ice cream (Witherly, 1987).

TEXTURE CONTRAST THEORY

Hyde and Witherly (1993) proposed a theory that can explain why a variety of different or contrasting sensory experiences in a meal are appealing to humans. They began by asking why a dessert such as ice cream should be so desirable, especially in the face of gross satiation. In other words, most people will eat ice cream after a meal even when they feel full. To explain this, they proposed that ice cream invokes several processes basic to highly palatable foods. These include central nervous system activation (arousal, in other words), the ability to evoke a variety of systems for enhancing pleasure, and the delivery of contrasting sensory experiences during consumption. The "variety of systems" includes temperature effects from cooling, viscosity effects from oromuscular activity like tongue movements, and of course, taste and other texture perception.

The contrasting or dynamic effects are of primary interest here. This single product—ice cream—can achieve some of the results of eating a variety of contrasting foods in a meal. Ice cream is basically a frozen foam, an emulsion of fat, ice crystals, sugar, and air. As it melts, the phase change provides a dual matrix, some semisolid parts floating in a fluid matrix. If we include various bits or chunks of additional ingredients such as chocolate chips, cookie parts, fruits, or nuts, we can enhance this perception of a multiphase product experience. As melting progresses, the semisolid parts give way to a totally fluid sensation, and swallowing is virtually mandatory. The added ingredients, if present, are dynamically destroyed by mastication. The thermal sensations may continue to grow for some time as the oral tissues are cooled by the product. So by the very nature of the way the product behaves once it enters the oral cavity, ice cream provides a changing array of sensory experiences. From one perspective this can be viewed as a mechanism to overcome sensory adaptation. "Ice cream melting in the mouth rapidly undergoes changes in texture, temperature, and flavor. This barrage of contrasting sensations gives orosensory afferents little chance to adapt" (Hyde & Witherly, 1993, p. 10).

Hyde and Witherly expanded this observation to the proposal that dynamic contrast provides a form of novelty or change in the flow of sensory experiences and that such novelty would in itself be rewarding. They go on to note that the tactile experiences are often those that provide the contrasting and dynamic elements. Foods that contain textural contrast, such as flaky pastry with a creamy filling or pizza with a variety of toppings, are most appealing. Aqueous sucrose (sugar in water) is not very appealing, and even with added aromatic flavors the mixture is rarely that enticing. However, lower the temperature and add carbonation to provide oral tactile sensation and you have a very popular beverage. Hyde and Witherly point out that the act of swallowing a carbonated beverage produces a temporary burst of tingling or even stinging sensations, which, when fleeting, are positive in nature.

Oral movements and the actions of the perceiver may be critical in maximizing the contrast or change and thus the pleasure. Foods that undergo changes in texture during mastication or that contain a variety of textural experiences are likely to be most appealing. Consider warm soggy breakfast cereal as opposed to crunchy cereal in cold milk. The latter requires action on the part of the eater and undergoes change in both texture and temperature before swallowing. Although Hyde and Witherly's notion of dynamic contrast applies to the consumption of a single food item, the principles they proposed can be extended to the kinds of sensory contrast and variety experienced in a meal.

It is important to view the experience of eating as a dynamic and interactive process. The observer is not a passive collection of receptor organs that are stimulated by an onslaught of unsought-after stimulus energies. Rather we seek out stimulation from food and interact with it dynamically. Some experiences, notably texture and viscosity, cannot be perceived unless a person is interacting with the food (oral movements are essential). This may also be true of some chemically derived sensations such as astringency. It is difficult to perceive the drying and roughing effects of astringent tannins on the surfaces of the mouth unless the tongue and lips move against other oral surfaces. Furthermore, it is likely that throughout our life span, we come to adopt optimal strategies to enhance our pleasure in eating by oral movements and chewing strategies that are personally gratifying. This might be one reason why people with new dentures complain that their food does not taste as good. In essence they have to learn to chew again. Such a disruption of the normal eating movements may also be one reason why astronauts in microgravity find food less appealing during space travel. The normal collection and manipulation of food in the mouth could be disrupted by weightlessness.

MEALS AS A MECHANISM FOR SENSORY VARIETY

Humans, like other omnivores, are seekers of variety (van Trijp, 1994). Thus we can view the meal as one way to add complexity and variety in our sensory combinations. Witherly (1987), reflecting on the importance of texture contrast, stated, "Texture contrasts (firm/crisp vs. soft/creamy) in a meal, on the plate or within a multiphased or uniphased food system enhance the eating experience *and reflect upon the excellence of food preparation*" (1987, p. 407, italics added). It is noteworthy that Witherly states that the achievement of texture contrast is one sign of the quality of the culinary art.

Foods tend to decrease in their appeal as we eat them. This is also true of things that are merely tasted without being consumed, so the effect apparently does not depend on any sense of fullness or satiation (Drewnowski, Grinker, & Hirsch, 1982). Rolls and coworkers have coined the term "sensory specific satiety" to refer to this decrease in the pleasantness of specific tastes or flavors over repeated exposure (Rolls, Rolls, & Rowe, 1983). Sensory specific satiety can be thought of as a decrease in the appeal of a food caused by exposure to the sensory qualities of that food (Johnson & Vickers, 1992). This concept is explored more fully in Chapter 6.

We can combine this effect with the notion that one purpose of having diverse foods and beverages in a meal is to provide variety. Of course, meals also provide contexts that can increase a food's appeal relative to items considered alone (Rozin & Tuorila, 1993). Milk and cookies are a lot more interesting than milk consumed alone or cookies consumed alone. So one way to view a meal is that it provides some gratification for our human tendency to seek variety in what we eat (van Trijp, 1994). A meal of diverse tastes and flavors also provides the opportunity for alternation of sensory experiences, a relief from sameness. Little is known about the specific parameters that affect the development of sensory specific satiety. However, it would seem reasonable to propose that the duration and continuity of exposure to an item should influence how dramatically the change in appeal comes about. One or two bites of a fried food or bites consumed several days apart should not produce much shift in liking. However, a big plateful of a fried item, consumed all at once and in isolation, can certainly change one's opinion. "Enough is enough" is a phrase I have heard at the all-you-can-eat fish fry.

The possibility of alternation and variety of sensory experiences, along with sensory specific satiety, puts the meal in a new light. It provides us the opportunity to maintain the appeal of a food that we might otherwise stop eating or grow tired of. Alternation of bitefuls of meat, vegetable, and sips of a beverage will produce a varied sensory array of flavors and differences in the momentary perception of that individual item, depending on what preceded it. Release-from-suppression effects tend to enhance taste and smell contrast. The meal with varied components in terms of flavors and textures can be seen as an antidote to our tendency to become bored or satiated to a specific sensory characteristic. Even in one-dish meals, there may be variety that increases the overall appeal. I no longer find a big bowl of oatmeal to be very interesting, and I am unlikely to finish a large portion if it is put in front of me. However, an oriental stir-fry with a variety of crunchy vegetables, juicy bits of meat, and chewy sauteed tofu is interesting from bite to bite for quite awhile.

REFERENCES

Acree, T.E., Barnard, J., & Cunningham, D.G. (1984). A procedure for the sensory analysis for gas chromatographic effluents. *Food Chemistry, 14*, 273–286.

Ayya, N., & Lawless, H.T. (1992). Qualitative and quantitative evaluation of high-intensity sweeteners and sweetener mixtures. *Chemical Senses, 17*, 245–259.

Bartoshuk, L.M. (1975). Taste mixtures: Is mixture suppression related to compression? *Physiology and Behavior, 14*, 643–649.

Bartoshuk, L.M., Murphy, C.L., & Cleveland, C.T. (1978). Sweet taste of dilute NaCl. *Physiology and Behavior, 21*, 609–613.

Brillat-Savarin, J.A. (1825/1972). *The physiology of taste or meditations on transcendental gastronomy* (p. 39). (1825 Trans. M.F.K. Fisher 1972). New York: Alfred A. Knopf.

Cagan, R.H. (1981). Recognition of taste stimuli at the initial binding interaction. In R.H. Cagan & M.R. Kare (Eds.), *Biochemistry of taste and olfaction* (pp. 175–204). New York: Academic Press.

Cain, W.S. (1975). Odor intensity: Mixtures and masking. *Chemical Senses and Flavour, 1*, 339–352.

Cain, W.S., & Drexler, M. (1974). Scope and evaluation of odor counteraction and masking. *Annals of the New York Academy of Sciences, 237*, 427–439.

Drewnowski, A., Grinker, J.A., & Hirsch, J. (1982). Obesity and flavor perception: Multidimensional scaling of soft drinks. *Appetite, 3*, 361–368.

Frank, R.A., & Byram, J. (1988). Taste-smell interactions are tastant and odorant dependent. *Chemical Senses, 13*, 455.

Frank, R.A., Wessel, N., & Shaffer, G. (1990). The enhancement of sweetness by strawberry odor is instruction dependent. *Chemical Senses, 15*, 576–577.

Gent, J.F. (1979). An exponential model for adaptation in taste. *Sensory Processes, 3*, 303–316.

Green, B.G., & Lawless, H.T. (1991). The psychophysics of somatosensory chemoreception in the nose and mouth. In T.V. Getchell, L.M. Bartoshuk, R.L. Doty & J.B. Snow (Eds.), *Smell and taste in health and disease* (pp. 235–253). New York: Raven Press.

Henick Kling, T., & Acree, T.E. (1994). Inappropriate wine flavors: Microbial origins; chemistry; sensory perception. In M.R. Dharmadhikari, C.E. Edson, & K.L. Wilker (Eds.), *Proceedings of the 9th Annual Midwestern Region Grape Wine Conference* (pp. 25–35), Department of Fruit Science: SW Missouri St. University.

Hyde, R.J., & Witherly, S.A. (1993). Dynamic contrast: A sensory contribution to palatability. *Appetite, 21*, 1–16.

Johnson, J., & Vickers, Z. (1992). Factors influencing sensory specific satiety. *Appetite, 19*, 15–31.

Karrer, T., & Bartoshuk, L.M. (1995). Effects of capsaicin desensitization on taste in humans. *Physiology and Behavior, 57*, 421–429.

Lavin, J., & Lawless, H.T. (1998). Effects of color and odor on judgments of sweetness among children and adults. *Food Quality and Preference, 9*, 283–289.

Lawless, H.T. (1977). The pleasantness of mixtures in taste and olfaction. *Sensory Processes, 1*, 227–237.

Lawless, H.T. (1982). Paradoxical adaptation to taste mixtures. *Physiology and Behavior, 29*, 149–152.

Lawless, H.T. (1986). Sensory interactions in mixtures. *Journal of Sensory Studies, 1*, 259–274.

Lawless, H.T. (1987). An olfactory analogy to release from mixture suppression in taste. *Bulletin of the Psychonomic Society, 25*, 266–268.

Lawless, H.T., & Stevens, D.A. (1984). Effects of oral chemical irritation on taste. *Physiology and Behavior, 32*, 995–998.

McBurney, D.H. (1966). Magnitude estimation of the taste of sodium chloride after adaptation to sodium chloride. *Journal of Experimental Psychology, 72*, 869–873.

Meiselman, H.L., & Halpern, B.P. (1973). Enhancement of taste intensity through pulsatile stimulation. *Physiology and Behavior, 11*, 713–716.

Murphy, C., & Cain, W.S. (1980). Taste and olfaction: Independence vs. interaction. *Physiology and Behavior, 24*, 601–605.

Perng, C.M., & McDaniel, M.R. (1989). Optimization of a blackberry juice drink using response surface methodology. In *Institute of food technologists*, Program and Abstracts, Annual Meeting (p. 216). Chicago: Institute of Food Technologists.

Piggott, J.R. (1990). Relating sensory and chemical data to understand flavor. *Journal of Sensory Studies, 4*, 261–272.

Rolls, E.T., Rolls, B.J., & Rowe, E.A. (1983). Sensory-specific and motivation-specific satiety for the sight and taste of food and water in man. *Physiology and Behavior, 30*, 185–192.

Rozin, E. (1983). *Ethnic cuisine: The flavor principle cookbook.* Brattleboro, VT: The Stephen Greene Press.

Rozin, P. (1982). "Taste-smell confusions" and the duality of the olfactory sense. *Perception and Psychophysics, 31*, 397–401.

Rozin, P. (1990). Getting to like the burn of chili pepper, biological, psychological and cultural perspectives. In B.G. Green, J.R. Mason, & M.R. Kare (Eds.), *Chemical senses, Vol. 2, Irritation* (pp. 231–273). New York: Marcel Dekker.

Rozin, P., & Tuorila, H. (1993). Simultaneous and temporal contextual influences on food acceptance. *Food Quality and Preference, 4*, 11–20.

Stevenson, R.J., Prescott, J., & Boakes, R.A. (1995). The acquisition of taste properties by odors. *Learning and Motivation, 26*, 433–455.

van Trijp, H.C.M. (1994). Product related determinants of variety seeking behavior for foods. *Appetite, 22*, 1–10.

Vickers, Z. (1991). Sound perception and food quality. *Journal of Food Quality, 14*, 87–96.

von Sydow, E., Moskowitz, H., Jacobs, H., & Meiselman, H. (1974). Odor-taste interactions in fruit juices. *Lebensmittel-wissenschaft und Technologie, 7*, 18–20.

Weil, A. (1975). Hot! Hot! — 1. Eating chilies. *Journal of Psychedelic Drugs, 8*, 83–86.

Wiseman, J.J., & McDaniel, M.R. (1989). Modification of fruit flavors by aspartame and sucrose. In *Institute of food technologists*, Program and Abstracts, Annual Meeting. Chicago, Institute of Food Technologists.

Witherly, S. (1987). Physiological and nutritional influences on cuisine and product development. In J. Solms (Ed.), *Food acceptance and nutrition* (pp. 403–415). London: Academic Press.

Yamaguchi, S. (1967). The synergistic taste effect of monosodium glutamate and disodium 5'inosinate. *Journal of Food Science, 32*, 473–475.

Chapter 6

Sensory-Specific Satiety and Variety in the Meal

Barbara J. Rolls

My most memorable meal was set in the mountains of central Italy. I was attending a conference that simply stated on the program that we would be attending a pasta party. We were bussed to a remote castle high in the mountains where, with no explanation, we were left for several hours to await our meal. Rumors started multiplying—it seemed that we were expecting the premier pasta chef in Italy. Finally, we were seated and the menu was presented. We were to have 14 courses of pasta! This apparently was an Italian joke. We started with gusto. The food was delicious, the wine flowed, and the company of colleagues could not have been better. All of these not only make for a convivial meal but are also known to improve the appetite and to increase food intake. But in this meal of pasta, which varied in shape and type of sauce, by the middle of the menu all but the most hearty eaters had given up trying to eat the pasta. Our host attempted to revive appetites by announcing that our next course was a special new pasta that all of Italy was clamoring to buy. Still no one wanted pasta. People got up and socialized throughout the last six or seven pasta dishes. Do we assume that they were completely satisfied and simply could not eat another bite? Perhaps, but why then did they rush back to the table and heartily eat the dolce or dessert? By the end of this chapter you should understand why the pasta had lost its appeal, but there was still an appetite for dessert.

The desire for variety in a meal is very basic, occurring across species and in human infants. Several types of mammals and birds, when presented exclusively with one palatable food for a long time, will avoid that food when additional, normally less preferred food is also provided. Human infants begin life consuming only milk. However on weaning, if given choices, they select a varied diet. In discussing the excellent nutritional status of these self-selecting infants, Clara Davis, who conducted studies in the 1920s, speculated that there must be some innate, automatic mechanism directing food selection (Davis, 1928). It is likely that a major factor ensuring that a varied diet is consumed is sensory-specific satiety (Hetherington & Rolls, 1996; Rolls, 1986, 1990).

SENSORY-SPECIFIC SATIETY

What is sensory-specific satiety? The next time you are eating a meal pay attention to how your responses to the foods you are eating change. Before eating, take a small bite of each of the foods that will make up the main course and also sample your dessert. Rate how much you like the taste of each food at the time of tasting.

When you have eaten as much as you want of the main course, pause before dessert and rate all of the foods again. If you experience sensory-specific satiety, you will have rated the foods you have just eaten as tasting less pleasant at the end of the course than at the start of the course, but the dessert should taste just as good or sometimes even better than on the first rating. Experimentally, we define sensory-specific satiety as the difference between the change in the pleasantness of the food that has just been eaten and the change in foods that were tasted but were not eaten.

What happens during a meal consisting of several different foods or courses is that satiety may be experienced for the foods already eaten but not for other foods, particularly those with very different sensory properties. Thus, satiety need not be for all foods but rather can be sensory specific. To maintain your enjoyment of a meal, as one food declines in pleasantness, you will switch to another food that remains appealing. If all the available foods have similar sensory properties, the satiety may generalize from one food to another. In our Italian meal, we had "pasta-specific" satiety.

By ensuring that we switch from food to food, sensory-specific satiety plays a fundamental role in maintaining nutritional balance. In this review, why sensory-specific satiety occurs and how it can influence what and how much we eat in a meal will be considered. Although we refer to the change in the hedonic response to a food as *sensory*-specific satiety, it is possible that the pleasantness of foods declines during a meal because eating has eliminated our need for both energy and specific macronutrients. We will consider how the energy and macronutrient content of foods affect the change in hedonic response before we consider the influence of the sensory properties of foods.

The Need for Calories and the Pleasantness of Food

Aristotle (cited 1973) suggested that the flavors and smells of foods are affected by our internal state: "these smells are pleasant when we are hungry, but when we are sated and not requiring to eat, they are not pleasant." This implies that the energy content of foods, which would satisfy hunger, is a critical determinant of the changing hedonic response to foods during a meal (Cabanac, 1971).

The way to determine the role of the energy content of foods in sensory-specific satiety is straightforward. The energy content of foods can be varied by substituting a high-calorie ingredient with one that is lower in calories. Such substitutions alter the energy density (i.e., the calories per gram) of a food. In several different studies, we have offered test participants foods that could be consumed freely and have assessed sensory-specific satiety. The energy density of the foods was varied by reducing the carbohydrate content of soup (Rolls, Hetherington, & Burley, 1988a), by replacing sugar with a noncaloric sweetener (aspartame) in pudding or gelatin dessert (Rolls, Laster, & Summerfelt, 1989), by substituting vegetables for pasta in casseroles (Bell, Castellanos, Pelkman, Thorwart, & Rolls, 1997), or by replacing fat with a noncaloric fat substitute (olestra) in potato chips (Miller, Hammer, Peters, & Rolls, 1996). In all these tests the participants consumed the same amount of a particular food type and therefore consumed more calories when the food was higher in energy density. The calories consumed of a particular food did not, however, affect the change in pleasantness of the foods after consumption. These studies suggest that sensory-specific satiety does not depend on the repletion of a physiologic need for energy.

Because the test participants stopped eating after a similar amount of a particular food type was consumed, it was hypothesized that sensory-specific satiety may be related to the amount or weight of food consumed, which in turn could relate to the amount of sensory stimulation associated with consumption. To confirm this hypothesis we independently varied the volume of a milk drink that subjects were required to consume while keeping the energy content constant, or we varied the energy content while keeping the volume consumed constant (Bell, Thorwart, & Rolls, 1998b). We found that sensory-specific satiety varied with the volume consumed so that the greater the volume the greater the decline in pleasantness of the drink. The energy content of the drinks did not affect sensory-specific satiety.

These findings indicate that during a meal the pleasantness of foods changes with the amount (weight or volume) of food eaten. People eat a food until it is no longer pleasant; that is, they stop eating just before it becomes unpleasant tasting. Because they stop after consuming a given weight of food, if the food is higher in energy density (more calories per gram), the greater the energy intake of that food will be. Likewise, if the food is lower in energy density (fewer calories per gram), the lower energy intake will be. This was clearly illustrated in all the studies reviewed earlier in which the effects of the energy content of foods on sensory-specific satiety were tested. An important implication of this finding is that one way to reduce caloric intake in a meal is to reduce the energy density of the available foods (Bell, Castellanos, Pelkman, Thorwart, & Rolls, 1998a). This can be achieved by decreasing the fat content or by increasing the proportion of foods low in energy density such as vegetables and fruits.

Macronutrients and Sensory-Specific Satiety

In our pasta feast, did we tire of pasta because we had eaten our fill of carbohydrates? Was the dessert still delicious because it was high in fat? Although the energy content of foods has not been found to affect sensory-specific satiety, it is possible that the macronutrient content is important. If this were the case, we would expect, for example, that consumption of a high-carbohydrate food would decrease the pleasantness of other high-carbohydrate foods. To determine the role of different macronutrients in sensory-specific satiety, we offered subjects equicaloric amounts of foods high in one macronutrient and low in other major nutrients (Rolls, Hetherington, & Burley, 1988b). To test for interactions due to similarities in composition, each test food was paired with a food—tasted but not eaten—that was similar in macronutrient composition but that had different sensory properties. For example, pasta was consumed as a test food and its effects on pretzels, which represent another high-carbohydrate food, were assessed. As expected, the largest change in the pleasantness of the taste occurred with the food eaten in the test meal (i.e., the pasta). No significant differences were found in the changes in the uneaten food with a similar macronutrient composition (i.e., the pretzels) and foods with different nutrient composition. None of the uneaten foods showed much change in palatability after the test meals. When a varied meal was offered 2 hours later, the type of macronutrient already consumed did not affect the proportion of macronutrients in the meal. Similar results were seen when either fat or protein was the main macronutrient in the test food. Thus, changes in the pleasantness of foods did not relate specifically to the macronutrient composition. Furthermore, no significant differences were found in the magnitude of the changes after the different macronutrients. Therefore, these results do not support suggestions (Cabanac, 1979; Guy-Grand & Sitt, 1976; Vandewater & Vickers, 1996) that some macronutrients are more effective than others in modifying the hedonic response to foods.

Sensory-Specific Satiety Depends on the Sensory Properties of Food

When we discussed the role of the energy content of foods in sensory-specific satiety, the studies indicated that the volume or weight of food consumed was a critical determinant of changes in the pleasantness of foods. This suggests that the amount of sensory stimulation associated with eating could be a critical influence on sensory-specific satiety. The timing of the development of sensory-specific satiety after eating further reinforces the importance of sensory stimulation. Measurement over the hour after eating of changes in the pleasantness of the taste, texture, appearance, and odor of a food such as cheese on crackers showed that the largest decrease in pleasantness was seen 2 minutes after eating stopped, with a gradual recovery in pleasantness over the hour (Hetherington, Rolls, & Burley, 1989). All the sensory properties of the eaten food were affected. The changes in ratings occurred before most of the food could have been absorbed and metabolized, again indicating that the energy and macronutrient content are not critical determinants of sensory-specific satiety. All these data support the hypothesis that the sensory stimulation associated with eating a food is a critical determinant of sensory-specific satiety.

Because the sensory properties of foods influence the development of sensory-specific satiety, it is of interest to ask whether the sensory properties of foods alone without actual food consumption can affect the hedonic response. In this way we can be sure that postabsorptive effects of foods have not influenced the results. In a recent study (Rolls & Rolls, 1997), chewing but not swallowing either fresh banana or cooked chicken for 5 minutes decreased the pleasantness of both the taste and the odor of the food that was chewed, but not of other foods. Furthermore, just smelling a food for 5 minutes decreased the pleasantness of the odor of that food, although the decline was of a smaller magnitude than when the food was either eaten or just chewed. Normally, the entire sequence of events associated with eating will contribute to satiety. The cues from the sensory properties of foods clearly provide rapid information about the amount consumed.

How Consumption of One Food Affects the Pleasantness of Other Foods

Because we are suggesting that the taste of foods is involved in sensory-specific satiety, we should question whether foods with similar basic tastes such as sweet or salty can interact so that a decrease in the pleasantness of one sweet food affects other sweet foods. We have found that this is what happens (Rolls, Van Duijvenvoorde, & Rolls, 1984). For example, eating chocolate pudding was associated with a decrease in the pleasantness of the taste of sweetened fruit yogurt and of a fresh ripe banana. Salty/savory foods were not affected by eating sweet foods. Eating a salty/savory food such as sausages decreased the pleasantness of the taste of other salty/savory foods such as cheese on crackers and potato chips, but sweet foods were unaffected. It seems likely that such interactions between foods of similar basic tastes, but not for foods with different tastes, explains why at the end of the pasta feast we were still eager for dessert. The dessert likely still tasted just as good as it did before the meal had begun.

When we eat a familiar food, our previous experiences with that food can influence how much we like it and the amount consumed. We learn to associate particular foods with the satiety that they produce (Booth, Mather & Fuller, 1982). We also learn to group foods by particular categories. These can be according to the type of food. For example, we may perceive that all soups are in the same category. Or we may group foods according to when they are eaten in a meal—appetizers, entrees, desserts. When we have assessed sensory-specific satiety, the consumption of some foods affected the pleasantness of other foods. For example, consumption of tomato soup affected the taste ratings of another soup, consommé, but not of fresh tomato (Rolls et al., 1988a) or of tomato-based casserole (Rolls, Fedoroff, Guthrie, & Laster, 1990). The taste of orange drink declined, perhaps because it was also a liquid. Thus, interactions take place between foods of the same type, but if they are very different in sensory properties such as texture, for example, such interactions may not occur. Understanding the basis of these interactions among foods is important for meal planning because the goal of a meal should be to maintain palatability at a high level throughout.

It is likely that at least part of the specificity of satiety stems simply from the knowledge that a particular food has been consumed. Because we learn how much of a food is appropriate to satisfy hunger in various situations (Booth et al., 1982), it is likely that when this limit is exceeded, the pleasantness of a food will be reduced. Thus, the changing hedonic response to foods is a complex response resulting from the sensory stimulation accompanying ingestion, the knowledge that a particular food has been consumed, and finally, after some delay, changes in the need for energy or nutrients.

Are Some Foods Resistant to Sensory-Specific Satiety?

Rogers (1994) has proposed that some foods are so pleasant that no matter how much we eat sensory-specific satiety will not develop. It is unlikely that any foods are so palatable that they will be resistant to sensory-specific satiety. Instead, the initial palatability of a food may influence how much of a food must be consumed before its pleasantness declines. Even chocolate, which is often rated as the most pleasant food in our tests, declines in pleasantness as it is eaten (Rolls, Rolls, Rowe, & Sweeney, 1981a).

If a food is frequently overeaten, this can alter the amount of food required to produce sensory-specific satiety. For example, individuals who habitually overeat chocolate show smaller changes in the pleasantness of chocolate after eating a given amount than do consumers who like chocolate as well but who do not overeat it (Hetherington & Macdiarmid, 1994).Thus, systematically overeating a food can weaken satiety signals for that food, including sensory-specific satiety.

Individual Differences in the Development of Sensory-Specific Satiety

It is possible that sensory-specific satiety could provide at least part of the explanation for the aberrant eating behavior observed in some clinical populations such as individuals with anorexia nervosa who eat too little to meet energy needs and those with bulimia nervosa who habitually binge. In a controlled study, we found that the persons with bulimia demonstrated sensory-specific satiety after a low-calorie salad but not after a high-calorie salad. In contrast, persons with anorexia reported sensory-specific satiety after a high-calorie salad and not after a low-calorie salad (Rolls et al., 1992). A possible interpretation of these results is that for the individuals with anorexia, the high-energy salad, which was loaded with cheese, meat, and high-fat dressing, was considered "dangerous" and therefore they reported a rapid decline in its pleasantness. The individuals with bulimia, on the other hand, may have experienced the high-calorie salad in the same way that they might experience a binge and accordingly recorded no change in response to the eaten food as a function of intake. This observation might explain, in part, why bulimics are capable of consuming very large amounts of food in a short period of time.

That sensory-specific satiety in these groups was dependent on the type of foods given demonstrates the power of cognitive processes over sensory characteristics of the foods in the expression of sensory-specific satiety. Because satiety is specific to certain foods and can be influenced by the type of food and the individual's cognitive appraisal of the food, it has been suggested that a more appropriate term for sensory-specific satiety is *food-specific satiety* (Booth, 1995). Certainly, this term conveys the importance of the features of the food, which are rooted in the sensory domain, but also gives attention to a more cognitive component. A person's appraisal of how safe (low-calorie, low-fat, etc.) or dangerous (high-calorie, high-fat) a food is may contribute to the development of sensory-specific satiety.

Thus, sensory-specific satiety can provide at least part of the explanation for the restrictive food practices associated with anorexia nervosa and the bingeing behavior characteristic of bulimia nervosa. Changes in sensory-specific satiety may also help us to understand changes in eating behavior seen with aging. Large-scale dietary surveys indicated that some elderly individuals have decreased variety in their diets (Fanelli & Stevenhagen, 1985). Because one role of sensory-specific satiety is to encourage consumption of a variety of foods, we predicted that aging would be associated with a decline in sensory-specific satiety. When we tested sensory-specific satiety in adolescents, young adults, older adults, and elderly individuals, we found that all groups but the elderly group showed sensory-specific satiety (Rolls & McDermott, 1991). The elderly participants showed little change in their ratings of the yogurt test food either when given a large amount that they were required to eat or when they ate as much yogurt as they wanted. One possible explanation for this is that sensory processing and acuity decline with age; however, we found no correlation between sensory impairment and loss of sensory-specific satiety. Other possible explanations are that the elderly are resistant to changes of any kind or that the impairments seen in the regulation of food intake in elderly individuals (Rolls, Dimeo, & Shide, 1995) are pervasive and extend to sensory-specific satiety. Because a varied diet is critical for good health, special care must be taken when planning meals for the elderly. These should incorporate a variety of foods with very different flavors, textures, and appearance (Duffy, Backstrand, & Ferris, 1995).

VARIETY AND FOOD INTAKE

If satiety is specific to particular properties of foods, more should be consumed during a meal if a variety of foods is available instead of just one food. Variety in a meal can enhance energy intake and the more different the foods are, the greater the enhancement is likely to be.

Variations in the flavor of food can enhance intake. Three successive courses of cream cheese–filled sandwiches with distinctive flavors (salt, curry, and lemon/

sweet) enhanced intake by 15% compared with another test when participants were offered only their favorite flavor. It is perhaps critical that the sandwiches were basically of different tastes because in another test no enhancement was seen with three different fruit flavors of yogurt or with three different flavors of chocolates when the other sensory properties of these foods such as texture and appearance were the same. The lack of effect could have been because all the foods were sweet and of the same type (Rolls, Rowe, & Rolls, 1982).

Variations in the shape of food, which altered both the appearance and the mouth feel, also affected intake. People ate 14% more when offered successive courses of three different shapes of pasta than when they had only the favorite shape. When just the appearance of a food was different, as in a test in which children were offered different colored candies, the variety of colors was not associated with a greater intake than when just one color was available (Rolls et al., 1982).

Thus, when just the flavor or shape of foods was varied, the enhancement of intake was around 15% over three successive courses. When more properties of the available foods differed, the enhancement was greater. For example, when four courses of sandwiches with very different foods were offered at lunch, intake was about one-third more than when only one type of sandwich was available (Rolls et al., 1981b). Even though the sandwiches differed in appearance, texture, flavor, and nutrient composition, they were still the same basic type of food—sandwiches. An even greater enhancement of intake was seen when very different foods were presented (Rolls et al., 1984). In a four-course meal of sausages, bread and butter, bananas, and chocolate dessert, intake was 60% more than when just one of the foods was available throughout the meal. These foods differed in basic type, nutrient composition, appearance, flavor, and texture. Thus, it is clear that the greater the differences between foods, the greater the enhancement of intake. The implications of these findings are that if you are choosing a meal to maximize a person's intake, the foods should be as varied as possible. On the other hand, if you want intake to be kept low, the foods should be similar.

Several studies indicate that intake is enhanced by variety because of sensory-specific satiety. During meals of different foods in successive courses, ratings of the pleasantness of the taste of foods not included in the meal declined little even though two or three courses had already been consumed (Rolls et al., 1984). However, if a food had been eaten in one of the courses, it became less pleasant. These changes in foods were related to how much of the food was eaten later in the meal. For example, if a food was eaten in a previous course and became less pleasant, little of it was eaten in the next course. In contrast, if a food had not been eaten in a previous course, it was likely to be eaten when offered later in the meal. The correlation between pleasantness changes and the amount eaten was statistically significant (Rolls et al., 1981a). These findings are consistent with the view that one way in which food intake is regulated is by a reduction in its pleasantness as eating takes place.

FOOD MONOTONY AND INTAKE

We have proposed that sensory-specific satiety is a short-lived phenomenon that would have its primary effect on food intake and selection within a single meal (Rolls, 1986). There are longer term effects in which changes in the palatability of foods persist from meal to meal and even from day to day. These are referred to as monotony effects. Such long-term effects are probably related to the knowledge that a particular food was recently consumed and a memory of its sensory properties. Because there is also a cognitive component to sensory-specific satiety, it is likely that sensory-specific satiety and monotony effects are a continuum rather than distinct phenomena.

It is difficult to distinguish sensory-specific satiety and monotony effects simply on the basis of the length of time they persist after consumption. It is not yet clear what affects the timing of the recovery of palatability after consumption. We have not seen complete recovery back to premeal ratings over the 1 to 2 hours after the meals that we have assessed (Hetherington et al., 1989). In other studies of repeated daily consumption of palatable snack foods such as potato chips or chocolate over several weeks, initial palatability of the snacks had returned to its previous level by the next test day (Hetherington, 1995; Miller et al., 1996). These highly palatable snacks were resistant to monotony effects over the several weeks they were consumed.

Some other types of foods are also spared from monotony effects. For example, staple foods such as bread, cereals, and dairy products, as well as tea and coffee, can be consumed every day without a loss of palatability. On the other hand, when foods of low initial palatability are eaten repeatedly, they become very unpalatable and are disliked for 3 to 6 months after consumption (Schutz & Pilgrim, 1958; Siegel & Pilgrim, 1958).

Moskowitz (1979) has described "time-preference curves" for different types of foods, which indicate that foods not consumed for about 3 months are highly desired, but those eaten the day before may not be desired at all. Foods such as meat and shellfish, foods with a high fat content, or foods that constitute the main component of a meal such as the entree have steep curves and are greatly desired if not eaten for a long period. Recent consumption, however, eliminates the desire for such foods. In contrast, items that are not a major component of a meal and do not have a high fat or protein content such as bread, salad, potatoes, and some desserts have a much flatter function on the time-preference curves, and thus a less significant impact on preference, and could be eaten every day.

Because repeat consumption of foods has been shown to lead to decreased acceptance, it is worth considering ways to avoid such monotony effects. Creating diets as varied as possible is recommended. Also, ensuring that the initial palatability of the foods offered is uniformly high will help to curtail the decline in pleasantness. One way of improving initial acceptability is to allow self-selection of foods when possible. Kamen and Peryam (1961) found that self-selection of foods to be included in a repetitive diet reduced dissatisfaction with the diet. Overall satisfaction with a 3-day, self-planned menu cycle was equivalent to a 6-day cycle chosen by someone else.

Requiring people to eat foods that they have not selected themselves can decrease the preference for those foods. This suggestion is supported by studies of young children. If the children were forced to eat a food to gain access to rewards such as watching television, the food decreased in preference (Birch, Birch, Marlin, & Kramer, 1982).

Understanding the effects of repeated consumption on the palatability of foods is a major challenge to the food industry. All companies want consumers to continue buying and eating their products. If they can determine why people tire of foods, potentially they can design foods that are more resistant to changes in palatability with repeat consumption. At this stage our understanding of critical determinants of human food intake behavior is rudimentary. Future communication between basic scientists and the food industry and other culinary experts could potentially improve the situation so that meals can be planned that maintain palatability while optimizing nutritional quality.

CONCLUSION

The findings from studies of sensory-specific satiety can be used in planning meals. Understanding what influences the development, magnitude, and persistence of sensory-specific satiety, as well as how different foods interact with one another so that consumption of one affects the pleasantness of another, can be critical for maintaining palatability at a high level throughout a meal. The goal would be to include foods that show little interaction. Meal patterns in many cultures already stress the importance of different sensory qualities of foods throughout a meal. A meal of soup or salad followed by meat with vegetables and ending with dessert, with an emphasis of contrasts in flavors, colors, and textures, would reduce the possibility of specific satieties diminishing appetite during the meal. Perhaps if at our pasta feast we had been served salad and sorbet between some of the courses, we could have continued eating pasta longer. Even so, it seems unlikely that many of us could have continued eating enthusiastically throughout all 14 pasta courses! Although we probably would have been tempted by that thin wafer mint at the end.

REFERENCES

Aristotle. (1973). *De Sensa* (G.R.T. Ross, Trans). New York: Arno Press.

Bell, E.A., Castellanos, V.H., Pelkman, C.L., Thorwart, M.L., & Rolls, B.J. (1997). Energy density and sensory-specific satiety. *Appetite, 29*, 388.

Bell, E.A., Castellanos, V.H., Pelkman, C.L., Thorwart, M.L., & Rolls, B.J. (1998a). Energy density of foods affects energy intake in normal-weight women. *American Journal of Clinical Nutrition, 67*, 412–420.

Bell, E.A., Thorwart, M.L., & Rolls, B.J. (1998b). Effects of energy content and volume on sensory-specific satiety. *FASEB Journal, 12*, A347.

Birch, L.L., Birch, D., Marlin, D.W., & Kramer, L. (1982). Effects of instrumental consumption on children's food preference. *Appetite, 3*, 125–134.

Booth, D. (1995). *The psychology of nutrition*. London: Taylor & Francis Publishers.

Booth, D.A., Mather, P., & Fuller, J. (1982). Starch content of ordinary foods associatively conditions human appetite and satiation, indexed by intake and eating pleasantness of starch-paired flavours. *Appetite, 3*, 163–184.

Cabanac, M. (1971). Physiological role of pleasure. *Science, 173*, 1103–1107.

Cabanac, M. (1979). Sensory pleasure. *Quarterly Review of Biology, 54*, 1–29.

Davis, C. M. (1928). Self selection of diet by newly weaned infants. *American Journal of Diseases in Childhood, 36*, 651–679.

Duffy, V.B., Backstrand, J.R., & Ferris, A.M. (1995). Olfactory dysfunction and related nutritional risk in free-living, elderly women. *Journal of the American Dietetic Association, 95*, 879–884.

Fanelli, M.T., & Stevenhagen, K.J. (1985). Characterizing consumption patterns by food frequency methodologies: Core foods and variety of foods in diets of older Americans. *Journal of the American Dietetic Association, 85*, 1570–1576.

Guy-Grand, B., & Sitt, Y. (1976). Origine de l'alliesthesie gustative: effets compares de charges orales glucosees ou protido-lipidiques. *Comptes-rendus de l'Academie des Sciences de Paris, 282*, 755–757.

Hetherington, M.M. (1995). Sugar and spice: The role of pleasure in food intake. *C-H-O, 6,* 1–3.

Hetherington, M.M., & Macdiarmid, J.I. (1994). Pleasure and excess: liking for and overconsumption of chocolate. *Physiology and Behavior, 57,* 27–35.

Hetherington, M.M., & Rolls, B.J. (1996). Sensory-specific satiety: theoretical frameworks and central characteristics. In E. D. Capaldi (Ed.), *Why we eat what we eat: The psychology of eating* (pp. 267–290). Washington, DC: American Psychological Association.

Hetherington, M.M, Rolls, B.J., & Burley, V.J. (1989). The time course of sensory-specific satiety. *Appetite, 12,* 57–68.

Kamen, J.M., & Peryam, D.R. (1961). Acceptability of repetitive diets. *Food Technology, 115,* 173–177.

Miller, D.L., Hammer, V.A., Peters, J.C., & Rolls, B.J. (1996). Influence of the fat content of a food on sensory-specific satiety. *FASEB Journal, 10,* A824.

Moskowitz, H.R. (1979). Mind, body, and pleasure: An analysis of factors which influence sensory hedonics. In J.H.A. Kroeze (Ed.), *Preference behavior and chemoreception* (pp.131–148). London: Information Retrieval.

Rogers, P.J. (1994). Mechanisms of moreishness and food craving. In D.M. Warburton (Ed.), *Pleasure: The politics and the reality* (pp. 38–49). Chichester, England: Wiley.

Rolls, B.J. (1986). Sensory-specific satiety. *Nutrition Reviews, 44,* 93–101.

Rolls, B.J. (1990). The role of sensory-specific satiety in food intake and food selection. In E.D. Capaldi (Ed.), *Why we eat what we eat: The psychology of eating* (pp. 197–209). Washington, DC: American Psychological Association.

Rolls, B.J., Andersen, A.E., Moran, T.H., McNelis, A.L., Baier, H.C., & Federoff, I.C. (1992). Food intake, hunger and satiety after preloads in women with eating disorders. *American Journal of Clinical Nutrition, 55,* 1093–1103.

Rolls, B.J., Dimeo, K.A., & Shide, D.J. (1995). Age-related impairments in the regulation of food intake. *American Journal of Clinical Nutrition, 62,* 923–931.

Rolls, B.J., Fedoroff, I.C., Guthrie, J., & Laster, L.J. (1990). Foods with different satiating effects in humans. *Appetite, 15,* 115–216.

Rolls, B.J., Hetherington, M.M., & Burley, V.J. (1988a). Sensory stimulation and energy density in the development of satiety. *Physiology and Behavior, 44,* 727–733.

Rolls, B.J., Hetherington, M.M., & Burley, V.J. (1988b). The specificity of satiety: the influence of foods of different macronutrient content on the development of satiety. *Physiology and Behavior, 43,* 145–153.

Rolls, B.J., Laster, L.J., & Summerfelt, A. (1989). Hunger and food intake following consumption of low-calorie foods. *Appetite, 13,* 115–127.

Rolls, B.J., & McDermott, T.M. (1991). Effects of age on sensory-specific satiety. *American Journal of Clinical Nutrition, 54,* 988–996.

Rolls, E.T., & Rolls, J.H. (1997). Olfactory sensory-specific satiety in humans. *Physiology and Behavior, 61,* 461–473.

Rolls, B.J., Rolls, E.T., Rowe, E.A., & Sweeney, K. (1981a). Sensory specific satiety in man. *Physiology and Behavior, 27,* 137–142.

Rolls, B.J., Rowe, E.A., & Rolls, E.T. (1982). How sensory properties of foods affect human feeding behavior. *Physiology and Behavior, 29,* 409–417.

Rolls, B.J., Rowe, E.A., Rolls, E.T., Kingston, B., Megson, A., & Gunary, R. (1981b). Variety in a meal enhances food intake in man. *Physiology and Behavior, 26,* 215–221.

Rolls, B.J., van Duijvenvoorde, P.M., & Rolls, E.T. (1984). Pleasantness changes and food intake in a varied four-course meal. *Appetite, 5,* 337–348.

Schutz, H.G., & Pilgrim, F.J. (1958). A field study of monotony. *Psychological Reports, 4,* 559–565.

Siegel, P.S., & Pilgrim, F.J. (1958). The effect of monotony on the acceptance of food. *American Journal of Psychology, 71,* 756–759.

Vandewater, K., & Vickers, Z. (1996). Higher-protein foods produce greater sensory-specific satiety. *Physiology & Behavior, 59,* 579–583.

Part III

The Meal and Culture

Chapter 7

Sociability and Meals: Facilitation, Commensality, and Interaction

Jeffery Sobal

INTRODUCTION

Meals are social events, as well as food events. Douglas and Nicod (1974) contrast a food event as an occasion at which food is ingested with a meal at which food is eaten in a structured format that is strongly bound by social rules. Analysts are often distracted by the food and overlook the eating, neglecting sociability as an important component of meals as social events.

Relationships between sociability (also referred to as sociality) and eating are an essential component of meals but have received only scattered attention from researchers in a range of disciplines using a variety of methods. This chapter will draw on diverse sources to examine several of the many dimensions of sociability and meals. The introduction will consider the role of sociability in meals, and later sections will explore three dimensions of sociability: facilitation (how people's eating is influenced by others), commensality (how eating partners are selected and excluded), and interaction (social interchanges that occur with others during meals).

SOCIABILITY AND MEALS

The sociologist Georg Simmel (1961) provided early insights about sociability and meals, suggesting that meals are intended to satisfy a need for interaction by a union with others in which talking is an end in itself. Simmel contrasted the conviviality of shared eating with the personal desires of the individual (Symons, 1994), which generated interest of the social aspects of meals among later analysts (e.g., Hirschman, 1996).

The concept of a meal is difficult to precisely define, with many complexities and conditions that must be considered in trying to specify what is and what is not a meal (Makela, 1991). Conceptualization of a meal is based on culturally learned examples that provide parameters for judging whether a particular episode of food ingestion constitutes a meal. A "proper" or "ideal" meal is typically eaten with others (Douglas, 1972; Murcott, 1982), with eating alone not considered a "real" meal for many people. Conceptions of the meal may be even more socially specific, including considerations of who is required to be present for food consumption to be defined as a meal. Eating without family may not be seen as an "actual" meal by some people, and eating with only a servant present may also not be considered a meal.

If the presence of others is required for a meal, what occurs when people eat alone? Dining alone is an abhorrent action in many cultures. Except when eating

119

for purely functional purposes, solo eating is generally seen as abnormal, undesired, and even unhealthy. People perceive stigmatization as lone eaters and feel a need to justify or rationalize eating alone. This may demand accounts (Orbuch, 1997; Scott & Lyman, 1968) to explain why their meal is being taken without company. The solitary diner in a restaurant is treated as an anomaly and often feels (and may be made to feel) awkward and out of place. Eating alone may be downgraded to the less significant social category of "snack" rather than being credited with being a meal, irrespective of the amount or type of food that is consumed.

Structural changes in society have increased the probability that people eat alone, even though the family meal may not actually be in decline (Murcott, 1997). Industrialization and subsequent social transformations are widely attributed to having led to major changes in residential, family, and employment arrangements that have influenced eating patterns, including a rise in solo eating both at home and at work (Rotenberg, 1981). Changes in marriage and family patterns are increasing the number of single-person households in postindustrial societies, with about one in four adults living alone in the United States (Bureau of the Census, 1998). Living by ones' self is an important contributor to eating alone because of the lack of availability of regular meal partners in the household. However, people who live alone do not necessarily eat alone (Torres, McIntosh, & Kubena, 1992), with living arrangements an imperfect proxy for social eating. Most solitary dwellers seek out others to eat with, either inviting them or being invited into their households or going out jointly to eating establishments for at least some meals with consumptive company.

Most research about the effects of eating alone on diet and nutrition focuses on the elderly, for whom living and eating alone are prevalent because of the high proportion of widows and widowers. The collective findings of these studies are mixed (Mahajan & Schafer, 1993). Many reports suggest that those who eat by themselves have unhealthy eating, diet, and nutrition patterns; other investigations report that those who eat solo are no different, and some others find lone diners healthier (Davis, Murphy, Neuhaus, & Lien, 1990; Davis & Randall, 1983; Davis, Randall, Forthofer, Lee, & Marger, 1985; Grotkowski & Sims, 1978; Lee, Templeton, & Wang, 1995; McIntosh & Shifflet, 1984, McIntosh, Shifflet, & Picou, 1989). Some elderly people living alone have similar nutritional status to those living with others, but their food preparation and consumption practices differ. One explanation for the variety of findings is that the elderly are such a heterogeneous population that no consistent overall patterns exist. Another explanation is that men, but not necessarily women, who live alone lack the motivation, knowledge, and skills for doing food and nutrition work and more often have unhealthy eating patterns (Donkin et al., 1998), with men frequently skipping meals and consuming foods out of convenience rather than making healthier food choices. A lack of shared meals leads people to invest less attention, time, and effort in food preparation and food choices (Falk, Bisogni, & Sobal, 1996). For example, elderly people who lived alone often made enough vegetables and potatoes for several meals rather than cooking every day (Westenbrink, Lowik, Hulshof, & Kistemaker, 1989).

Frequency and acceptability of eating alone vary by meal. Breakfast is the meal most often eaten alone in modern societies because it is the least formal meal (not counting snacks), and joint eating is the most easily abandoned early in the day to accommodate differing schedules of various household members. Many people skip breakfast entirely, not being available to eat with others (Haines, Guilkey, &

Popkin, 1996). Dinners are least likely to be consumed alone because they are the most socially significant and largest meals of the day. Holiday meals and ritual feasts are often structured to involve others in shared eating (e.g., Wallendorf & Arnould, 1991; Williams, 1997) and are rarely consumed alone except in the most extreme circumstances.

FACILITATION

Social influence is the effect of the presence of others on a person's behavior (Zajonc, 1965). Social facilitation is the positive enhancement of performance when others are present (Guerin, 1993). Eating is often facilitated in the company of others, with many species eating more together than alone.

The presence of others during eating facilitates many human behaviors, including increasing food consumption (Berry, Beatty, & Klesges, 1985; Clendenen, Herman, & Polivy, 1994; de Castro 1990, 1994; de Castro, Brewer, Elmore, & Orozco, 1990; Klesges, Bartsch, Norwood, Kautzman, & Haugrud, 1984; Redd & de Castro, 1992; Shide & Rolls, 1991). A series of analyses of group size and eating behaviors reported that social facilitation occurs across all mealtimes during the day, including breakfast, lunch, dinner, and snacks (de Castro et al., 1990); on both weekdays and weekends (de Castro, 1991); in many settings, including home, restaurants, and elsewhere (de Castro et al., 1990); and for food consumed with and without alcohol (de Castro et al., 1990).

Several mechanisms for social facilitation of eating have been proposed, including modeling, exposure, and disinhibition (Guerin, 1993). Eating with another person who consumes a large amount of food serves as a behavioral model, and co-eaters consume significantly more themselves (Polivy, Herman, Younger, & Jaeger, 1979). Social modeling also facilitates drinking (Engell, Kramer, Malafi, & Salomon, 1996). The duration of mealtimes has been advanced by de Castro and Brewer (1992) as the primary mechanism for social facilitation of eating, in which the number of others directly increases as a power function of the amount of time spent in meals by offering increased opportunities to consume more food. Other investigations support the primacy of the meal-duration effect (Sommer & Steele, 1997), along with the enhanced perception of social atmosphere (Feunekes, de Graaf, & Staveren, 1995).

Social facilitation applies to drink as well as food, with sociability often a more important aspect of drinking alcoholic beverages than eating food. Alcohol is deeply enmeshed in norms about social consumption (e.g., Carlson, 1992; Dietler, 1990). Solitary drinking is often perceived as a marginal and antisocial practice, although it occurs among a substantial proportion of the population, typically at home and as part of a meal eaten alone (Demers & Borugault, 1996; Klein & Pittman, 1990). Duration of exposure in drinking settings is an important mechanism in social facilitation. Beer drinkers who are in groups remained for a longer time in drinking settings than those who were alone, with people who started drinking alone but were later joined by others remaining in drinking settings the longest (Sommer, 1965). However, large groups may not facilitate alcohol consumption in all cultures (Sykes, Rowley, & Schaefer, 1990). Duration is directly associated with social facilitation of nonalcoholic, as well as alcoholic, beverages as seen in coffeehouses (Sommer & Sommer, 1989). Alcohol consumption increases the rate of eating, meal duration, and caloric intake (Westerterp-Plantenga & Verwegen, 1999).

Social facilitation of eating varies according to who is present at meals. Gender and familiarity are two influences on the relationship between facilitation and consumption. Men produce increased social facilitation for eating among women, but not other men (de Castro, 1994). Both women and men attempt to eat in a socially desirable way when eating with someone of the opposite gender who is perceived as a desirable partner (Pliner & Chaiken, 1990). Women who eat less are perceived as being more feminine (Chaiken & Pliner, 1987; Mori, Chaiken, & Pliner, 1987), both when they are with others and when they are eating alone (Basow & Kobrynowicz, 1993). People eat more with family than strangers (Clendenen et al., 1994; de Castro, 1990, 1994), possibly because of exposure, relaxation, or disinhibition. Future research is needed to further elaborate social facilitation effects for other important demographic characteristics, such as age, socioeconomic status, ethnicity, and other social attributes.

Social facilitation operates as a mechanism by which social networks offer social support to individuals during meals. Social networks are the system of social ties that individuals are involved in, including the web of family, friends, neighbors, coworkers, and acquaintances (Cohen & Syme, 1985). Sharing meals provides important opportunities for people to interact with others in their networks. Social networks provide social support through the exchange of information, exertion of social control, and expression of emotional attachments that facilitate relationships and food consumption. Food and eating support tend to be offered and sought in times of need. Support often operates in a hierarchy beginning with family, moving to friends, and then involving neighbors (Ahluwaila, Dodds, & Baligh, 1998). Spousal partner's eating norms tend to exert the strongest influence on eating patterns (Oygard & Klepp 1996).

COMMENSALITY

Commensality is eating food together (Mars, 1997), and who someone shares the intimacy of eating with is an important dimension of the sociability of meals. Commensal partners are people chosen to share a meal. Commensality is an understudied topic (Grignon, 1996), despite its importance as a conceptual tool for understanding meals and eating. Many examples of social analysis of commensality exist in the anthropological literature but are typically not labeled or explicitly recognized as considering commensality, with some exceptions (Anigbo, 1987; Feinman, 1979; Gefou-Madianou, 1992; Mars, 1997; Morrison, 1996; Seremetakis, 1993; van Esterik, 1997).

Commensal eating may be a risky act, both physiologically and socially (Bloch, 1999; Miller, Rozin, & Fiske, 1998). Physiologically, ingestion includes a risk of taking in poisons or contaminants (Rozin & Vollmecke, 1986). Compared with the shelter of eating alone, eating with others (especially strangers) increases the possibility of exposure to physical risks in a meal. The risk of transmission of disease is shared with co-consumers, and people take more care in hygiene and cleanliness when eating with others than when eating alone. Socially, eating communicates information about what, how, when, where, and with whom a person ate. Such knowledge may constitute potentially risky biographical information. People often guard social access to themselves when eating, consuming foods "backstage" (Goffman, 1959) rather than in more public settings where they are observable and available for interactions with others who may not be seen as desirable or discreet meal partners. People differentiate formal eating settings (restaurants, weddings, etc.) and informal eating settings (home, friends' home, work), although both are still considered "public eating" because other people are present (Zdrodowski, 1996).

Commensality involves many levels and types of sharing food and is governed by a cultural code that provides a moral order for guiding food exchange and eating patterns by mediating between desire and restraint, selfishness and generosity, possessiveness and sharing (Hamer, 1994). Gittelsohn (1991) classified preferential household food distribution that occurs by gender, age, economic contribution, birth order, body size, health, and other factors. Food serving allocates foods in relationship to who serves, the order of serving, the quality of foods served, the quantity of foods served, and the method of serving (automatic, on request of the server or recipient, or self-serving). Second helpings and refusals by a server may express favoritism and neglect. Serving patterns of channeling desirable foods to some people but not others and substitutions of less desirable foods can influence food consumption. Gittelsohn (1991) identified four major food allocation patterns: (1) egalitarian, (2) favoritism in quality, (3) favoritism in quantity, and (4) shifts in allocation caused by transitory states like pregnancy or illness.

Meals are embedded in social life, reflecting affiliations and boundaries of acceptable interaction within social systems and the social structure of networks and relationships that establish commensal relationships (Grignon, 1996; Tapper & Tapper, 1986). People are defined and socially situated by their commensal behaviors under the concept that "You are who you eat with." Commensal relationships are not set or fixed but rather are continually considered, assessed, evaluated, negotiated, and managed. Commensal relationships are revealed by who is present or not invited, who is seated at the same table or elsewhere, who is seated adjacently or opposite, and who is seated on the right or left. Propinquity during meals provides a social map of roles, reference groups, status, and social networks.

The concept of commensal circles can be used for classifying actual and potential eating partners into various levels of inclusion in meals that reflect their location in larger social structures. Commensal circles represent the frequency of eating with others, with commensal partners ranging in a hierarchy from usual to occasional, rare, or absent. Drawing commensal circles may be based on preference for partners that extends from those who would be desirable, acceptable, avoided, and prohibited from sharing a meal. Various types of food have differential significance in commensal sharing, with some foods better "social conductors" than others (Bloch, 1999).

People are embedded within social networks, and their commensal circles reflect their social affiliations. Who someone shares food with locates them in the social world, revealing the types of people they are willing or not willing to include in their commensal circle. Commensal circles demarcate "us" versus "them" by coconsumption of "our foods" and not "their foods." Decisions about commensality are metaphors for the acceptance and rejection of others and the groups they represent (Hamer, 1994). Just as there are core, secondary, and peripheral foods (Bennett, Smith, & Passin, 1942; Jerome, 1975,1980; Koehler et al., 1980), commensal hierarchies demarcate core, secondary, and peripheral commensal partners.

The caste system of India and other societies is the classic exemplar of asymmetrical commensal relationships. Caste commensality is based on the application of the concepts of purity and pollution of food transactions (Appadurai, 1981; Gupta, 1975; Hollup, 1993). The institutionalized caste hierarchy uses commensality to maintain separation between hereditary caste groups through a system of rules about who can eat together that reflects the boundaries of each caste's place in society (Kumar, 1983). Food can be exchanged between people in equal status castes, given to those in lower status castes, and received from those in higher status castes (Achaya, 1998). Members of higher status castes have taboos not only against eating with people from lower status castes but also often extend this to prohibit symbolic commensality by avoiding foods prepared by lower status castes. The concept of pollution is so strongly observed that no method of cleaning would be deemed adequate to sufficiently purify food vessels used by the lowest status castes so that high-status caste members would be willing to eat from them (Achaya, 1994). The specific implementation of the caste hierarchy in food transactions is highly complex in its application of these general rules (Appadurai, 1981; Marriott, 1968).

The concept of commensal unit (Ashkenazi, 1991; Powers & Powers, 1984) represents the individuals eating together for a particular meal. Commensal units are established in line with commensal circles and vary by occasion and setting. The most common (and most symbolically idealized) commensal unit is the family, with other routine commensal units being composed of coworkers or peer groups and less routine ones including distant contacts or strangers. Important eating rituals typically involve careful selection, recruitment, and mobilization of appropriate commensal units. Feasting is rarely a solitary act, and decisions about who is involved are often at least as important as what is eaten. Even intentional fasting and dieting for weight loss are often done with others as a demonstration of social solidarity and a means for invoking support (Allon, 1975).

Commensal interactions may occur with people who are not actually present during the meal. Many people regularly watch television during meals (Locher, Burgio, Yoels, & Ritchie, 1997), drawing commensal companions from the mass media by interacting with television characters in "artificial relationships" (Caughey, 1984). Similarly, ceremonial meals in some cultures may "set a place" at the table to honor deceased ancestors by including them as commensal partners. Many cultures engage in commensal exchanges with deities (Haley, 1980). In a parallel vein, commensal privileges are also extended to nonhuman animals as people share food and company with pets or livestock as a part of a meal.

Commensality is widely recognized as an important component of initiating, building, confirming, deepening, and reaffirming social connections (Anigbo, 1987,1996; Mennell, Murcott, & von Otterloo, 1992; Van Den Berghe, 1984). Social institutions like pubs or clubs and food events such as festivals or holidays use food to communicate and enhance common identities and social ties (Brown & Mussell, 1985; Camp, 1989; Humphrey & Humphrey, 1988). Sharing of foods in commensal eating builds and reinforces mutual bonds of reciprocity that express shared sociability. Rites of passage and social transformations are often marked with special consumption rituals and particular foods, such as wedding banquets and feasts (Charsley, 1991,1992) and holiday foods like Thanksgiving turkey in the United States (Wallendorf & Arnould, 1991). The romanticizing of communal commensality is a prevalent and powerful social rhetoric (Bell & Valentine, 1997).

In contrast to the integrative aspects of commensality, Mennell et al. (1992, p. 117) emphasize the need to examine the obverse of inclusiveness and solidarity, pointing out that "inclusion implies exclusion." They observe that the construction of commensal communities also establishes xenophobic norms and ideals that exclude and even stigmatize others with respect to food and eating. Nationalities may be typified by derogatory epithets on the basis of their cuisines, such as "Frogs" or "Krauts," who avoid eating together (Murcott, 1996), and vegetarians and nonvegetarians cast aspersions on each other for their food choices and avoid sharing meals (Beardsworth & Keil, 1992).

Feasts are an important component of the commensal food life of many cultures and are used for many purposes such as to enhance social bonding, provide mutual assistance, display status, recruit labor, and foster economic gain (Dietler, 1996; Hayden, 1996; Young, 1971). Sharing food in feasts is often done to advance the social position or power of individuals or groups in what has been called "gastropolitics" (Appadurai, 1981) or "commensal politics" (Dietler, 1996).

Commensality influences the amount and type of foods that are consumed at meals. People tend to eat more with friends than strangers (Clendenen et al., 1994; de Castro, 1990,1994). Gender is an important modifier of commensality that shapes food choices, with men tending to eat more with women but not with other men, and women tending to eat less with men and more with other women (de Castro, 1994; Pliner & Chaiken, 1990). To the extent that people can choose their meal partners, commensality functions as a precursor to the operation of social facilitation.

Neophobia is fear of what is new. Psychologists have identified and extensively examined food neophobia, in which people are reluctant to consume unfamiliar foods (Crandall, 1985; Martins, Pelchat, & Pliner, 1997; Pliner & Hobden, 1992; Pliner & Pelchat, 1991; Prescott & Khu, 1995). The concept of neophobia can be extended to apply to commensality as a reluctance to eat with new people. What can be termed "commensal neophobia" influences people to avoid eating and to eat less with strangers (Clendenen et al., 1994; de Castro, 1990,1994). This suggests that changing with whom a person eats may be as difficult as changing what they eat.

INTERACTION

Interaction at meals provides opportunities for a myriad of types of communication, including sociability and socialization. Meals are structured social events that offer a major arena for social interaction, occurring regularly and having normative social rules that set mealtime apart from other activities. Meal conversation norms specify appropriate and inappropriate topics, typically excluding subjects that are controversial or disgusting on the basis of the justification of avoiding stress to facilitate completion, enjoyment, and digestion of a meal.

Research by psychologists shows that meals are important interactional settings with standing patterns of behavior. At dinner parents talk to each other more than with children (Dreyer & Dreyer, 1973; Lewis & Feiring, 1982), although larger families are more child centered. The amount of conversation does not increase with family size, and each person speaks less at dinner in larger families (Lewis & Feiring, 1982). Meal interactions in larger families are also less structured and noisier. Mothers tend to have responsibility for both the food itself and the interaction at meals (DeVault, 1991; Feiring & Lewis, 1987).

Sociolinguistic research provides an understanding of meal interactions as a part of its larger goal of examining how language is used in society and how language varies in social groups (Goffman, 1981; Hymes, 1974; Labov, 1972). Mealtimes, particularly longer and more structured dinner meals, are seen as an "opportunity space" for families to come together and have a captive audience for a joint interchange (Ochs, Smith, & Taylor, 1989). At dinner families engage in narratives about many topics, with discourse about food and other issues (Ochs et al., 1989).

Personal narratives are stories that a speaker tells an audience about a topic and form an important part of dinnertime discourse (Erickson, 1990). Interaction norms at a meal prescribe that family members present narratives and comment about what is presented by others in a set of "story rounds," although these norms may not be strictly followed in many families (Schultz, Florio, & Erickson, 1982). Mealtime discourse is constructed by the participants into conversations that maintain congruity through continuity of topic, consistency in tone, group participation in maintaining coherence, and persistence in the prominence of particular individuals at the meal (Erickson, 1990).

Family table talk operates to transmit cultural and family ideals and norms to children (Bossard, 1943; Bossard & Boll, 1950). Mothers talk to their children more frequently at meals than many other settings (Hoff-Ginsberg, 1991). Blum-Kulka (1997) provides an extensive sociolinguistic analysis of dinner talk as a site for understanding family discourse. She portrays dinners as communication events, describing the specific discourse roles for men and women, adults and children, and family members and guests. Discourse at dinner provides cultural and pragmatic socialization for children. Children develop cultural skills through dinner talk, telling tales in family narrative events. Pragmatically, social control is taught by means of dinner talk through the stating and enforcing of rules about politeness. Meals are intergenerational social events in which children participate but are controlled and constrained in the form and amount of participation that is permitted by adults (Blum-Kulka, 1994).

Family food interaction is the particular discourse that involves exchange and interchange between family members related to decisions and activities involved in making food choices (Gillespie & Acterberg, 1989; Schafer, Schaefer, Dunbar, & Keith, 1999). Family food interaction involves a range of relationships and behaviors, including discussions about healthy eating, suggestions about food consumption, requests for specific foods, and eating foods together. Marital partners who engage in more family food interaction tend to have healthier diets (Gillespie & Acterberg, 1989; Schafer et al., 1999).

In addition to being major sites for socialization of children, meals are also central settings for acculturation of adults from other societies. The rituals and procedures of eating are a crucial part of learning a culture, and the meal is a fundamental setting that needs to be learned in the assimilation into a new culture (Blum-Kulka & Sheffer, 1993).

Meal arrangements vary along several dimensions, including temporal, spatial, and activity foci (Ochs et al., 1989). Temporal arrangements may be staggered, in which household members may eat at several times, or synchronous, when eating occurs together. Spatial arrangements may be dispersed, when people eat in different places in the household, or assembled, when everyone is in one place. Activity focus may be diverse, in which some people talk while others watch television and still others eat separately, or may be shared, in which everyone is engaged in common interaction. The combination of these arrangements provides the social setting within which mealtime interactions occur (Feiring & Lewis, 1987). Families with more centralized meal arrangements include interaction rituals in which members can share stories that promote solutions to problems and reinforce dynamics of social control in the family (Ochs et al., 1989).

Meals include routines that structure which interactions take place, with culturally recognized standards for beginning, enacting, and ending meals (Wilhite, 1983). Who may be present, where they are located, what order they eat in, who can offer and request food, who receives how much of what types of food, what may be talked about and how the talk is structured, when people may leave, and what is involved in departure all have norms and rules associated with them. Variation may occur in mealtime interaction routines, but clear boundaries exist that prescribe what is and is not acceptable interaction. Children are exempt from some interaction rules at meals until they are judged to be sufficiently mature to be held accountable to adult standards (Wilhite, 1983).

Nonverbal communication is important at meals. Simply being present at a meal communicates participation. Much communication that occurs at meals is nonverbal, with kinesics (also termed "body language") an important aspect of communication during meals (Birdwhistell, 1970). For example, posture and clothing are used by commensal participants to communicate messages during mealtime.

Environmental situations also shape the social activities involved in meals (Bell & Meiselman, 1995; Marshall, 1993; Marshall & Bell, 1996; Meiselman, 1996). Proxemics (spatial positioning), such as who sits next to whom and how far each person is from others, communicate important messages at the meal (Almond & Esser, 1965; Hall, 1969).

Social interaction may distract from eating activities, and some social institutions attempt to suppress sociability to make eating faster, meals shorter, and consumption more efficient. This often occurs in total institutions (Goffman, 1961; Mennell et al., 1992), which are organized to control their inmates. For example, school lunchroom supervisors often discourage children from talking so they will finish their meals during the fixed lunch period (Morrison, 1996). Some utopian sects that emphasized efficiency enforce a code of silence during mealtimes, such as the Shakers in the United States (Andrews, 1953). In this way, institutions not only standardize and control food, they also regiment mealtime interactions.

Social interaction is perhaps the most emphasized and structured at banquets and feasts that are organized to promote conviviality. Jeanneret (1991) explains that such events order and codify convivial discourse at the meal in both speech and the language of food. The vocabulary of celebration includes ideals enacted in rules for good conversation and nonverbal expressions of shared civility embodied in table manners. Food and affability are also central components of visiting and the hospitality that accompanies social interactions with friends and neighbors in both public and private meals and other social gatherings (Julier, 1997). Much remains to be examined about social interaction at meals, with a need for investigation of many aspects of that dimension of sociability.

CONCLUSION

Meals are social events that provide great sensory pleasures, divide up the activities of the day, and delineate appropriate times when everyday life must or can be interrupted. Specific ceremonial meals are often used to demarcate social cycles over years and lifetimes. The interruptive and communal aspects of meals provide opportunities and obligations for sociability, where people who consider each other's company to be important meet and share food. Sociability at meals can provide a useful window into broader social patterns.

This chapter examined sociability and meals with the intent to examine selected past work to raise issues rather than present definitive conclusions about this underinvestigated topic. Many questions in this area remain unresolved. Is a meal really a meal when eaten alone? Under what conditions does the presence of different types of others exert various influences on eating? How are commensal partners identified and delineated, and how are commensal boundaries enforced? How are social interchanges at meals initiated, constructed, negotiated, and managed? Other questions about sociability and meals abound and should not remain unattended.

Meal sociability deserves multidisciplinary and multimethod examination. This chapter attempted to bring together ideas and information that are widely scattered across disciplines, fields, and professions that used a variety of investigative techniques. Three dimensions of sociability in meals (facilitation, commensality, and interaction) were used to focus the discussion, but others exist, such as etiquette and manners (e.g., Carson, 1990; Elias, 1978; Visser, 1993). The focus here was on contemporary postindustrial societies (although examples from other times and places were included), which was determined more by accessibility and availability of information than a decision to be bound by culture or history. Future attention to meals as social events needs to further explore and examine the influences, processes, and consequences of sociability. To fully understand meals requires knowing about the composition and activities of those who eat as well as what was eaten.

REFERENCES

Achaya, K.T. (1994). *Indian food: A historical companion* (pp. 63–64). Delhi: Oxford University Press.

Achaya, K.T. (1998). *A historical dictionary of Indian food* (pp. 63–64). Delhi: Oxford University Press.

Ahluwalia, I.B., Dodds, J.M., & Baligh, M. (1998). Social support and coping behaviors of low-income families experiencing food insufficiency in North Carolina. *Health Education and Behavior, 25*(5), 599–612.

Allon, N. (1975). Latent social services in group dieting. *Social Problems, 23,* 59–69.

Almond, R., & Esser, A.H. (1965). Tablemate choice of psychiatric patients: A technique for measuring social contact. *Journal of Nervous and Mental Disease, 141,* 68–82.

Andrews, E.D. (1953). *The people called Shakers: A search for the perfect society.* New York: Dover Publications.

Anigbo, O.A.C. (1987). *Commensality and human relationship among the Igbo.* Nsukka: University of Nigeria Press.

Anigbo, O.A.C. (1996). Commensality as cultural performance: The struggle for leadership in an Igbo village. In D. Parkin, L. Caplan, & H. Fisher (Eds.), *The politics of cultural performance* (pp. 101–114). Oxford: Berghahn Books.

Appadurai, A. (1981). Gastropolitics in Hindu South Asia. *American Ethnologist, 8,* 494–511.

Ashkenazi, M. (1991). From Tachi Soba to Naoria: Cultural implications of the Japanese meal. *Social Science Information, 30*(2), 287–304.

Basow, S.A., & Kobrynowicz, D. (1993). What is she eating? The effects of meal size on impressions of a female eater. *Sex Roles, 28*(5–6), 335–344.

Beardsworth, A., & Keil, T. (1992). The vegetarian option: Varieties, motives, conversions, and careers. *Sociological Review, 40,* 252–293.

Bell, D., & Valentine, G. (1997). *Consuming geographies: We are where we eat.* New York: Routledge.

Bell, R., & Meiselman, H.L. (1995). The role of eating environments in determining food choice. In D.W. Marshall (Ed.), *Food choice and the consumer* (pp. 292–310). London: Blackie Academic.

Bennett, J.W., Smith, H.L., & Passin, H. (1942). Food and culture in southern Illinois: A preliminary report. *American Sociological Review, 7,* 645–660.

Berry, S.L., Beatty, W.W., & Klesges, R.C. (1985). Sensory and social influences on ice cream consumption by males and females in a laboratory setting. *Appetite, 6,* 41–45.

Birdwhistell, R.L. (1970). *Kinesics and context: Essays on body motion communication.* Philadelphia: University of Pennsylvania Press.

Bloch, M. (1999). Commensality and poisoning. *Social Research, 66*(1), 133–149.

Blum-Kulka, S. (1994). The dynamics of family dinner-talk: Cultural contexts for children's passages to adult discourse. *Research on Language and Social Interaction, 27,* 1–51.

Blum-Kulka, S. (1997). Dinner talk: Cultural patterns of sociability and socialization in family discourse. Hillsdale, NJ: Lawrence Earlbaum.

Blum-Kulka, S., & Sheffer, H. (1993). The metapragmatic discourse of American-Israeli families at dinner. In G. Kasper, & S. Blum-Kulka (Eds.), *Interlanguage pragmatics* (pp. 196–224). New York: Oxford University Press.

Bossard, J.H.S. (1943). Family table talk: An area for sociological study. *American Sociological Review, 8*(3), 295–301.

Bossard, J.H.S., & Boll, E.S. (1950). *Ritual in family living: A contemporary study.* Philadelphia: University of Pennsylvania Press.

Brown, L.K., & Mussell, K. (Eds.), (1985). *Ethnic and regional foodways in the United States: The performance of group identity.* Knoxville, TN: University of Tennessee Press.

Bureau of the Census. (1998). *Statistical abstract of the United States—1998.* Washington, DC: U.S. Government Printing Office.

Camp, C. (1989). *American foodways: What, when, why and how we eat in America.* Little Rock, AK: August House.

Carlson, R.G. (1992). Symbolic mediation and commoditization: A critical examination of alcohol use among the Naya of Bukoba Tanzania. *Medical Anthropology, 15*(1), 41–62.

Carson, B.G. (1990). *Ambitious appetites: Dining, behavior, and patterns of consumption in federal Washington.* Washington, DC: American Institute of Architects Press.

Caughey, J.L. (1984). *Imaginary social worlds: A cultural approach.* Lincoln, NE: University of Nebraska Press.

Chaiken, S., & Pliner, P. (1987). Women, but not men, are what they eat: The effect of meal size and gender on perceived femininity and masculinity. *Personality and Social Psychology Bulletin, 13*(2), 166–176.

Charsley, S.R. (1991). *Rites of marrying: The wedding industry in Scotland.* Manchester: Manchester University Press.

Charsley, S.R. (1992). *Wedding cakes and cultural history.* New York: Routledge.

Clendenen, V.I., Herman, C.P., & Polivy, J. (1994). Social facilitation of eating among friends and strangers. *Appetite, 23*(1), 1–13.

Cohen, S., & Syme, S.L. (Eds.), (1985). *Social support and health.* New York: Academic Press.

Crandall, C.S. (1985). The liking of foods as a result of exposure: Eating doughnuts in Alaska. *Journal of Social Psychology, 125*(2), 187–194.

Davis, M., & Randall, E. (1983). Social change and food habits of the elderly. In M.W. Riley, B.B. Hess, & K. Bond (Eds.), *Aging in society: Selected reviews of recent research* (pp 199–217). Hillsdale, NJ: Lawrence Earlbaum.

Davis, M., Randall, E., Forthofer, R.N., Lee, E.S., & Marger, S. (1985). Living arrangements and dietary patterns in older adults in the United States. *Journal of Gerontology, 40,* 434–438.

Davis, M.A., Murphy, S.P., Neuhaus, J.M., & Lien, D. (1990). Living arrangements and dietary quality of older U.S. adults. *Journal of the American Dietetic Association, 90,* 1667–1672.

de Castro, J.M. (1990). Social facilitation of duration and size but not rate of the spontaneous meal intake of humans. *Physiology and Behavior, 47*(6), 1129–1135.

de Castro, J.M. (1991). Social facilitation of the spontaneous meal size of humans occurs on both weekdays and weekends. *Physiology and Behavior, 49*(6), 1289–1291.

de Castro, J.M. (1994). Family and friends produce greater social facilitation of food intake than other companions. *Physiology and Behavior, 56*(3), 445–455.

de Castro, J.M., & Brewer, E.M. (1992). The amount eaten in meals by humans is a power function of the number of people present. *Physiology and Behavior, 51*(1), 121–125.

de Castro, J.M., Brewer, E.M., Elmore, D.K., & Orozco, S. (1990). Social facilitation of the spontaneous meal size of humans occurs regardless of time, place, alcohol or snacks. *Appetite, 15*(2), 89–101.

Demers, A., & Bourgault, C. (1996). Changing society, changing drinking: Solitary drinking as a non-pathological behaviour. *Addiction, 91*(10), 1505–1516.

DeVault, M. (1991). *Feeding the family: The social organization of caring as gendered work*. Chicago: University of Chicago Press.

Dietler, M. (1990). Driven by drink: The role of drinking in the political economy and the case of early Iron Age France. *Journal of Anthropological Archaeology, 9*, 352–406.

Dietler, M. (1996). Feasts and commensal politics in the political economy: Food, power, and status in prehistoric Europe. In P. Weissner, and W. Schiefenhovel (Eds.), *Food and the status quest: An interdisciplinary perspective* (pp. 87–125). Oxford: Berghahn Books.

Donkin, A.J.M., Johnson, A.E., Lilley, J.M., Morgan, K., Neale, R.J., Page, R.M., & Silburn, R.L. (1998). Gender and living alone as determinants of fruit and vegetable consumption among the elderly living at home in urban Nottingham. *Appetite, 30*, 39–51.

Douglas, M. (1972). Deciphering a meal. *Daedalus, 101*, 61–82.

Douglas, M., & Nicod, M. (1974). Taking the biscuit: The structure of British meals. *New Society, 19 Dec*, 744–747.

Dreyer, C.D., & Dreyer, A.S. (1973). Family dinnertime as a unique behavior. *Family Process, 12*, 291–301.

Elias, N. (1978). *The civilizing process. Volume I: The history of manners*. Oxford: Basil Blackwell.

Engell, D., Kramer, M., Malafi, T., & Salomon, M. (1996). Effects of effort and social modeling on drinking in humans. *Appetite, 26*(2), 129–138.

Erickson, F. (1990). The social construction of discourse coherence in a family dinner table conversation. In B. Dorval (Ed.), *Conversation organization and its development* (pp. 207–239). Norwood, NJ: Ablex.

Falk, L.W., Bisogni, C.A., & Sobal, J. (1996). Food choice processes of older adults. *Journal of Nutrition Education, 28*, 257–265.

Feinman, S. (1979). Food sharing and sociability. *Free Inquiry in Creative Sociology, 7*(2), 123–127.

Feiring, C., & Lewis, M. (1987). The ecology of some middle class families at dinner. *International Journal of Behavioral Development, 10*(3), 377–390.

Feunekes, G.I.J., de Graaf, C., & Staveren, W.A. (1995). Social facilitation of food intake is mediated by meal duration. *Physiology and Behavior, 58*(3), 551–558.

Gefou-Madianou, D. (1992). Alcohol commensality, identity transformations and transcendence. In D. Gefou-Madianou (Ed.), *Alcohol, gender, and culture* (pp. 1–34). New York: Routledge.

Gillespie, A.H., & Acterberg, C.L. (1989). Comparison of family interaction patterns related to food and nutrition. *Journal of the American Dietetic Association, 89*, 509–512.

Gittelsohn J. (1991). Opening the box: Intrahousehold food allocation in rural Nepal. *Social Science and Medicine, 33*(10), 1141–1154.

Goffman, E. (1959). *The presentation of self in everyday life*. New York: Anchor.

Goffman, E. (1961). *Asylums: Essays on the social situation of mental patients and other inmates*. New York: Anchor.

Goffman, E. (1981). *Forms of talk*. Philadelphia: University of Pennsylvania Press.

Grignon, C. (1996). Commensality and social morphology: An essay of typology. Unpublished manuscript, INRA/CORELA, Paris.

Grotkowski, M.L., & Sims, L.S. (1978). Nutrition knowledge, attitudes and dietary practices of the elderly. *Journal of the American Dietetic Association, 72*, 499–506.

Guerin, B. (1993). *Social facilitation*. New York: Cambridge University Press.

Gupta, S.P. (1975). Changes in the food habits of Asian Indians in the United States: A case study. *Sociology and Social Research, 60*(1), 87–99.

Haines, P.S., Guilkey, D.K., & Popkin, B.M. (1996). Trends in breakfast consumption in US adults between 1965 and 1991. *Journal of the American Dietetic Association, 96*, 464–470.

Hall, E.T. (1969). *The hidden dimension*. New York: Anchor.

Hamer, J. (1994). Commensality, process and the moral order: An example from Southern Ethiopia. *Africa, 64*(1), 126–144.

Hayden, B. (1996). Feasting in prehistoric and traditional societies. In P. Weissner, and W. Schiefenhovel (Eds.), *Food and the status quest: An interdisciplinary perspective* (pp. 127–147). Oxford: Berghahn Books.

Hayley, A. (1980). A commensal relationship with God: The nature of the offering in Assamese Vaishnavism. In *Sacrifice* (pp. 107–125). New York: Academic Press.

Hirschman, A.O. (1996). Melding the public and private spheres: Taking commensality seriously. *Critical Review, 10*(4), 533–550.

Hoff-Ginsberg, E. (1991). Mother-child conversation in different social classes and communicative settings. *Child Development, 62*(4), 782–796.

Hollup, O. (1993). Caste identity and cultural continuity among Tamil plantation workers in Sri Lanka. *Journal of Asian and African Studies, 28*(1–2), 67–87.

Humphrey, T.C., & Humphrey, L.T. (Eds.), (1988). *"We gather together": Food and the festival in American life*. Ann Arbor, MI: UMI Research Press.

Hymes, D. (1974). *Foundations in sociolingusitics: An ethnographic approach*. Philadelphia: University of Pennsylvania Press.

Jeanneret, M. (1991). *A feast of words: Banquets and table talk in the Renaissance*. Chicago: University of Chicago Press.

Jerome, N.W. (1975). On determining food patterns of urban dwellers in contemporary United States society. In M.L. Arnott (Ed.), *Gastronomy: The anthropology of food habits* (pp. 91–111). The Hague: Mouton.

Jerome, N.W. (1980). Diet and acculturation: The case of Black-American immigrants. In N.W. Jerome, R.F. Kandel, & G.H. Pelto (Eds.), *Nutritional anthropology: Contemporary approaches to diet and culture* (pp. 275–325). Pleasantville, NY: Redgrave Publishing Company.

Julier, A. (1997). *The expression of hospitality: Shared meals and the construction of friend and family relationships*. Madison, WI: Association for the Study of Food and Society.

Klein, H., & Pittman, D.J. (1990). Drinker prototypes in American society. *Journal of Substance Abuse, 2*(3), 299–316

Klesges, R.C., Bartsch, D., Norwood, J.D, Kautzman, D., & Haugrud, S. (1984). The effects of selected social and environmental variables on the eating behavior of adults in the natural environment. *International Journal of Eating Disorders, 3*, 33–41.

Koehler, K.M., Harris, M.B., & David, S.M. (1980). Core, secondary, and peripheral foods in the diets of Hispanic, Navajo, and Jemez Indian children. *Journal of the American Dietetic Association, 89*, 538–540.

Kumar, A. (1983). Religion, politics and social stratification: A case study of Satnamis of Chhattisgarh. *Indian Journal of Social Research, 24*(1), 63–69.

Labov, W. (1972). *Language in the inner city: Studies in Black English vernacular*. Philadelphia: University of Pennsylvania Press.

Lee, C., Templeton, S., & Wang, C. (1995). Meal skipping patterns and nutrient intakes of rural southern elderly. *Journal of Nutrition for the Elderly, 15*(2), 1–14.

Lewis, M., & Feiring, C. (1982). Some American families at dinner. In L.M. Laosa & I.E. Sigel (Eds.), *Families as learning environments for children* (pp. 115–145). New York: Plenum Publishing.

Locher, J.L., Burgio, K.L., Yoels, W.C., & Ritchie, C.S. (1997). The social significance of food and eating in the lives of older recipients of Meals on Wheels. *Journal of Nutrition for the Elderly, 17*(2), 15–33.

Mahajan, K.H., & Schafer, E. (1993). Influence of selected psychosocial factors on dietary intake in the elderly. *Journal of Nutrition for the Elderly, 12*(4), 21–41.

Makela, J. (1991). Defining a meal. In E.L. Furst, R. Prättäla, M. Ekstrom, L. Holm, & U. Kjarnes, (Eds.), *Palatable worlds: Socio-cultural food studies*. Oslo: Solum Sorlag.

Marriott, M. (1968). Caste ranking and food transactions: A matrix analysis. In M. Singer & B.S. Cohn (Eds.), *Structure and change in Indian society* (pp. 133–171). Chicago: Aldine.

Mars, L. (1997). Food and disharmony: Commensality among Jews. *Food and Foodways, 7*(3), 189–202.

Marshall, D. (1993). Appropriate meal occasions: Understanding conventions and exploring situational influences on food choice. *International Review of Retail, Distribution, and Consumer Research 3*, 279–301.

Marshall, D., & Bell, R. (1996). The relative influence of meal occasion and situation on food choice. In A. Worsley (Ed.), *Multidisciplinary approaches to food choice* (pp. 99–102). Australia: University of Adelaide.

Martins, Y., Pelchat, M.L., & Pliner, P. (1997). "Try it; it's good and it's good for you": Effects of taste and nutrition information on willingness to try novel foods. *Appetite, 28*, 89–102.

Meiselman, H.L. (1996). The contextual basis for food acceptance, food choice, and food intake: The food, the situation, and the individual. In H.L. Meiselman & H.J.H. MacFie (Eds.), *Food choice, acceptance, and consumption* (pp. 239–263). London: Blackie Academic.

Mennell, S., Murcott, A., & van Otterloo, A.H. (1992). *The sociology of food: Eating, diet and culture*. Newbury Park, CA: Sage Publications.

McIntosh, W.A., & Shifflett, P.A. (1984). Influence of social support systems on dietary intake of the elderly. *Journal of Nutrition for the Elderly, 4*(1), 5–18.

McIntosh, W.A., Shifflett, P.A., & Picou, J.S. (1989). Social support, stressful events, strain, dietary intake and the elderly. *Medical Care, 27,* 140–153.

Miller L., Rozin, P., & Fiske, A.P. (1998). Food sharing and feeding another person suggest intimacy: Two studies of American college students. *European Journal of Social Psychology, 29,* 423–436.

Mori, D., Chaiken, S., & Pliner, P. (1987). "Eating lightly" and the self-presentation of femininity. *Journal of Personality and Social Psychology, 53,* 693–702.

Morrison, M. (1996). Sharing food at home and school: Perspectives on commensality. *Sociological Review 44*(4), 648–74.

Murcott, A. (1982). On the social significance of the 'cooked dinner' in South Wales. *Social Science Information, 21*(4/5), 677–695.

Murcott, A. (1996). Food as an expression of national identity. In S. Gustavsson & L. Lewin (Eds.), *The future of the nation-state: Essays on cultural pluralism and political integration* (pp. 49–77). Stockholm: Nerenius and Santerus.

Murcott, A. (1997). Family meals: A thing of the past? In P. Caplan (Ed.), *Food, health, and identity* (pp. 32–49). New York: Routeledge.

Ochs, E., Smith, R., & Taylor, C. (1989). Detective stories at dinnertime: Problem solving through co-narration. *Cultural Dynamics, 2,* 238–257.

Orbuch, T.L. (1997). Peoples' accounts count: The sociology of accounts. *Annual Review of Sociology, 23,* 455–478.

Oygard L, & Klepp K. (1996). Influences of social groups on eating patterns: A study among young adults. *Journal of Behavioral Medicine, 19*(1), 1–14.

Pliner, P., & Chaiken, S. (1990). Eating, social motives, and self-presentation in women and men. *Journal of Experimental Social Psychology, 26*(3), 240–254.

Pliner, P., & Hobden, K. (1992). Development of a scale to measure the trait of food neophobia in humans. *Appetite, 19,* 105–20.

Pliner, P., & Pelchat, M.L. (1991). Neophobia in humans and the special status of foods of animal origin. *Appetite, 16,* 205–218.

Polivy, J., Herman, C.P., Younger, J.C., & Jaeger, J. (1979). Effects of a model on eating behavior: The induction of a restrained eating style. *Journal of Personality, 47*(1), 100–117.

Powers, W.K., & Powers, M. (1984). Metaphysical aspects of the Oglala food system. In M. Douglas (Ed.), *Food in the social order: Studies of food and festivities in three American communities* (pp. 40–96). New York: Russell Sage Foundation.

Prescott, J., & Khu, B. (1995). Changes in preference for saltiness within soup as a function of exposure. *Appetite, 24,* 302.

Redd, M., & de Castro, J.M. (1992). Social facilitation of eating: Effects of social instruction on food intake. *Physiology and Behavior, 52*(4), 749–754.

Rotenberg, R. (1981). The impact of industrialization on meal patterns in Vienna, Austria. *Ecology of Food and Nutrition, 11,* 25–35.

Rozin, P., & Vollmecke, T.A. (1986). Food likes and dislikes. *Annual Review of Nutrition, 6,* 433–456.

Schafer, R.B., Schafer, E., Dunbar, M., & Keith, P.M. (1999). Marital food interaction and dietary behavior. *Social Science and Medicine, 48,* 787–796.

Schultz, J., Florio, S., & Erickson, F. (1982). Where's the floor?: Aspects of the cultural organization of social relationships in communication at home and at school. In P. Gilmore & A. Glatthorn (Eds.), *Ethnography and education: Children in and out of school* (pp. 88–123). Washington, DC: Center for Applied Linguistics.

Scott, M.B., & Lyman, S.M. (1968). Accounts. *American Sociological Review, 33,* 46–62.

Seremetakis, C.N. (1993). The memory of the senses: Historical perception, commensal exchange, and modernity. *Visual Anthropology Review, 9*(2), 2–18.

Shide, D.J., & Rolls, B.J. (1991). Social facilitation of caloric intake in humans by friends but not strangers [abstract]. *International Journal of Obesity, 15*(Suppl. 3), 8.

Simmel, G. (1961). The sociology of sociability. In T. Parsons, E. Shils, K.D. Naegele, & J.R. Pitts (Eds.), *Theories of society: Foundations of modern sociological theory* (pp. 157–163). New York: Free Press.

Sommer, R. (1965). The isolated drinker in the Edmonton beverage room. *Quarterly Journal of Studies on Alcohol, 26,* 95–110.

Sommer, R., & Sommer, B.A. (1989). Social facilitation effects in coffeehouses. *Environment and Behavior, 21*(6), 651–666.

Sommer, R., & Steele, J. (1997). Social effects on duration in restaurants. *Appetite, 29,* 25–30.

Sykes, R., Rowley, R.D., & Schaefer, J.M. (1990). Effects of group participation on drinking behaviors in public bars: An observational survey. *Journal of Social Behavior and Personality, 5*(4), 385–402.

Symons, M. (1994). Simmel's gastronomic sociology: An overlooked essay. *Food and Foodways, 5*(4), 333–351.

Tapper, R., & Tapper, N. (1986). "Eat this, it'll do you a power of good": Food and commensality among Durrani Pashtuns. *American Ethnologist, 13*(1), 62–79.

Torres, C.C., McIntosh, W.A., & Kubena, K.S. (1992). Social network and social background characteristics of elderly who live and eat alone. *Journal of Aging and Health, 4,* 564–578.

Van den Berghe, P.L. (1984). Ethnic cuisine, culture in nature. *Ethnic and Racial Studies, 7*(3), 387–397.

van Esterik, P. (1997). Women and nurture in industrial societies. *Proceedings of the Nutrition Society, 56,* 335–343.

Visser, M. (1993). *The rituals of dinner: The origins, evolution, eccentricities, and meaning of table manners.* New York: Penguin USA.

Wallendorf, M., & Arnould, E.J. (1991). "We gather together": Consumption rituals of Thanksgiving day. *Journal of Consumer Research, 18,* 13–31.

Westenbrink, S., Lowik, M.R., Hulshof, K.F., & Kistemaker, C. (1989). Effects of household size on nutritional patterns. *Journal of the American Dietetic Association, 89*(6), 793–799.

Westerterp-Plantenga, M.S., & Verwegen, C.R.T. (1999). The appetizing effect of an aperitif in overweight and normal-weight humans. *American Journal of Clinical Nutrition, 69,* 205–212.

Wilhite, M. (1983). Children's acquisition of language routines: The end-of-meal routine in Cakchiquel. *Language in Society, 12*(1), 47–64.

Williams, J. (1997). "We never eat like this at home": Food on holiday. In P. Caplan (Ed.), *Food, health, and identity* (pp. 151–171). New York: Routeledge.

Young, M. (1971). *Fighting with food.* Cambridge, England: Cambridge University Press.

Zajonc, R.B. (1965). Social facilitation. *Science, 149,* 269–274.

Zdrodowski, D. (1996). Eating out: The experience of eating in public for the "overweight" woman. *Women's Studies International Forum, 19,* 665–654.

Chapter 8

The Role of Flavor
in the Meal and the Culture

Elisabeth Rozin

"Animals feed, men dine," proclaimed the nineteenth century gastronome Brillat-Savarin. The aphorism neatly defines the fundamental difference between the human food experience and the animal, its elaboration from an instinctual drive designed to nourish the individual and propagate the species into a complex behavior rich with significance and associations and marked with a wealth of unique products and practices.

Although much of human food behavior comes out of ideational considerations that use food as a way of expressing or illuminating social structure, religious belief, festive, or celebratory occasions, much is driven by the gustatory, that peculiarly human conjunction between the nutritional and the aesthetic, the need to nourish ourselves in an appropriate, gratifying, and pleasurable fashion. Whatever the tradition, the occasion, or the dish we may undertake to examine, we are faced with the powerful, primary, and universal need of people to produce food that is palatable and enjoyable, that satisfies in the pot, on the table, and, most especially, in the mouth.

The means by which humans accomplish this goal is the behavior we call cooking, the deliberate manipulation or transformation of basic foodstuffs into appropriate edibles. All people cook and seem to have done so from the very earliest times; indeed, cooking, like language or art, is one of those accomplishments that defines our species and distinguishes it from its animal brethren. And although cooking, as it occurs in cultures throughout time and across the world, seems to exist in an almost limitless variety of forms, much of the enterprise is common and universal (Rozin, 1996).

In earlier works I outlined what seem to be the universal components of culinary systems or cuisines (Rozin, 1973, 1982). The first of these components is that of basic foods, the substances selected by any group for preparation and consumption. All cultures make such selections that, in their formative stages at least, are largely shaped by geography, by what is more or less locally available. These choices, whether of animal or vegetable foods, tend to be conservatively maintained, even when new ingredients are introduced into the system: lamb is still, as it has been for thousands of years, the most esteemed red meat of the Middle East; corn remains the favored staple grain of Mexico even though rice and wheat have made significant inroads since the arrival of Europeans some 500 years ago. The selection of basic foods depends on a wide variety of factors: availability; environmental variables such as climate, soil, and precipitation; ease of production; cost of importation; nutritional benefit; palatability; and religious or social sanction. All foods, whether meat, fish, or fowl; vegetables; grains; tubers and roots provide unique and distinctive qualities of flavor, texture, aroma, and appearance, and the preference for these qualities tends to remain strong.

The second component of cuisine is that of culinary techniques, those practices or manipulations selected by any culture to transform basic foodstuffs into appropriate edibles. Basic cooking techniques can be classified into three general categories: (1) processes that change the physical size, shape, or mass of the food, including such techniques as chopping or grinding, processes of separation or extraction, and processes of incorporation; (2) processes that alter the water content of foods, by either adding or removing liquid, including such techniques as soaking, marinating, drying, smoking, and so forth; (3) processes that change foods chemically, including the most pervasive and widespread of all techniques, the application of heat. Heat techniques include dry heat (baking, roasting), cooking with liquid (boiling, steaming), and cooking with fat (frying, sautéing). The other great category of chemical alteration is that of fermentation, the use of various molds, bacteria, and enzymes under controlled conditions of time and temperature for the culturing of dairy products; the production of wines, beers, and condiment sauces like soy sauce; and the leavening of bread with yeast. From these basic categories of cooking techniques all cultures make a selection based initially, at least, on such environmental factors as the kinds and availability of fuel, the nature of the ingredients to be processed, and the general level of technology. Every technique provides unique experiences of flavor, texture, aroma, and appearance; to preserve fish, for example, the natives of the American Northwest smoke their salmon, whereas the Portuguese salt and dry their codfish, and the Scandinavians pickle their herring.

The third component of culinary structure is flavor, the deliberate modification of the taste of cooked food in addition to whatever flavor is provided by the foodstuffs themselves and the cooking techniques by which they are prepared. It seems to be, in the end, that part of culinary practice that is most capable of evoking a particular ethnicity, that is most crucial in providing a sensory and cultural label for the food of any group. Flavor is, of course, a complex phenomenon involving the interaction between the basic taste system—the chemical receptors for sweet, sour, salty, and bitter; the pain receptors stimulated by temperature and irritants such as those found in chile peppers, ginger, and horseradish—and smell, which conveys a great deal of what most people ordinarily experience as flavor.

Most cultures go to great lengths to achieve what is regarded as good and pleasing flavors for food and, despite differences in ingredients, do so in remarkably similar fashions. A clear tendency exists in cultures throughout the world to use combinations of flavoring ingredients consistently and pervasively; these seasoning compounds, which I call "flavor principles," provide powerful and characteristic flavor profiles that are familiar and pleasing to those within the system, recognizable, and replicable to those from without (Rozin, 1973). Think, for example, of two widely divergent cuisines, Greek and Indonesian, which both grill chunks of lamb over charcoal. The Greek souvlaki is characteristically flavored with lemon and oregano, whereas the Indonesian satay is typically seasoned with a sweet and spicy blend of soy sauce, coconut, chiles, and ground peanuts. The same basic food and the same cooking technique here results in two wholly different dishes whose distinctiveness resides in the flavoring ingredients. Similarly, we can contemplate two varieties of boiled wheat noodles, one Chinese and the other southern Italian (which, incidentally, seem to have had separate and independent origins, despite the persistent myth of Marco Polo's responsibility). The Chinese dish is likely to have a sauce based on soy sauce, gingerroot, and sesame oil, whereas the Neapolitan dish is almost certain to carry sauce made from olive oil, garlic, tomatoes, basil, and oregano. The two dishes cannot be confused because the sensory experience each offers by way of flavoring ingredients is characteristic and unique.

The notion of flavor principles describes persistently recurring combinations of flavor ingredients, without specifying the kinds or categories of substances typically used. A closer analysis reveals the kinds of ingredients most commonly selected by cuisines for the clear purpose of providing flavor, categories that once again seem to be shared by cultures throughout the world. The structure that encompasses the varieties of seasoning substances is called "flavor systems."

The first category of flavor systems is that of fats or oils, which have many functions. They can be used as an essential part of the cooking process (as in frying and sautéing); as a component of spreads, sauces, and emulsions (as in salad dressings or mayonnaise); or as simple condiments (butter on bread, sour cream on a baked potato, olive oil on pasta). Fats and oils are among the most effective flavoring agents because they both carry their own unique flavors and are powerful conveyors of other flavors. In kitchens all over the globe the flavor of cooked food is initiated by a process of browning meat, bones, and aromatics in hot oil or fat, a technique that caramelizes the sugars in vegetables and converts meat proteins into savory flavor-rich compounds. Oils of vegetable origin (sesame, olive, peanut, etc.) and fats of animal origin (lard, chicken fat, butter, etc.) are frequently preprocessed to provide desirable flavor (the roasting of sesame seeds to obtain the fragrant Asian sesame oil, the cooking and clarifying of butter to produce the nutty flavored Indian ghee).

The second category of flavor systems is the liquid component, typically used in soups, stews, and sauced dishes. Liquids are of both animal and vegetable origin and can be further classified as fresh or cultured (fermented) products. Fresh animal-based liquids include meat, poultry, and fish or seafood stocks; milk and cream; cultured animal liquids include fish sauce, commonly used throughout southeast Asia and once a ubiquitous feature of ancient Roman cuisine; sour cream; and yogurt. Familiar fresh vegetable–based liquids are tomatoes, coconut milk, vegetable juices, and infusions made from, for example, dried mushrooms soaked in water. Cultured or fermented vegetable liquids include such widely used products as wine, beer, soy sauce, and sauerkraut juice. Each of these liquids, as opposed to plain water, offers a unique and distinctive flavoring element to the food in which or with which it is prepared.

The third category consists of a wide variety of ingredients that are typically added in relatively small amounts and that, unlike fats or oils and liquids, seem to serve no function other than to provide additional flavor. These ingredients are used in both fresh and processed forms; they include aromatics (members of the allium family such as onion, leek, garlic, scallion, and shallot), sweet peppers and pungent chiles, carrots, celery, and gingerroot. A large array of fresh green herbs such as parsley, dill, basil, mint, and cilantro are widely used, and an astounding assortment of spices is provided by bark, leaves, berries, and seeds. Also included in this category is a number of acidic ingredients such as citrus juice, vinegar, and tamarind, and flavored sweeteners such as honey, maple syrup, fruit juices, and fruit concentrates.

This outline of flavor systems is obviously a simplified classification designed to serve as a useful guide in discovering and describing the flavoring practices of any cuisine. It does not necessarily account for the inevitable complexity and subtlety of many traditions, in which the three components of culinary structure—basic foods, cooking techniques, and flavor—are constantly manipulated and played with to produce a variety of dishes that are familiar and pleasing, and yet not always exactly the same. Still, it is clear that across cultures the notion of flavor principles is a viable one, providing a kind of thematic ground on which any number of variations is played (Rozin & Rozin, 1981). It is also obvious that the three categories of flavor systems are frequently flexible and capable of shift, so that ingredients can serve multiple or overlapping functions. Wine may be used as the primary liquid in a dish, but it also contributes an acidic note that might be inappropriate if provided by lime juice, for example, or yogurt. The unique flavor of rendered chicken fat is critical to the final flavor of certain European Jewish dishes, but the accompanying richness, lubricity, and unctuous mouth-feel are also an essential part of the food's appeal.

However we may attempt to understand and analyze flavor, we cannot under-estimate its value as a central and crucial part of the human food experience, and this has apparently been the case from the earliest times of our evolution as *homo culinarius*. Whatever other vital functions our need and ability to cook may serve, increasing the nutritional value of foods, improving digestibility, preventing spoil-age, or eliminating dangerous or unpleasant elements, the application or produc-tion of appropriate and pleasing flavor is primary in our understanding of food and in our understanding of ourselves. It provides sensory labels that enable us to iden-tify ourselves as members of a culture, a tribe, or a family, a process that may begin in utero and that is surely enhanced by breastfeeding; it makes boring foods more palatable and unfamiliar foods more acceptable (recall the American GI, who rou-tinely doses all camp rations with a comforting coating of ketchup); it lifts food out of the doldrums of the merely nutritional and transports it into a realm to which no animal (except my Boston terrier) has ever aspired—that of the aesthetic.

There is good reason to understand the deliberate production of flavor as a uniquely human response to imperfect food (Rozin, 1994). All cultures have beliefs about what the best is: the purest oil, the firstborn lamb, the tenderest baby veg-etables. The best can be the youngest or the rarest or the purest or the costliest or the most nutritionally valuable. No matter how a culture defines the best, the fact is that most ordinary people do not have easy or frequent access to it, and it is season-ing that often functions as a means of making less-than-perfect food more accept-able and more satisfying. The Chinese, for example, categorize flavor in two basic ways (Lin & Lin, 1969). The first, and best, is the natural flavor inherent in the finest of food substances—the choicest meats and seafoods, the most delicate new veg-etables—flavor that is coaxed and nurtured with consummate skill, care, time, and expense. This kind of flavor—deceptively simple, subtle, refined—is different from the other kind, which is called "cheap" flavor. It relies not on the finest ingredients or the most skillful technique but rather on strong, salient seasoning substances that are quickly and easily fashioned into savory, mouth-filling sauces and condi-ments, which in turn convert imperfect or unsatisfying foods into palatable and pleasurable eating experiences.

And it is, of course, "cheap" flavor that most of the world, of necessity, enjoys. When we look at the broad spectrum of human flavoring practices, we see one very curious correlation: the heavier the dependence on plant or vegetable foods, the more pronounced the seasonings; the heavier the consumption of animal foods, the less pronounced the seasonings. Those cuisines that clearly demonstrate a highly spiced or complex seasoning profile—southeast Asia, India, Africa, Mexico—have all long relied on high-plant, low-meat diets. Conversely, the heavy meat, fat, and dairy cultures of northern and western Europe and of central Asia show less salient, more underplayed flavoring practices. It looks as though the heavier seasoning of vegetable foods is a way of compensating for the lack of animal foods, and, more specifically, of red meat, with its savory juices and heavy load of fat. This compensation is minimally nutritional, however; rather, it is primarily gustatory.

When we come to assess the role that flavor plays in the meals of traditional cultures, we can see clearly how important the gustatory has become in the human experience. The Bemba (Richards, 1939), a traditional Bantu people from northern Rhodesia, have long subsisted on a thick millet porridge called *ubwali*, a bland stiff mixture that is scooped from the bowl and rolled with the fingers into a ball. The ball is then dipped into a tasty sauce or relish called *umunami*, a liquid seasoned stew made from a variety of ingredients that may include bits of meat, ants, caterpillars, chiles, peanuts, and vegetables. Although the Bemba regard the millet porridge as essential to the meal and do not believe they have been adequately fed unless they have eaten *ubwali*, they also feel they cannot eat the *ubwali* without the accompanying *umunami*. The seasoned sauce makes the porridge balls "slide down the throat," allowing the thick glutinous starch to go down easier. And there is no question that it also makes the food taste better, adding zest and variety to an otherwise bland and monotonous diet. It is a practice echoed throughout sub-Saharan Africa, where the basis of the diet and the focus of the meal is a starchy mush or porridge made from cassava or plantains or cornmeal eaten with an enhancing savory sauce or stew.

The *ubwali-umunami* system of the Bemba is strikingly similar to a dietary principle long held by the Chinese, although accounted for in very different terms. Chinese world view rests on a strong belief in balance and harmony, the yin-yang; a meal that presents a balance of elements is not only pleasing and enjoyable for the palate but is also appropriately nourishing, creating a harmony between body and spirit. The dietary principle that describes this balance is called *fan-t'sai*, and it divides food into two basic categories: *fan*, which is the basic grain or starch, and *t'sai*, which is a variety of dishes made from other ingredients that are a necessary complement to *fan*. *Fan* is usually rice but can also be noodles or porridge, whereas *t'sai* comprises an extraordinary variety of preparations made from meat, seafood, bean curd, or vegetables cooked and flavored with such ingredients as soy sauce, rice wine, gingerroot, garlic, and hoisin. Although *fan* is the central and more important component without which no meal is a meal, *t'sai* is necessary to balance or complement, to round out the whole (Anderson, 1988; Chang, 1977). Like the Bemba, the Chinese system illustrates the importance, both dietary and aesthetic, of providing variety and richness to basic grain or carbohydrate-focused meals by means of flavored sauces and garnishes. From the nutritional point of view a meal consisting solely of grains or starch represents an imbalance; similarly, from the aesthetic point of view monotony is an imbalance. Once again, imperfect foods require some additional supplement or embellishment to make them more acceptable and more pleasing to their human consumers.

The pattern is repeated in another cuisine more familiar to western diners. Pasta, in a stunning variety of shapes and sizes, has for centuries been the beloved focus of southern Italian meals, serving frequently as the single dish in a meal and at times as one course of a multi-tiered meal. It is a crucial and central food without which a meal is not considered complete. Yet pasta, exquisitely appreciated for its own flavor and texture, is never brought to the table wholly unadorned; it is always served with an appropriate *condimento*, a flavorful coating or savory sauce that may be as simple as olive oil and garlic or as complex as a sauce Bolognese, a rich mixture of ground meat and tomatoes, cheese, herbs, and cream. Pasta, like cornmeal mush or rice, is the valued staple that fills the belly and sticks to the ribs, but it is its alliance with additional enriching, highly seasoned ingredients that elevates it to a lip-smacking experience, a proper meal.

The preceding examples from Africa, Asia, and Europe illustrate a practice that is widespread and pervasive enough to be called universal, one that accounts for a great number of representative dishes that are produced in cuisines everywhere. In many cases, these dishes are the central feature of the meal or, in fact, define the meal. As important as these preparations are, however, they are not the only ones to use flavor in consistent and predictable ways; indeed, when we look at the total output of any culinary tradition, we see that the parameters of any particular flavor system are manipulated and varied to produce a number of other categories of prepared food.

Salads or relishes, for example, are a common adjunct to the central dish that frequently defines the meal, and they too are constructed with their own appropriate flavors, ingredients selected from the flavor system of the culture in question. Typically seasoned with acid and salt (the word *salad* comes from the Latin word for salt, as do the words for *sauce* or *salsa*), salads individualize themselves culturally by the ingredients that carry the acid and salt, as well as by additional flavoring substances. A Thai salad is typically flavored with acid from lime juice or rice vinegar, salt from fermented fish sauce, and fresh coriander leaf and chopped peanuts; an Italian salad is more characteristically seasoned with wine vinegar, olive oil, and perhaps fresh basil or garlic. Other such repetitive flavoring practices can be observed in a number of categories of prepared foods, such as sweets, preserved foods, snacks, and street foods.

One more category of prepared foods is of particular interest in any consideration of flavor in the meal and the culture. Condiments are substances that are separately prepared and designed for no other purpose than to add a final layer of flavor to an already cooked and seasoned dish. They tend to be concentrated in flavor with strongly salty, sweet, acidic, or irritant properties, providing little dollops of stimulation that are not necessarily pleasant by themselves but are titillating as zesty enhancements to the rest of the meal. They function by exciting the palate with complement and contrast, perking up bland foods with pungent flavors, refreshing fatty foods with cleansing acidic tastes, highlighting salty or smoky flavors with sweetness. Their structure and function are remarkably similar across cultures and they illustrate eloquently the role of flavor as a gustatory agent. Imagine sushi without pickled gingerroot and pungent wasabi. Boring! A hot dog without mustard and sauerkraut. Naked! Fish and chips without a good splash of malt vinegar. A surfeit of grease! Condiments are typically not eaten by themselves but only in conjunction with the foods they are designed to enhance, yet without them our meals are unfinished and incomplete. Their use, however, is frequently optional; it is left to the discretion of the diner to decide whether to use them and in what amount. From the simplest squeeze of a lemon wedge or a sprinkle of pungent chopped chiles to complex combinations of flavor-rich ingredients—chutneys and sambals, relishes and salsas, dips and spreads—condiments are the meal's fashion accessories, whose only purpose is to embellish and intensify the aesthetic experience of food.

The flavoring practices described to this point, informed and characterized by unique individual ethnic selections of ingredients, are based on stable traditions that evolved more or less gradually over generations and that retain their identity as recognizable and reproducible systems. Traditional cuisines tend to be conservative, maintaining familiar and long-established parameters even as new elements are introduced from without. But cuisines, although fundamentally conservative, are rarely if ever completely static; novel ingredients and practices are accepted and, indeed, sought out for a number of reasons: enhanced nutrition, convenience, social status, the need for increased palatability, and variety. The tension between the novel and the familiar varies from culture to culture and from time to time, but the twentieth century has witnessed an unprecedented swing toward novelty. And nowhere is that tendency more pronounced than in the use of flavor.

The reasons for this are obvious enough: the explosion of technology and the consequent burgeoning of an aggressive food industry; ever-increasing urbanization and sophistication, with a subsequent diminishing of strong ethnic family and communal ties; growing ease of transport and exportation, so that unique products once available only locally can now be found in almost every corner of the world; and finally, a growing awareness, disseminated globally by an unprecedented system for information transmission, of the medical, nutritional, and prestige value of certain foods and diets.

The United States is at the forefront of shifting flavor practices in the contemporary food scene. Although it is a nation of immigrants, its political and cultural mainstream has been historically almost exclusively grounded in western Europe, a reality made palpable in its food traditions. The traditional American diet reflected the European taste for red meat, animal fats, and dairy products, with a corresponding minimal appreciation of vegetables and strong, salient seasoning ingredients. Despite the steadily growing nonwestern European populations along the borders, in rural pockets, and in urban ghettos, it was not really until after World War II that "ethnic" food began to have some significant impact on the mainstream; until then, ethnic was a vaguely derogatory term designating unfamiliar and distasteful food (and the people who ate it) that reeked of garlic or stung the mouth with hot chile peppers.

The end of the war brought greater opportunities for travel, a sense of expanding horizons, and a willingness to sample novel foreign foods; the seventies saw a surge of interest in ethnic cooking, including cuisines once denigrated as coarse or lower class. Then in the eighties came the burst of medical data indicating the unhealthy correspondence between the typical American diet and an increasingly sedentary lifestyle and the clear health benefits of traditional ethnic cuisines that were high in vegetable foods and low in red meat, animal fats, and dairy products. Ethnicity in food was now perceived as a positive attribute and the high seasoning associated with many of these cultures became a necessary part of the "good" life, now esteemed and fashionable. Garlic and soy sauce were "in"; olive oil knocked butter out of the larder; salsa gave ketchup a run for its money as Americans began to be aware that chile peppers offered not merely a generic hotness but a truly remarkable range of flavors and pungencies in a rich tradition that had existed on these shores long before Europeans arrived.

This emerging desire for novelty, heightened stimulation, and what is perceived as a better way to eat has not yet run its course; greater levels and varieties of flavor are constantly being sought in the grand American (or perhaps more generally human) notion that if some is good, more is better. With most of the world's ingredients now so widely available, not only can the dishes of almost any ethnic cuisine be reproduced with reasonable authenticity but the ingredients and flavors can be mixed and matched, combined, and reapplied in wholly new and unexpected ways. Asian aromatics like lemongrass and gingerroot lend their exotic perfume to French cream sauces; spicy chiles add unexpected highlights to sweet fruits and ice creams. The products of this "fusion" cuisine are as exotic and titillating to the contemporary urban palate as spicy tomato sauces must surely have been to the citizens of sixteenth century Naples. For, of course, this need for ever-increasing kinds and intensities of flavor has shown itself at times throughout history, but most commonly only for those of wealth and stature. The cuisines of both ancient Rome and aristocratic medieval England were distinguished by a lust for exotic spices, rare and costly ingredients that were frequently combined in bizarre and unprecedented ways. The recipes for those ancient dishes are hauntingly similar to the menu selections from today's trendiest restaurants, their names and descriptions a hodgepodge of foods and flavors from all the world's kitchens.

However we may use, or misuse, flavor, one fact is abundantly clear: flavor is a critical and defining part of human eating behavior, and for a number of reasons. It is a nutritional "enabler," encouraging a greater consumption of bland grains and starchy carbohydrates, the basic staples that nourish most of the world's population. It provides sensory labels that identify food as a recognizable part of family, tribe, or ethnic tradition—familiar, nurturing, and safe. At the same time it can also accommodate the apparently deep-rooted human need for variety, novelty, and excitement, satisfying on a level wholly apart from the nutritive. For in the end it is the gustatory that flavor serves most crucially for the human consumer, gratifying that unique inclination to experience pleasure for its own sake.

Although this can be seen most clearly in the production of individual dishes, it operates in a more complex fashion in the context of a meal, where flavor functions more specifically to define and illuminate the purpose and process of the event. The meal, of course, can be understood in many different ways by different traditions under a variety of circumstances. It can range from a simple, quick, periodic, mouth-stuffing satiation of hunger to lengthy, leisurely, multicoursed food events; the most notorious of these were the legendary banquets of ancient Rome, in which intermissions of ritualized bulemia permitted the prolongation of the gastronomic experience well beyond physiological capacity. A meal can consist of a single central dish served with or without accompanying dishes like condiments, or as a single structure in which many elements are organized as a whole, as, for example, in a cheeseburger, where bread, meat, cheese, vegetables, and condiments are orchestrated into a single layered composition. Or, those individual elements can be provided temporally as a succession of different dishes, as in the typical Italian meal, in which a plate of varied cold hors d'oeuvres is followed by a pasta, then by the meat, the vegetable, the salad, and the sweet, each one a separate and distinctive experience. In this kind of meal presentation flavor is frequently manipulated into deliberately alternating patterns of contrast and complement, intensity and subtlety, richness and refreshment, resulting in complex temporal experiences designed to gratify on many levels, serial as well as simultaneous.

Flavor can also be an essential means of defining the occasion or the significance of a meal. Although all cultures select from the ingredients designated by their particular flavor systems, not all of those foods or flavors are acceptable or appropriate with every dish or every meal. Bitter green herbs dipped into salty water and thick, sweet, spicy fruit and nut pastes might not be considered desirable for a schoolbox lunch or a backyard barbecue, but they are an indispensable part of the Passover seder, ritual foods whose flavor highlights and enforces the commemorative meaning of the occasion. Similarly, although ketchup may provide the constant and beloved flavor enhancement for an American family's weekly round of meatloaf and burgers and hash browns, it is not likely to be considered an appropriate addition to the oldest daughter's wedding feast. At such a celebratory meal a whole other set of preparations comes into play—a hollandaise sauce, for example, or a béarnaise—defining the occasion with special flavors. All kinds of meaning, significance, and import are richly conveyed by the sensory attributes of flavor and the foods that carry it.

It may not be easy to define the meal, and it is not always possible to define flavor, to separate it absolutely from the foods that carry it, the techniques that shape it, and the characteristics of texture and aroma and temperature that invariably accompany it. But as both individuals and participants in a variety of social networks, we all know a meal when we see one. And we all know what makes it taste good.

REFERENCES

Anderson, E.N. (1988). *The food of China*. New Haven, CT: Yale University Press.

Chang, E.C. (Ed.) (1977). *Food in Chinese culture*. New Haven, CT: Yale University Press.

Lin, H., & Lin, T. (1969). *Chinese gastronomy*. New York: Hastings House.

Richards, A.I. (1939). *Land, labour and diet in Northern Rhodesia*. London: Oxford University Press.

Rozin, E. (1973). *The flavor-principle cookbook*. New York: Hawthorn Books.

Rozin, E. (1982). The structure of cuisine. In L.M. Barker (Ed.), *The psychobiology of human food selection*. Westport, CT: Avi.

Rozin, E. (1994). *The primal cheeseburger*. New York: Penguin Books.

Rozin, E. (1996). *The universal kitchen*. New York: Viking Penguin.

Rozin, E., & Rozin, P. (1981). Culinary themes and variations. *Natural History 90*, 2.

Chapter 9

Holiday Meals:
Rituals of Family Tradition

Lucy M. Long

INTRODUCTION

Popular media tends to portray family holiday meals as times of togetherness, as warm gatherings full of affection and extended family. Such meals, however, are multivalent events, reflecting the complexity of family structures and interpersonal relationships. In reality, they often include conflict, distancing, and discord. This is partially because such meals represent the intersection of food with two other symbolically rich domains—those of "family" and "holidays." These domains offer a sense of identity and a pool of aesthetic traditions for expressing and constructing those identities. In family holiday meals, various identities can be enacted and negotiated, allowing for affirmation, challenging, or reversal of established or emerging identities. As such, these meals can be deeply personal and emotive and can hold intense symbolic meaning for individuals.

In this chapter, I take a folkloristic perspective to the study of such meals and to the domains they represent. Folklore as an academic discipline attends to the personal and the aesthetic or expressive in cultural activities and productions. It explores the social construction of meaningfulness, recognizing culture as a manipulatable resource for artistic expression, social interaction, identity construction, and the display and even the imposition of power (see Bauman, 1984; Ben-Amos, 1971; Glassie, 1983; Oring, 1986b; Paredes & Bauman, 1972; Toelkin, 1996). Holidays and rituals, foodways, and family are all subdisciplines within the field of folklore. The concept of the meal is both subject and scholarly construct in each one.

Focusing on the American holiday of Thanksgiving, this chapter explores the intersection of these constructs in understanding family holiday meals. Thanksgiving provides an excellent example of the multiple layers of activities and meanings involved in such meals. Far from being a simple celebration of unity, Thanksgiving meals can be arenas for the construction of family identity and individual roles within that identity. Furthermore, they enact the specific meanings being given to the holiday by the larger culture.

The American Thanksgiving is an official national holiday set aside in 1865 by President Abraham Lincoln purportedly to commemorate the Pilgrims' settling of Plimouth Colony in Massachusetts in the 1620s, the thanks they gave to God for their first harvest in the New World, and their appreciation of the assistance given them by Native Americans (Humphrey, 1996; Santino, 1994a). The actual origins of the holiday are disputed—Plimouth is the official origin, but some historians contend that an earlier settlement, Berkeley Plantation, in Virginia actually predated the Plimouth celebration (C. Geist [personal communication, 1998]. Also, the nature of the first celebrations has been challenged. Far from being the reverent, religious ceremony of thanks portrayed in many of the history books, the settlers of Plimouth held a rather boisterous and rowdy extended harvest festival similar to what they would have experienced in their English homeland (Deetz & Anderson, 1972). Venison, rather than the commonplace turkey, would probably have been the centerpiece meat because it would have carried associations with English aristocracy for the settlers. The Virginia celebration, on the other hand, was a somber, religious service of prayer and thanksgiving. The role of the Native Americans in the survival of the Pilgrims is also questioned, as is the appropriateness of purportedly celebrating cultures that were later dominated, if not exterminated, by the European newcomers. Be that as it may, contemporary American Thanksgiving is promoted as a time of national unity commemorated by family gatherings centering around a festive meal of meat (turkey), starch (potatoes, stuffing, cornbread), vegetables and fruits associated with the harvest season (corn, apples, green beans, squash) and with the assumed New England origins (cranberries), and plentiful traditional "all-American" desserts (pumpkin pie, apple pie, and, for some, egg custards and molasses puddings).

Although American in its history and practice, the Thanksgiving meal provides a model for study that can be applied to any celebratory meals that involve family gatherings, particularly those that draw on both seasonal imagery and religious ritual foods and food activities. Some sort of harvest festival seems to occur in every culture, and most of these have a strong feasting component, although not all of them emphasize family gatherings or include religious overtones. Halloween and Harvest Home in Ireland are comparable to the American Thanksgiving as is the Canadian Thanksgiving. The autumn festival of *Chusok* in Korea celebrates the harvest and is a time of feasting and visiting with friends and family. *Eid*, in Islamic cultures, is a period of sharing sweets and delicacies with neighbors and relatives after the daily fasting during the holy month of *Ramadan* and includes giving thanks to *Allah* for enabling the fast. Festivals in Italy and Chile celebrate the grape harvest with feasting and thanks for the harvest. These celebrations, and many more, share aspects of the American Thanksgiving.

FOODWAYS

Meals are the culmination of numerous activities surrounding their conceptualization and production. As the visible focal point of a range of activities, they offer a starting point for examining this range. Furthermore, meals are the tangible expression or enactment of the beliefs, customs, history, and aesthetics surrounding a culture's eating habits (Brown & Mussell, 1984; Camp, 1989; Jones, Giuliano & Krell, 1981; B.S. Turner, 1982). Folklorist Don Yoder introduced the term "foodways" to refer to this extended network of activities surrounding the procurement, preservation, preparation, presentation, performance, and consumption of food (Yoder, 1972). His model also involves the beliefs, aesthetics, economics, and politics involved in food behaviors. "Foodways" suggests the full meaning of food in our lives and why it carries such significance; the activities surrounding food are integrated into all aspects of our lives. A brief overview of the foodways of an American Thanksgiving meal demonstrates the pervasiveness of foodways in our daily activities and is applicable to any celebratory feast.

Procurement involves obtaining the ingredients and items needed for a meal. Historically, it would have necessitated hunting for wild turkey and using the harvest crops, but for most Americans it now involves placing a special order at a butcher shop or turkey farm, shopping at the local grocery, or perhaps visiting a pumpkin patch or apple orchard. It is also possible to procure the entire Thanksgiving meal by ordering it ready-made from a grocery store, an act that can call into question a parent's commitment to family by more traditionally minded cooks.

Preservation involves strategies used for keeping foods frozen or fresh and storing them until needed. Turkeys, because of their size, frequently present a problem for storage. Due to recent publicity about the dangers of salmonella, some families have given up the traditional whole turkey and substituted parts—legs and breasts.

Preparation usually begins days in advance of the consumption of the meal and can be a communal or individual activity, involving chopping, marinating, and otherwise readying foods to be cooked. Before anything can begin, however, the menu must be selected and recipes decided on. Many holidays have special dishes associated with them, offering formulas for constructing holiday meals. These are frequently special dishes found only at that holiday or are elaborations of a regular meal; however, every family will construct their own meal on the basis of their individual tastes, bank accounts, traditions, and numerous other factors. This construction can be an extensive and highly social process, involving communication and discussion over issues such as the appropriateness of dishes and the accuracy of recipes. The availability of ingredients, abilities of the cook, and the tastes of the cook and intended consumers would also be taken into account, and cooking equipment and methods would also be selected. For example the more traditional roasting of the turkey has given way recently to smoking in a covered outdoor grill (probably a West Coast innovation) or deep-frying in large vats of boiling oil (popular this past year in the American Southeast).

Preparation leads directly to the product, the food itself, and includes the actual recipes and ingredients used. This is the most visible aspect of foodways and the one most readily accessible to study. Most holiday meals, including Thanksgiving, are a mixture of traditional components, variations on those components, and complete innovations. This mixture offers a key to the identities of the producers and consumers and the ways in which their meal is expressive of their culture, their values, their food aesthetics, and their immediate contexts for eating. Also, certain forms of food carry emotive and symbolic connotations. Garnishes signify a special occasion, as do elaborate dishes requiring sauces and extensive preparation. Raw fruits and vegetables may be carved into artistic designs—radishes into roses, carrots into curlicues, cucumbers into geometric patterns. Everyday foods may be dressed up by adding little extras, such as nuts (almonds to green beans), spices (ginger to carrots), and other sauces (pecan butter to brussels sprouts). Gelatin molds can be varied to fit the holiday by using appropriate colors (orange or yellow for Thanksgiving, red for Christmas, green for St. Patrick's Day) or fillings (apples and nuts for Thanksgiving molds). Margaret Visser points out that " . . . joints and whole birds (whole turkeys, for example) have become especially festive in connotation: they are for relatively rare large gatherings, when almost all the meat will probably be consumed at a sitting" (1986, p. 17). A 20-pound turkey almost demands extended family and friends to be in attendance, and its size reflects the size of one's familial and social universe—an observation quickly understood by anyone separated from family or friends during Thanksgiving.

Presentation of food refers to how it is physically displayed, brought to the table, served to consumers. It includes the artistic placement of dishes and the laying out of the various food items. For Thanksgiving, for example, a whole roasted turkey is usually the centerpiece, both spatially and visually. Romantic images of the holiday portray a large turkey surrounded by vegetables (usually root vegetables such as potatoes and onions) and garnishes (parsley) sitting on a platter in the center of the table. In reality, this arrangement proves unwieldy for carving and serving the turkey, so once the meal has formally started, the turkey is frequently moved to another spot on the table or returned to the kitchen to be carved. Similarly, celebratory meals are frequently marked by special plates and silverware, and dishes with holiday motifs are commercially available. Thanksgiving platters, plates, napkins, and tablecloths come in autumn colors and may include turkeys, leaves, sheaves of grain, cornucopias, pumpkins, or pilgrims.

Performance in foodways includes the place of those foods within the usual meal system and cuisine. Judith Goode and others have clearly demonstrated that a meal must be understood within the larger framework of the overall daily, weekly, annual, and celebratory meal cycles (1984). The social roles and functions of that meal emerge partly from its genre. For example, we have different expectations of breakfast and lunch, not only of what foods will be offered and at what times but also of whom those meals can be shared with, the kinds of interactions that might occur, and the formality of the event. The Thanksgiving meal, for example, usually occurs around midday, anywhere from noon to late afternoon (but before the annual televised football games). This allows families to gather and gives enough time for preparation and clean-up. It also means that the abundant meal can be digested before bedtime, allowing for a secondary meal based on leftovers (usually hot or cold turkey sandwiches). Performance also includes how people interact with each other through food, using it for conversation, for bonding, for competition. All holiday meals, including Thanksgiving, are rich resources for such interaction because they are frequently the central activity in a celebration and tend to be consciously constructed to set a particular tone and to evoke emotional associations, helping to define the nature of that celebration.

Consumption refers to how people eat the meal: what utensils they use, what mixtures they create, and the order in which items are consumed. Formal celebrations, such as Thanksgiving tends to be, usually require more attention to one's eating style and the use of appropriate manners. Thanksgiving meals tend to be "sit-down" affairs in which the food is served at the table, and the consumers are expected to use the full range of cutlery and napkins. Many families banish young children to a table of their own until such manners have been acquired. One of the appeals of the supper or late-night snack that usually follows the mid-day meal is that it is less formal; turkey sandwiches are eaten with the hands, paper plates might be used, other foods might be eaten straight from storage containers (olives and pickles from the jar, stuffing or dressing taken from the refrigerator rather than from a serving dish).

Last in the foodways model is the clean-up after food preparation and consumption. Although usually ignored in popular portrayals of holiday meals, this aspect of the meal, as anyone carrying responsibility for it knows, can affect the selection of recipes and preparation methods and the experiencing of the meal itself. Cooks may limit the number of dishes or the elaborateness of their cooking to minimize the amount of cleaning up necessary. Innovations in cooking equipment and techniques are marketed with clean-up in mind—large roasting bags for turkeys is one timesaver that has proved popular recently because it means there is no roasting pan to scrub. Time constraints may also require that the meal be consumed in a particular amount of time, so that clean-up can occur before another scheduled event (returning home, watching football on television, visiting). Clean-up includes questions of storage and the use of leftovers. Women's magazines commonly carry recipes using turkey, addressing the problem of, in Visser's words ". . . a whole aspect of household management—one hesitates to call it a cuisine . . .: that of What to Do with Left-Overs, how to present them at another meal in a form which the family will accept" (Visser, 1986, p. 17).

With food involving so much of our lives, it is easy to see how it would be significant to family structure and tradition. Meals, as the culmination of a range of activities, involve much more than simply eating. They serve as a manifestation not only of a cultural system; they are the realization of foodways and the role of the foodways traditions of individual families. As such, they can play a crucial role in the construction of family roles and relationships and in the maintenance of a felt connection with that family's past.

FAMILY AS FOLK GROUP AND TRADITION

Family can be thought of as a biological category, a social institution, or a political rhetorical tool, but, in terms of food, it is useful to speak of it as a folk group. A folk group is a group in which common experiences are shared and expressive traditions have emerged out of those shared experiences (Oring, 1986b, p. 1). The structure of the family in western societies is such that numerous experiences occur in common among family members. These can be mundane or everyday kinds of experiences, as well as more celebratory and special ones. Families develop histories of shared experiences, and these histories both define that particular family and provide subjects and forms for expressing family membership (Boatright, 1958; Sherman, 1988). For example, offering birthday cakes at a meal celebrating an individual's birthday is an expected component of such meals in the United States, and the type and size of cake (or the lack of one) can be the subject of conversations and memories in the future.

Families use a variety of aesthetic forms—oral, material, customary—to express their being and identity, and different families use different forms. Sometimes differences are due to ethnicity or culture, religion, region, or socioeconomic level, but the temperament and personality of family members also come into play. One family may intentionally use storytelling as a self-conscious expression of tradition, whereas another family may use recreational activities. Because the acquisition, production, and consumption of food are necessary activities in every family, foodways provide extensive opportunities for shared experiences. As such, they become invested with the imprint of family and can become highly expressive of family identity. This expression frequently occurs in unself-conscious ways, and sometimes it is only in looking back that we realize that we have developed patterns of behaviors and expectations that seem to embody our history. With my own children, for example, bedtime snack was elevated to the importance of a meal and had to consist of milk with either graham crackers or cereal. I was unaware of the meal's significance and structure until I tried to change it one evening and discovered that the toast and hot chocolate I offered was an unacceptable substitute for family tradition.

Family folklore contains several themes, all of which can be related to food and suggest the pervasiveness of food in family traditions (Zeitlin, Kotkin, & Cutting Baker, 1982). Foodways can be used to both express these themes and to construct them, that is, to create, shape, and define them. These themes include family history, heroes and rogues in the family gallery, relationships among family members, and the broad theme of identity.

Family history refers specifically to events in the past believed to be connected to the family, as well as the actual genealogy of the family. Like many southern highlanders, my own family claims an Ulster Scots heritage, ignoring the German, Welsh, English, and possibly Dutch blood that entered in at various points. Events in Scottish history, particularly those that have been romanticized, were selected as part of our family mythology, and foodways were occasionally related to these events, usually in the form of drinking tea and eating shortbread at "Highland Games," annual festive gatherings of Scottish descendants. In a more intentional bonding of foodways and history, the Jewish Passover feast re-enacts the historical event of Moses leading the Israelites out of Egypt. On a smaller scale, particular foods may be associated with events or images in the past—cornbread baked in an iron skillet re-creates the cooking methods of my Appalachian ancestors in the past; preparing flattened soda bread farls on a stove-top griddle reminds us of the year we spent living in Northern Ireland.

A significant part of family history involves *characters* from the past: heroes, rogues, and ordinary family members. My maternal ancestors trace (probably erroneously) their lineage to William Bradford, the first Governor of Plimouth Plantation in Massachusetts. Every Thanksgiving, my children eat their meal with the sense that one of their ancestors was responsible for the holiday in the first place. In a more accurate rendition of history, they also joke that the meal should include peanut butter because their grandfather likes to eat it with almost every food. I also usually make a cornbread stuffing and a gelatin mold that my mother always serves, reminding me of the Thanksgiving tables of my own childhood. Similarly, my favorite recipe for birthday cakes comes from my mother. In this way, particular foods can remind us of other family members.

Relationships between family members can also be enacted through holiday meals. The attendance of a member at a family meal can be understood as a statement of commitment to the family: conversely, nonattendance may signal that loyalties lie elsewhere. Similarly, the lack of an invitation extended to a family member can be seen as a rejection of that member. Animosities or rivalries between members may be expressed through competitive cooking, or, as I remember from my own childhood, competitive consumption. Roles within the family can also be expressed through holiday meals. The eldest female hosting the Thanksgiving meal affirms the nurturing role of women in the family. Similarly, taking responsibility for preparing a dish to bring to a meal signifies the maturity and willingness to take on other responsibilities in the family. Gender and age roles can also be affirmed or challenged.

Identity, a dominant theme in family folklore and tradition, referrs to the characteristics seen as defining a family. Components of identity include ethnicity, region, socioeconomic class, occupation, religion and ethos, and recreational interests. Every holiday meal reflects a number of these components, frequently with the participants unaware of that reflection.

Ethnicity is one of the more obvious components of family identity and a dominant shaper of holiday foodways. A process-oriented approach to culture defines ethnicity as a process in which one culture is contextualized as subordinate to another culture (Oring, 1986a, p. 24–28). Dishes from the home culture will frequently be added to a more stereotypical "American" menu for such meals. These are frequently celebratory foods requiring a certain amount of expense, time, or expertise to prepare—for example, eastern European and Italian sweet breads and braided breads for Easter, Scandinavian dried cod for Christmas Eve; tamales included with turkey dinner for Christmas and Thanksgiving by Mexican-Americans. Sometimes ethnic foods are included that represent one member of the family or draw on the unique history of that family. These are usually foods known to be favored by the more American tastes of other family members: Korean barbecued meat added by a Korean in-law to a Thanksgiving dinner; Vietnamese eggrolls included by military veterans in a July Fourth picnic. Ethnic ingredients and spices can also be added to the menu: olives and goat's cheese in salads may reflect a Mediterranean heritage, chili peppers may reflect a Central American or Asian heritage, whereas soy sauce reflects an Asian heritage. Entire meals from the home culture can be reproduced as well: the Italian seafood dinner for Christmas Eve; a Chinese dim sum brunch for a birthday meal.

Region is a less obvious factor of identity and is one that participants are frequently unaware of until they discover contrasting foodways customs. On the basis of the physical geography of a culture and a consciousness of place as a resource of tradition (Allen, 1990, p. 1; Lockwood & Lockwood, 1991), region can be expressed in the menu selected for a meal or in the ingredients for special dishes. The question of whether the food cooked inside a turkey is "dressing" (southern nomenclature) or "stuffing" (northern and New England) is a regional one. Similarly, cornbread and pecans tend to be the southern basis for dressing, whereas crumbs from wheat (either white or brown) bread are common in New England and have become the commercialized national form. Oyster stuffing, an older dish harking back to our colonial past, tends to be found in the Mid-Atlantic and scattered pockets in the Midwest.

In my immediate family, we find not only differences in menu and ingredients but also in the overall aesthetics of the Thanksgiving meal table and the rules for whom to include. My husband is from New England and expects a meal of the basics: turkey, bread stuffing, mashed potatoes, gravy, cranberry sauce, mashed turnips, mashed winter squash, green beans, and pies (pumpkin, apple, custard). This is eaten with immediate family members. My family background is southern and Appalachian, and my expectations are of a meal with numerous dishes, each dressed up to be celebratory with the addition of sauces or "extras": the largest turkey that will fit into the oven; perhaps a ham; cornbread stuffing; sweet potatoes mashed with pecans and brown sugar; mashed potatoes with giblet gravy; biscuits and honey; corn; glazed carrots; creamed baby onions; a large variety of pickles; cauliflower in cheese sauce; brussels sprouts in pecan butter; green beans almondine; apple waldorf salad; a Jell-O mold containing cottage cheese, pineapple, and pecans; and endless desserts, such as mince pie, pecan pie, apple crisp, chocolate cake, all potentially topped with whipped cream and ice cream. The table is overloaded with much more food than could ever be consumed by a single family, but extended family and guests are expected to join the meal. It is easy to see how I would come away from my husband's meal feeling unsatisfied with the sparseness and mundaneness of the dishes, whereas he would feel surfeited by the overabundance of food and the superfluousness of the sauces and condiments. Both meals reflect our regional heritages, and in the meal that we have constructed for our children, we select from those traditional meals to give them a sense of their combined regional backgrounds.

Socioeconomic level is an aspect of family that has obvious effects on holiday foodways because it shapes access to particular food items and ingredients. Free public meals are common at Thanksgiving and Christmas and usually offer the more standard menus. Furthermore, socioeconomics may affect the preparation and preservation methods that can be used. A whole turkey, the usual centerpiece of winter holiday meals, is an inexpensive meat but requires refrigerators and ovens large enough to hold it. Also, the lengthy cooking time may be an extravagance that some people can ill afford. One individual, in response to my question of his favorite part of the Thanksgiving meal, stated that his rural Louisiana family always ate "fancy" sandwiches for Thanksgiving. The 4 to 6 hours of cooking time for a turkey required too much propane gas and was too expensive for the family. It was only when he left home for college that he discovered that the sandwiches were usually called "subs" and were not standard fare for Thanksgiving. Conversely, a family of a higher socioeconomic class may use more expensive ingredients to make the occasion special: artichokes or palm hearts added to salads, real cream and butter instead of substitutes, macadamia nuts in place of almonds.

Occupational identity is frequently but not necessarily intertwined with class. Only a few occupations impact directly on the choice of foods used—fresh eggs, meat, dairy products, or produce for farmers; seafood for fishermen. Occupation, however, can shape the procurement and preparations of foods. Those involved in the purchasing and selling of foods may have access to fresher or less-expensive foods and may have more choices available. Jobs that do not allow time off for holidays can dramatically shape holiday meals. Quicker cooking methods or preprepared ingredients may have to be used. Dishes requiring extensive preparation may simply be left out. Or, as is becoming more and more common, the family holiday meal may be held in a restaurant, where the preparation and clean-up are conducted by nonfamily. For similar reasons, many supermarkets now offer an entire precooked Thanksgiving meal that can be taken home and simply reheated.

Religious affiliation is a significant feature of family identity and frequently overlaps with ethnicity and region. Religions that include proscriptions for foodways have an obvious impact on family holiday meals. Perhaps one reason the turkey has become the national symbol of Thanksgiving foods is that it is not banned by any of the major religions. The ham frequently included in southern Easter and Christmas dinners and the pork eaten for southern and German-American New Year's Day meals would not be welcomed by those practicing Judaism or Islam. The steaks and hamburgers grilled outside for summer holiday meals would not be partaken of by Hindus. Similarly, the special alcoholic beverages associated with specific holidays (eggnog and wassail with Christmas; green beer and Irish coffees with St. Patrick's Day; champagne with New Year's Eve) would not be enjoyed by observant Mormons, Jehovah's Witnesses, and those of certain southern Protestant denominations.

Ethos is similar to religion but refers to a worldview and belief system apart from an organized religion. The most obvious example is vegetarianism, an ethos that challenges every major American holiday meal because these meals usually focus on meat. Sometimes vegetarians simply leave out the meat component of the meal and as with the vegetarians in my family, the individuals not wanting meat fill their plates with foods that are acceptable. A meat substitute can also be offered. Vegetarian cookbooks and magazines suggest such centerpieces for Thanksgiving as tofu or nut loaves, stuffed pumpkin or squash, breads or molded casseroles shaped like the meat they are displacing. Middle Eastern and Asian foods that tend to be vegetarian anyway (couscous, tabouli, falafel, tofu, tempeh) are frequently added to the meal. Ethos can also include values. Purposely containing the extravagance of the meal can be done as a means of challenging the consumerism and commercialization that tends to dominate many American holidays. Inviting individuals without family ties to join the family meal can reflect a concern with reaching out to the larger community, as would helping to prepare or serve a public meal for the homeless.

A final component of family identity is *recreational interests*. The recreations most clearly related to Thanksgiving would be hunting, fishing, and gardening, all of which can provide items for a meal. I know our Thanksgiving pumpkin pies are especially meaningful to us because they are made with the pumpkins my husband has cultivated during the past summer, and the leeks and onions used in the stuffing frequently come from our backyard garden. A friend with a rural background in Pennsylvania related how her father and uncles would provide the venison for their Christmas Day dinner, and the bonding that occurred in the procuring of the venison made the eating of it all the more celebratory.

Meals—and foodways, in general—obviously play an integral role in family tradition. The nature of holiday meals is such that food's power is intensified and magnified. When these meals function as rituals, they provide tools for affirming and maintaining the family as a folk group, and when we recognize the power of food to evoke family, we also realize the power of food to construct family.

HOLIDAYS AND RITUALS

Holidays are days that are formally set aside for celebration. Coming originally from "holy day," holidays are times meant to be set apart from the ordinary, secular, and mundane as somehow sacred or special—literally a "time out of time," to quote Allessandro Falassi in reference to festivals (1987a, p. 1). This specialness is not necessarily of a religious nature, although many holidays do have a religious basis; it is more a matter of participating in out-of-the-ordinary activities and holding expectations for those activities as being special (Fabre, 1995, pp. 1–9; Santino, 1994a, 1994b, 1996). Holidays tend to be complex events that incorporate many aspects of our lives, allowing for a sense of integration and holistic engagement of self (V.W. Turner, 1982). Foodways, because they are so intertwined into other aspects of our lives, offer a way for holidays to permeate our sense of self.

Foodways, thus, tend to be a significant part of most holidays (Humphrey & Humphrey, 1988; Manning, 1983; Powdermaker, 1932; Gutierrez & Fabre, 1995). Specific dishes are associated with particular holidays, and the consumption of those dishes is frequently a central activity in holiday celebrations. For example, turkey is associated with Thanksgiving; ham, boiled eggs, rabbit-shaped cakes sprinkled with coconut with Easter; highly decorated cakes with candles with birthdays; Matzo crackers with Passover; seafood dishes with Italian-American Christmas

Eve meals; black-eyed peas and rice with southern New Year's Day lunch. Not only the foods themselves but also the foodways surrounding these particular foods—specifically, the procurement, preparation, and consumption of them—are significant to holiday celebrations. These activities can be looked at as rituals, and the types and functions of these rituals help define the character of a holiday.

Rituals can be defined as recurring activities with a symbolic reference (Rappaport, 1992, pp. 249–254; Smith, 1972; V.W. Turner, 1969). The repetition sets up expectations for future performances and allows for the emergence of rules for defining and evaluating the event. No set number of repetitions is required to make an activity a ritual. As most parents know, doing something once is frequently enough for children to define an activity as a ritual. Most often, however, we think of rituals as those activities having an extensive history and continuity with the past. Furthermore, ritualistic activities have emotional or intellectual associations beyond the immediate activity. Folklorist Jack Santino points out that although some scholars use the term "ritual" to refer to any repeated action, it is more useful to specify as rituals those "repeated and recurrent symbolic enactments, customs, ceremonies that are often carried out with reference to the sacred, or at least to some overarching institution or principal: the state, the government, the alma mater. . ." (Santino, 1994a, p. 10.) Food consumption, although a daily recurring activity for most people, is not necessarily a ritual; it is only when it occurs with the intentional referencing to meanings larger than the immediate meeting of physical and nutritional needs that it becomes so.

Anthropologists have categorized a number of rituals, but I discuss here those varieties most relevant to family holiday meals. Rituals tend to be typed according to their sacred or secular nature and to the immediate purpose of the ritual (see Falassi, 1987a). The nomenclature is loosely defined, and scholars have adapted the models to their own data and concerns.

Rites of spectacle, also called rites of conspicuous display (Falassi, 1987a, p. 4), are obvious ones for foodways. Involving the display of symbolic objects, these rituals allow for those objects to be publicly recognized, even worshipped and adored. Translated to foodways, rites of spectacle involve the aesthetic display of foods, usually in large amounts and usually presented in a formally artistic manner. Thanksgiving tables, for example, overloaded with numerous foods, display the abundance of the harvest, signaling an optimistic outlook for the future. A July Fourth picnic table filled to overflowing with barbecue foods demonstrates the hearty, enthusiastic involvement of the participants in recognizing the nation's birthday and in celebrating the season of summer. Rites of spectacle also involve rituals of conspicuous consumption involving eating and drinking in abundance (Falassi, 1987a, p. 4). "Traditional meals or blessed foods are one of the most frequent and typical features of festival, since they are a very eloquent way to represent and enjoy abundance, fertility, and prosperity" (Falassi, 1987a, p. 4). Not only are the foods visually admired, they are also consumed, usually in specially designated meal forms, such as feasts, banquets, barbecues, and receptions.

Rites of season celebrate the annual cycle, connecting individuals and cultures to the natural world. Santino points out that many holidays become associated with the season in which they occur and holidays having a different intended function can serve to celebrate a season (see Santino, 1994a). Christmas, for example, is associated with winter and "operates as a midwinter festival as much as, if not more than, a religious one, and so snow and other symbols of the season are regularly associated with it, even in such sunny places as Florida and California" (Santino,

1994b, p. 11). Similarly, foods associated with specific seasons tend to be used with particular holidays that fall during that season. Holidays occurring during the summer use summer foodways (picnic foods, cold dishes, outdoor grilling, barbecues, watermelon, strawberry shortcake, garden salads); those in the spring use foods associated with rebirth (eggs, flower blossoms, lamb, spring crops—asparagus, new potatoes, peas).

Rites of passage acknowledge transitions: individuals passing from one stage of life to another, as in graduations, birthdays, weddings (van Gennep, 1960). Foods and meals usually play a central role in these transitions and sometimes help define the new stage of being. For example, the inclusion of alcohol at a twenty-first birthday or a graduation can refer to the individual's new status as an adult. Similarly, a child being allowed to sit at the "grown-ups' table" for a Christmas or Thanksgiving meal can be a rite of passage, signifying that child's maturation. Being allowed to carve the turkey, grill the steak, or open the champagne bottle are signals that a child is now capable of taking on adult responsibilities. Acknowledging changes in the natural cycle (seasonal, agricultural, death, birth) are also rites of passage, and food is frequently included. Funeral meals acknowledge the transition from life to death but also affirm the community of the living and the significance for that community of the memories held of those who have died.

Rites of affirmation can be similar to rites of passage in that they tend to focus on individuals, literally affirming their presence and existence. In American custom, birthday parties are obvious rites of passage, but the choice of foods (cake and frosting type, beverages, other snacks) is frequently left up to the birthday celebrant, allowing for them to be affirmed as an individual. Similarly, many families include a birthday meal aside from a party, in which the celebrant selects the menu. Such choices can include the venue for eating as well. Particular restaurants may offer a favorite cuisine or a special birthday rate. A Mexican-American restaurant chain in the Midwest, for example, provides free fried ice-cream (choice of strawberry, chocolate, or vanilla ice cream deep fried in batter and topped with whipped cream and fudge sauce) for customers celebrating their birthdays there. The ice cream is served with the staff singing a version of the "Happy Birthday" song, while the customer stands and wears a special hat.

Rites of unity celebrate the togetherness of a group. Foods are central in such rituals because they frequently represent family and community. Foods used tend to be inclusive and familiar rather than exotic, encouraging full participation of all members. The foods also tend to be ones that can be easily shared—large cuts of meat, such as turkeys or hams, casseroles, and soups. Similarly, meals are structured to be inclusive—everyone eats at the same time and sits together—and they are usually formally framed with blessings at the beginning and toasts at the end.

Rites of reversal turn the normal social order upside down, a process that ironically affirms that order. Meals in such celebrations are frequently haphazard events, with individuals helping themselves as they find it necessary or convenient to eat. Foods included are easily served and consumed—finger foods, cold cuts, cheeses and breads, chili, or soup that can be left on the stove to warm. Also, the foods used can be the antithesis of the nutritional family meal—candy and sweets for Halloween (Santino, 1996). Here the food is part of a marked "solicitation ritual" and is used as barter in a negotiated relationship between adults and children rather than a commonly shared meal that suggests community or communion.

Although these types of rituals have different immediate purposes, they overlap in terms of the functions they ultimately serve. Four main functions of rituals exist: *social*, in that they bring people together, clarify and affirm social roles, delineate hierarchies, affirm a group's identity and define its boundaries; *teleological*, in that they connect individuals with the larger cycles of nature, birth and death, offering a spiritual depth to people's lives; *cultural*, in that they celebrate, clarify, or challenge the ethos, values, and identity of a culture; and *psychological*, in that they potentially provide release from stress or worry and offer opportunities for playfulness and entertainment, for engagement with something outside oneself. Foodways play a significant role in defining the function of a ritualistic meal. As carriers of identity and memory, foods bring emotional associations to a ritual and enable participants to interpret the ritual's meanings. For examples, foods having sacred or religious symbolic uses will point to a teleological function, as might foods associated with specific seasonal cycles.

As rituals, family holiday meals can serve all of these functions. In this sense they are a reflection or enactment of the family, but they can also critique the family in that they can intensify, idealize, or invert the family structure. Through such acts, meals can also be used to construct family and the relationships constituting it.

Intensification is a magnification of the existing structure. The everyday—the ordinary—is concentrated and enlarged, thus clarified. For example, if the usual gender roles in a family prescribe that cooking and cleaning are done by women while men eat and rest, then a holiday meal in which the cooking and cleaning are conducted by the women in the family can be seen as an intensification of those roles. Not only are the women performing their usual chores, but they are doing much more than usual. Similarly, a grandmother hosting a holiday meal emphasizes the nurturing role of women in organizing and producing the family meals. The foods used can be intensifications of the everyday menu and foodways. If meat and potatoes are the staples of a culture's diet, they are usually the centerpiece of a holiday meal as well, probably with a larger quantity and better quality of meat and the potatoes ornamented or prepared in a marked way. The use of the turkey and mashed potatoes as the focal point for Thanksgiving is a good example.

Holiday meals can also be idealizations of the family. "Like all rituals, they [feasts] express idealized concepts, that is the way people believe relations exist or should exist rather than how they are necessarily manifested in daily activities" (Dietler, 1996, p. 89). These meals offer an occasion for articulating the ideal of the family. Perhaps the women in the family normally do not cook extensive family dinners but wish they had the time or skills to. This is an occasion for them to act on their wishes, to enact their idealized vision of their family. Similarly, husbands helping in the preparation or clean-up may also be acting on a vision of the cooperative family, as might cooperation in the kitchen among siblings. Individual members may also be idealized—children are told to be on their best behavior, to act how their parents wish they would always act, and individuals are expected to ignore animosities in the name of family peace.

Rituals can also be inversions of the family structure. Depending on the interpretations given by the participants, men helping in the kitchen may be not only an idealization of family structure but also a reversal of the usual roles. Women refusing to cook a holiday meal invert the expectations of the family and society. For example, the mother of a friend insisted on being treated to a restaurant meal every Thanksgiving, claiming that it was no holiday for her to spend 3 days in the kitchen preparing a large family meal because she did that every other day of her life. Similarly, allowing children to select the menu for their birthday meal gives them a role normally reserved for adults.

To further complicate matters, a holiday meal can act simultaneously to intensify, idealize, and invert the family structure. A father preparing and cooking the Thanksgiving turkey, for example, can be intensification in that it is common for men to take responsibility for large cuts of meat, but it may be idealization in that some men would take more responsibility for nurturing their families through food if their circumstances were different. And finally it can be reversal in that some men would never consider the kitchen their normal place and their entry into it is marked as unusual.

As rituals, then, holiday meals have enormous power to make people feel connected to the ideals being celebrated and to the other participants. They can affirm the individual's place within the family and the validity of the family as institution. Furthermore, they remind us of our place within the natural cycle of the seasons and of birth, growth, and death. Yet they can serve many purposes and offer variations in interpretation, allowing individual families to make their meals speak for themselves. Ultimately, meals and the foodways surrounding them are integral to the success of a holiday celebration.

THE POLITICS OF FAMILY HOLIDAY MEALS

With foodways permeating so many aspects of our lives, it can be a powerful carrier of identity, social relationships, emotive associations, and personal histories. As such, it also carries the potential for conflicts. If we use "politics" in its broader sense as referring to the display and manipulation of power and the relationships based on issues of power, the conflicts surrounding foodways can be seen as political issues and maneuvers (Goody, 1982). These conflicts over power occur on a number of overlapping levels.

Arjun Appadurai uses the term, "gastro-politics," to refer to the role of food in statements of power in class struggles or the wielding of economic status (1981, 1986). The questions of who has access (geographical, physical, and financial) to certain food items or to the means of production and distribution would fall under this category (Mintz, 1985). For example, large supermarkets tend to be located on the outskirts of towns and suburbs, necessitating travel by car. Those individuals lacking access to a private car are forced to shop at neighborhood grocers, which in urban areas frequently offer either high-quality foods and neighborly service but high prices or, in less financially secure neighborhoods, poorer quality foods and limited selections but still high prices. In some rural areas and suburbs, convenience stores, often attached to gas stations, would be the only other shopping venue. Such stores tend to offer narrow selections of goods, usually less nutritious, packaged foods with very high prices, and the lack of space in these outlets means they would rarely stock large items such as frozen turkeys. Also such venues rarely offer the special sales that larger supermarkets advertise. For individuals without personal transportation, obtaining a turkey for Thanksgiving is a logistical difficulty not addressed in the popular imagery of the holiday. Gastropolitics can also be the basis of the development of special cuisines. Folklorist Mario Montano demonstrated how the *fajaita* tradition among the Mexican population in South Texas developed originally out of lack of economic access to better cuts of meat (1997). Similarly, the fresh tripe (calf's stomach) used in the specialty dish, *menudo*, is no longer easily available because of sanitation laws that ban backyard butchering and the purchase of tripe directly from local slaughterhouses. The frozen tripe now sold at grocery stores is not only more expensive but also of poorer quality (Montano, 1997).

Another level of the politics of food refers to the use of food in defining, manipulating, and performing social roles and relationships. As Mary Douglas and others have demonstrated, meals are the enactment of social relationships and cultural roles and, as such, can be used to critique those relationships (Douglas, 1982; Sobal, Chapter 7). Dietler suggests the phrase "commensal politics" to refer to the politics of eating together (1996). In his discussion of food and power in prehistoric Europe, he demonstrates that . . . "food is a pervasive and critical element in the articulation and manipulation of social relations" (1996, p. 91). Because meals are the performance of foodways—the actual event that foodways leads up to and surrounds—it is at specific meals that commensal politics occurs, celebratory meals particularly so because it is at such meals that people tend to recognize the symbolic potential of foodways activities.

In reference to family holiday meals, commensal politics ultimately refers to the obligations and roles of family members. The manner and extent of participation can point to commitment to the family and to family roles, roles that may reflect age, gender, sibling relationships, and may be based on past familiar roles rather than present personal identity. A common complaint heard about family gatherings is that when adults return home for reunions, they return to their childhood roles as well: a successful doctor becomes once again the pesky little brother; the published author becomes the quiet middle child.

Furthermore, every role pertaining to the meal—procuring the foods, preparing them, presenting them, and cleaning up—carries the potential for defining and affirming current familial relationships. Hosting a meal for extended family, for example, infers a position of authority and leadership in that family. Although this authority may be traded off with the added expenses and energy that go into hosting, it does give the host some control over who is invited, when and how the meal will occur, and what dishes will be included. Similarly, carving a turkey or serving the main dish can be a responsibility signaling a status of maturity and authority within the family. In the United States, slicing the turkey is commonly thought of as a male chore, and one that can act as a rite of passage to demonstrate recognition of adulthood or even as a sign of acceptance into the family. Rivalry between siblings or in-laws may display itself through individuals vying for recognition within any of the components of foodways. Seating arrangements, as every child knows, can reveal status within the family and the categories with which that family organizes itself—gender, age, competence of eating techniques. Finding the best prices for ingredients, using the most traditional or the most interesting recipe, offering the most bounteous table can be ways of contesting status. Similarly, competitive cooking, demonstration of the most skillful use of preparation techniques, is a frequent enactment of rivalries among women.

A final level of the manipulation and display of power, usually referred to as "cultural politics," involves the issues surrounding who has the authority to define the traditions and institutions enacting identity and heritage (see Whisnant, 1983). It is at this level that conflicts emerge over control of the ethos and aesthetics of a foodways activity. Different menus can reflect different paradigms of the meal, and those paradigms can reflect aspects of identity. Through meals, then, individuals can ultimately contest, negotiate, and affirm the identity of the family.

As discussed previously, family identity can involve regional and ethnic heritage, religious beliefs and affiliations, socioeconomic status, occupation, recreational interests, and ethos or values. Ideally, those various aspects are brought together in celebratory rituals, but the actual enactment of family through a meal can force recognition of diversity through the inclusion or prominence of particular aspects. For example, a Thanksgiving meal held by a family with in-laws from varied ethnic backgrounds may emphasize its inclusivity and openness to those individuals by including dishes from those cuisines, symbolically extending a welcome to the diversity represented. On the other hand, a family may self-consciously limit their menu to standard fare to present a cohesive, unified identity. Furthermore, conflicts can arise over the prominence given to some aspects of identity over others in the enactment of family identity through meals. Ultimately, cultural politics in family meals refers to whose vision of the family will dominate.

Similarly, family holiday meals enact and allow for the working out of the meaning of that holiday—its purpose, theme, and character. Because different interpretations of holidays are possible, politics can enter into whose interpretation will be enacted. For example, Thanksgiving is promoted in contemporary mass media as a time for family gathering. Such gatherings, however, can range from the nuclear family to grandparents and siblings, to estranged blood relatives, and friends thought of as "family." Some families feel that the meal should be limited to immediate family, whereas others feel that the holiday should be a time of generosity, so guests should be included. Similarly, individuals may differ over the prioritizing of activities associated with a holiday. Some family members may feel that the food component should be the center of the Thanksgiving holiday, whereas others may highlight recreational components, such as watching football games on television.

The use of politics in reference to food emphasizes that eating together is not always a socially positive or pleasant event. Meals meant to be bonding affirmations of unity actually offer the potential for conflicts over status, for displays of power, and for challenging authority and familial roles. This potential for contested domains within the meal, however, attests to the power of food to carry meaning, to communicate, and to construct social relationships.

CONCLUSION

In conclusion, family Thanksgiving meals can be complex, multifaceted events that are the culmination of numerous activities, reaching into every niche of everyday and celebratory life. As family tradition, they have the potential to evoke our pasts and our visions of the future, and as rituals they can be tools for constructing the present. The feasts associated with this particular holiday are not unique in this potential. All family holiday meals draw on richly emotive and symbolic domains of our lives. In these meals, we see how food is used to attach meaningfulness to the mundane, to turn social interactions into social relationships, to enable individuals to transcend their individuality to forge ties of belonging to a collective. Through these meals we establish ourselves as social beings, as having meaningful connections to cycles and institutions larger than ourselves. In this sense, family holiday meals—and the conflicts that may arise with them—are not a trivial matter at all but are the very essence of what makes life worthwhile, memorable, and enjoyable.

REFERENCES

Allen, B. (1990). Regional studies in American folklore scholarship. In B. Allen & T.J. Schlereth (Eds.), *Sense of place: American regional cultures* (pp. 1–13). Lexington: The University Press of Kentucky.

Appadurai, A. (1981). Gastro-politics in Hindu South Asia. *American Ethnology, 8*, 494.

Appadurai, A. (Ed.). (1986). *The social life of things: Commodities in cultural perspective.* Cambridge: Cambridge University Press.

Bauman, R. (1984). *Verbal art as performance.* Prospect Height, IL: Waveland Press, Inc.

Ben-Amos, D. (1971). Toward a definition of folklore in context. *Journal of American Folklore 84*, 3–15. Reprinted in A. Paredes & R. Bauman, (Eds.) *Toward new perspectives in folklore*, American Folklore Society Bibliographical and Special Series, Vol. 23. Austin: University of Texas Press.

Boatright, M. (1958). The family saga as a form of folklore. In M. Boatright (Ed.), *The family saga and other phases of American folklore.* Urbana: University of Illinois Press.

Brown, L.K., & Mussell, K. (Eds.). (1984). *Ethnic and regional foodways in the United States: The performance of group identity.* Knoxville: University of Tennessee Press.

Camp, C. (1989). *American foodways: What, when, why, and how we eat in America.* Little Rock, AK: August House.

Deetz, J. & Anderson, J. (1972, November 25). The ethnogastronomy of Thanksgiving. *Saturday Review of Science.* 29–38.

Dietler, M. (1996). Feasts and commensal politics in the political economy: Food, power, and power in prehistoric Europe. In P. Weissner & W. Schiefenhovel (Eds.), *Food and the status quest: An interdisciplinary perspective* (pp. 87–126). Providence, RI: Berghahn Books.

Douglas, M. (1982). Food as a system of communication. In *The active voice.* London: Routledge & Kegan Paul.

Fabre, G. (1995). Feasts and celebrations: Introduction. In R.A. Gutierrez, & G. Fabre, (Eds.), *Feasts and celebrations in North American ethnic communities* (pp. 1–9). Albuquerque, NM: University of New Mexico Press.

Falassi, A. (1987a). Festival: Definition and morphology. In A. Falassi (Ed.), *Time out of time: Essays on the festival* (pp. 1–10). Albuquerque, NM: University of New Mexico Press.

Falassi, A. (Ed.). (1987b). *Time out of time: Essays on the festival.* Albuquerque, NM: University of New Mexico Press.

Glassie, H. (1983). Folkloristic study of the American artifact: Objects and objectives. In R.M. Dorson (Ed.), *Handbook of American folklore* (pp. 376–383). Bloomington, IN: Indiana University Press.

Goode, J., Theophano J., & Curtis, K. (1984). A framework for the analysis of continuity and change in shared sociocultural rules for food use: The Italian-American pattern. In L.K. Brown & K. Mussell (Eds.), *Ethnic and regional foodways in the United States: The performance of group identity* (pp. 66–88). Knoxville, TN: The University of Tennessee Press.

Goody, J. (1982). *Cooking, cuisine and class: A study in comparative sociology.* Cambridge, UK: Cambridge University Press.

Gutierrez, R.A., & Fabre, G. (Eds.). (1995). *Feasts and celebrations in North American ethnic communities.* Albuquerque, NM: University of New Mexico Press.

Humphrey, L.T. (1996). Thanksgiving day. In J.H. Brunvand (Ed.), *American folklore: An encyclopedia* (pp. 705–706). New York: Garland Publishing, Inc.

Humphrey, T.C., & Humphrey, L.T. (Eds.). (1988). *"We gather together": Food and festival in American life.* Ann Arbor, MI: UMI Research Press.

Jones, M.O., Guiliano, B., & Krell R. (Eds.). (1981). *Foodways and eating habits: Direction for research.* Glendale, CA: Folklore Society.

Lavenda, R.H. (1983). Family and corporation: Celebration in Central Minnesota. In F.E. Manning (Ed.), *The celebration of society: Perspectives on contemporary cultural performance* (pp. 51–64). Bowling Green, OH: Bowling Green University Popular Press.

Lockwood, Y.R., & Lockwood, W.G. (1991). Pasties in Michigan's upper peninsula: Foodways, interethnic relations, and regionalism. In S. Stern & J.A. Cicala (Eds.), *Creative ethnicity: Symbols and strategies of contemporary ethnic life* (pp. 3–20). Logan, UT: Utah State University Press.

Manning, F.E. (1983). *The celebration of society: perspectives on contemporary cultural performance.* Bowling Green, OH: Bowling Green University Popular Press.

Mintz, S.W. (1985). *Sweetness and power: The place of sugar in modern history.* New York: Penguin.

Montano, M. (1997). Appropriation and counterhegemony in South Texas: Food slurs, offah meats, and blood. In T. Tuleja (Ed.), *Usable pasts: Traditions and group expressions in North America* (pp. 50–67). Logan, UT: Utah State University Press.

Oring, E. (1986a). Ethnic groups and ethnic folklore. In E. Oring (Ed.), *Folk groups and folklore genres* (pp. 23–44). Logan, UT: Utah State University Press.

Oring, E. (1986b) On the concepts of folklore. In E. Oring (Ed.), *Folk groups and folklore genres* (pp. 1–22). Logan, UT: Utah State University Press.

Paredes, A., & Bauman, R. (Eds.). (1972). *Toward new perspectives in folklore.* Austin, TX: University of Texas Press.

Powdermaker, H. (1932). Feasts in New Ireland: The social foundation of eating. *American Anthropologist 34*, 236–247.

Rappaport, R.A. (1992). Ritual. In R. Bauman (Ed.), *Folklore, culture performances, and popular entertainments* (pp. 249–260). New York: Oxford University Press.

Santino, J. (1996). *New old-fashioned ways: Holidays and popular culture.* Knoxville, TN: The University of Tennessee Press.

Santino, J. (1994a). *All around the year: Holidays and celebrations in American life.* Urbana, IL: University of Illinois Press.

Santino, J. (Ed.) (1994b), *Halloween and other festivals of death and life.* Knoxville, TN: The University of Tennessee Press.

Sherman, S.R. (1988). The Passover seder: Ritual dynamics, foodways, and family folklore. In T.C. Humphrey & L.T. Humphrey (Eds.), *"We gather together": Food and festival in American life* (pp. 27–42). Ann Arbor, MI: UMI Research Press.

Smith, R.J. (1972). Festivals and celebrations. In R.M. Dorson (Ed.), *Folklore and folklife* (pp. 159–172). Chicago: University of Chicago Press.

Toelken, B. (1996). *The Dynamics of folklore* (Rev. ed.). Logan, UT: Utah State University Press.

Turner, B.S. (1982). The discourse of diet. *Theory and Culture in Society, 1*, 2–32.

Turner. V.W. (Ed.). (1982). *Celebration: Studies in festivity and ritual.* Washington: Smithsonian Institution Press.

Turner, V.W. (1969). *The ritual process.* Chicago: Aldine Publishing Co.

van Gennep, A. (1960). The rites of passage. Chicago: University of Chicago Press.

Visser, M. (1986). *Much depends on dinner.* New York: Macmillan Publishing USA.

Weismantel, M.J. (1988). *Food, gender, and poverty in Ecuadorian Andes.* Philadelphia: University of Pennsylvania Press.

Whisnant, D.E. (1983). *All that is native and fine: The politics of culture in an American region.* Chapel Hill, NC: University of North Carolina Press.

Yoder, D. (1972). Folk cookery. In R.M. Dorson (Ed.), *Folklore and folklife* (pp. 325–350). Chicago: University of Chicago Press.

Zeitlin, S.J., Kotkin, A.J., & Cutting Baker, H. (Eds.). (1982). *A celebration of American family folklore.* New York: Pantheon/Schocken Books.

Part IV

The Meal
and Cuisine

Chapter 10

Chinese Meals

Jacqueline M. Newman

With three good meals a day, be content.

A good breakfast doesn't take the place of a good dinner.

The more you eat, the less flavor; the less you eat the more flavor.

Better a man should wait for his gruel, than the gruel should wait for him.

Over a bowl of congee or rice, one should remember the trouble it has cost to produce it.

—Some Chinese proverbs about eating (Lamb, 1935).

Food is an important aspect of culture. This dictum is particularly true for the Chinese because they and no other nationality, with the exception perhaps of the French, take such an interest in and derive such enjoyment from eating. The Chinese food culture is, however, quite different from the French. It is also quite different from other cultures in the Western world because it aims at creating and enjoying outstanding dishes at all levels of society (Simoons, 1991). To people of this culture, eating is one of the rare joys of living, with food playing a central role in the protocols and ceremonies of life. It should be noted that the culture of the Han population, who comprise more than 93% of the people in China and most Chinese outside of it, is more than one quarter of the world's population. The Han Chinese had major influences on many other cultures, particularly those of the Asian world, including the meals and the foods of countries such as Japan, Korea, Vietnam, Cambodia, Laos, and Thailand.

The food traditions of these Han Chinese are not new; theirs is an ancient culture with a food heritage documented for at least 3000 years (Chang, 1977). Although not exactly as it was thousands of years ago, changes have occurred, but they have not altered the fundamental character of Chinese food (Hom, 1989). To understand the food and meals of the Chinese, knowledge of the physical aspects of the country and its history and traditions, as well as the nature of the Chinese people, is needed. Included should be information about the range of individual foods consumed, which is perhaps the most extensive in the world; the number of dishes that can be and have been prepared is in the thousands, and the number of Han Chinese people who think about and discuss nuances in the food culture includes almost everybody who considers himself or herself Chinese.

SOME BACKGROUND

China, known by its inhabitants as the Middle Kingdom, is a vast land of 9.6 million kilometers (3.7 million square miles). It descends from the Himalayan highlands to the sea in three steps from the Tibetan plateau to the western uplands and then on to a vast agricultural plain. The land mass of China, slightly larger than that of the United States, is about the same size as all of Western Europe, with arable land accounting for approximately 11% of the total. This is less land than that which can be cultivated in the United States and less than in all of Western Europe, and yet it needs to feed more than one billion people, the largest population by far of any country worldwide.

In excavations at Ban Po Village in Shaanxi Province in China, circa 5000 BC, it has been determined that this culture knew about the importance of and how to domesticate grains; at Ban Po the main grain found was millet. Rice was domesticated almost as early in Ho Mu-tu village in the Yangtze Delta. In these times, China was a semiagricultural society consuming those grains and others, leaf crops, root vegetables, and meats including pigs, water buffalo, sheep, game, and all manner of fish and creatures from the sea. Soybeans, conspicuously absent, were probably introduced about 1000 BC, and they and the other foodstuffs were eaten raw or cooked, plain, salted, fermented, dried, or pickled.

In the middle of the Chou dynasty, 1122 BC to 425 BC, recipes were found recorded on bamboo slips, wooden panels, and silk sheets advising how to prepare food (Anderson, 1988; Chang, 1977). The importance of food and variety in victuals is expressed in these and in other written words such as in a poem circa 600 BC that discusses more than 10 vegetables. They are also expressed in another poem written 200 years later that entices the soul to return to earth for all kinds of good food (Lai, 1984).

Another importance of food can be seen as it moves from one place to another and as it is adopted, expanding the variety of items consumed. For example, during the Han Dynasty 206 BC to AD 220, China began acquiring edibles from Turkestan. Within the country, wheat moved from north to south and rice did the opposite, from south to north. Some foodstuffs such as yogurt and sour milks were consumed everywhere, probably from Mongol influence during the Tang Dynasty from AD 618 to 907, examples of an influence that remained in Beijing and some northern regions but is no longer popular elsewhere (Latourette, 1964).

Some expressions also show the depth of food concerns. One common one is: *To the people, food is heaven;* another used instead of hello, asks: *Have you eaten?* Still other indicators of the importance of food can be found in restaurant names such as: Eating Comes First, Restaurant of Rich and Fine Viands, or House of Exquisite Beauty. The importance of meals can be seen in the number of people devoted to food preparation. One Imperial household devoted almost 60% of its staff to preparing food and drink, namely 2271 persons; another employed more than 6000 cooks (Chang, 1977).

IMPACT ON EATING HABITS

Perhaps, the most important influence on what people serve at meals is availability. In China, scarcity of food, ever-present hunger, and a series of famines plagued a country where status and prestige among the well-to-do meant that a gentleman ranked a cook equal to or higher than the chief physician. From these and other influences, this culture developed a healthy regard for food with respect for the inherent qualities of the ingredients (Koo, 1982); appropriate use of wok, steamer, clay-pot casseroles, and other utensils; and ability to implement a plethora of different cleaver techniques be they for cutting, chopping, or slicing. This latter item, making large pieces of food into smaller ones, conserved fuel and reduced cooking times, making the foods available at more meals and to the common man. Cooking wet or dry, in oil or water, and which culinary preparation to use was serious business to be discussed before and during meals by people of all social classes (Anderson, 1988; Chang, 1977; Simoons, 1991).

In addition to the items already discussed, many philosophies influenced the eating habits and the meals of the Han Chinese. For example, in the *Li Ji*, a classic translated as the Book of Rites, there is discussion of appropriate sacrificial and feast foods. In another volume, the *Book of Songs* edited by Confucius, there are rules to be followed in recipes and eating behaviors. Taoism, another major influence, made an impact on other things, including hygiene, nutritional aspects of food, and what they said was *the way* to do things (Chang, 1977).

As in other cultures, most Chinese eat three meals a day, in the morning, at mid-day, and in the evening. What they eat varies considerably from other cultures because the bulk of their diet, particularly at mid-day and in the evening, is carbohydrate. The amount of carbohydrate they consume contributes more than 60% to their total caloric intake (Newman & Linke, 1982). Contrary to popular belief, the most common grain consumed is not rice. Only approximately 40% of the Chinese population eat this particular carbohydrate. Others depend on wheat, barley, millet, *kaoling* (sorghum), and more recently, corn. Rice is most favored in the south, wheat and lesser grains more popular in the north. The main carbohydrate is called *fan*, a word that translates as both *rice* and *a meal*. The dishes that accompany the *fan* are referred to as *cai*. That word translates as "vegetable" but really means all the dishes that accompany the *fan* and help down the rice (Chang, 1977).

Besides grain differences, mountain ranges and rivers tend to make some foods and culinary techniques region specific. After building the Grand Canal, AD 581 to 618, foods and cookery styles moved easily from place to place, making them available everywhere. Despite this mobility, some region-specific differences still exist in taste preferences (Hom, 1989). They remain as the way people associate origins of dishes and tastes of meals.

Ancient travelers said that Chinese foods in the south tasted mostly sweet, in the north they were salty, in the west the taste was frequently hot, and in the east it was sour (Newman, 1981). Generalizing, southeastern regions still favor meals with mild and sweet tastes, plenty of seafood, and fresh vegetables, and dishes are cooked quickly to preserve both color and taste. People from the southwest, where the climate is often hot and humid, prefer their meals with many piquant foods. Seafood is not easily available in this region, so it is reserved for special occasions (Newman, 1997). In the north around Beijing, winters can be bitter, necessitating salting or drying vegetables and meats. This has accustomed people of northern Chinese heritage to saltier foods (Zhu, 1998). People in the east around Shanghai have a love affair with sweet tastes. Their foods are often red, cooked in soy sauce with ample amounts of sugar (Newman, 1997).

FOOD TRADITIONS PRACTICED AT MEALS

From an early age, children are taught to leave the table 70% full (Chang, 1977), to eat *cai* foods in lesser quantities than *fan* or grain foods, and to listen and learn from adult table conversation (Fan, 1998). Chinese people believe that daily meals alleviate hunger and know that Confucius himself ate sparingly. Gorging is considered a sin, freshness of foods a delight, and maintaining the body's energy or *qi* and its balance of *yin* and *yang* a necessity. These concepts maintain good health, and lack of *qi* or balance brings on ill health and disease (Fan, 1998). In addition to these health considerations, the Chinese believe that each meal should include five basic taste sensations, sweet, briny, sour, hot, and bitter, and that it should have the Taoist duality of *yin* and *yang* foods at each meal (Newman, 1985).

The Chinese pay attention to how they feel and take actions at mealtimes to keep their bodies in harmony. Diseases are most often referred to as "conditions" that are either *yin* or *yang*. Various foods, also considered either *yin* or *yang*, are prescribed to treat them. For example, hot foods, not in temperature but in the quality of *yang,* need cooling foods known as *liang* foods to keep the body in balance. Thus, *yin* conditions need hot or *jeh* foods for a counterbalance. Meals are made up of lots of grains because they are neutral and a smaller amount of hot and cold foods is needed to maintain a person's *yin-yang* balance. People who believe their bodies are out of balance or are told so by a health practitioner select the alternate *cai* foods to assist in bringing their bodies back into harmony. For example, a pregnant woman wants to eat foods in the opposite category as her "condition" of being pregnant. Therefore, she will eat *yang* foods because her pregnancy is a *yin* condition. A person with hypertension, which is considered a *yang* condition, would avoid eating *yang* foods and, therefore, select almost all *yin* foods to return the body to a harmonious balance (Anderson, 1988). When one is healthy and in balance, all meals need to have both *liang* and *jeh* foods so that the diner can select some of each of them. It is only when they do not feel right or a practitioner tells them to that they choose more of a needed food or foods to counterbalance their particular health condition; otherwise, everyone eats both *yin* and *yang* foods although they may not even be aware of doing so (Newman, 1985).

The Chinese also use food and meals as markers of social status. They mark life-cycle events such as birthdays, marriage, or funerals, and celebrate and culminate business deals with food (Newman, Ludman, & Lynn, 1988). These events are social gatherings that revolve around a meal. They are planned to be both pleasant and festive occasions. Most frequently held outside the home, these special events are preferred at noisy restaurants not quiet eateries because the Chinese words for "pleasant time" in English translates to "hot and noisy" (Tiger & Wolf, 1985).

REGIONAL VARIATIONS

Regional differences may be decreasing, but they are still important matters of discussion and almost all Chinese know that the *east,* with foods and preparations from Jiangxi, Anhui, Zhejiang, Fujian, and Jiangsu along the Yangtze River, is a sophisticated region with many climates, many foods, many varieties of rice and tea, and many gastronomes. They know that dishes from this region are cooked longer, have a rich flavor, often a sweet taste, are refined, exquisite in appearance, and have some delicacy (Newman, 1984).

The cuisine of the people from the *southeast,* including Guangdong and Guangxi provinces and in and around the city of Guangzhou (Canton), is known as one with many outside influences such as some fruits and vegetables from the New World, including potatoes, corn, tomatoes, and green pepper. People from this region use many different kinds of rice and are known to cook with very little oil. They prefer light soy sauce, a pinch of sugar, and gingerroot to enhance and not mask the original flavors of foods. Fermented sauces are used frequently in this region; oyster sauce, hoisin sauce, and fermented black bean sauce are popular. An early day's meal, called a tea lunch or a *dim sum,* or *yum cha,* meal originated in this region. It is the most popular breakfast or lunch meal eaten here, usually in restaurants because the dishes require labor-intensive preparation and are enjoyed with others when discussing business, family, news, and mostly food (Newman, 1984).

The *southwest,* including Sichuan, Yunnan, Guizhou, Chongging, and Hubei, is an agriculturally self-sufficient land-locked area with few seafoods. It is a region that likes tea with a smoky flavor and foods dry, chewy, twice cooked, and piquant. Foods are cooked longer than in other regions, and people enjoy many different brine-preserved vegetables. They make liberal use of casserole dishes, other smoked foods, meat puddings, and dishes with many flavors. Foods that are preferred are salty, sweet, sour, and with *fagara,* which is a Sichuan peppercorn or *Xanthoxylum* and mealtime foods cooked with tangerine peel and sesame oil (Newman, 1984).

The *northeast* and the provinces of Henan, Shandong, Hubei, Shanxi, Shaanxi, and Jiangsu are ancient China's birthplace. The land is flat and can be dry and windy. This area needs lots of irrigation for the wheat, *kaoling* (sorghum), sweet potatoes, and corn grown to feed people and the pigs, lamb, and sheep that they eat, as well as the turnips, squash, and other vegetables they adore. The current capital, *Beijing,* has many restaurants that prepare regional foods and complete meals using cooking styles originally brought to the capital to please the rulers, be they Han, Mongol, Tibetan, Manchu, or Moslem. The basic starches in this region are wheat, millet, sorghum, soy, and corn flours. They are used alone or with other ingredients to make noodles (*mein*), pancakes, and steamed breads. Grilling and roasting are popular cooking techniques in the region, and rich sauces are beloved; some are made with wine and liquor and others with a dark or a flavored soy sauce such as mushroom soy. Also popular is a hot-pot meal, in which diners simmer their own meats, vegetables, and noodles in broth and then consume the liquid with or without an egg broken into the leftover stock at the end (Newman, 1984).

MEAL PATTERNS

The three meals per day and the southern love of a *dim sum* meal are not always common to everyone. Some people in the north, particularly in winter, eat only two meals each day. Less affluent people, from the north or south, might make a meal of one or two meat-filled dumplings or breads steamed, baked, or fried, or they might have just rice or noodles plain or in a bowl of boiled water with or without other foods to flavor them. These dumplings or soups are often purchased and eaten on the street, in a tiny food facility, at a work site, at one's desk or work table, or taken home to eat. Many people consider them a snack, others are eating them in place of a complete meal. The beverage of choice with any of the above or at regular meals can be plain boiled water—the Chinese rarely drink water that has not been boiled— the water that vegetables were cooked in, tea, or the soup(s) served during the meal.

Meals have their own character and rules. At family meals, every person has a large bowl of *fan* with at least two cupfuls of cooked rice or noodles in it; a pair of chopsticks; perhaps a separate soup bowl and soup spoon; and three, four, or as many dishes to flavor the *fan* as can be afforded or as there are people at the table. At family meals at home or fancier ones such as banquets in restaurants, each diner may also have a very small saucer for soy or another sauce and a mid-sized flat plate. In years past, between-meal snacks were rarely consumed; now they have become popular. Some people eat them frequently, others rarely, and if consumed they might be anything from a dumpling to a noodle dish. Where and when to stop for them, for tea, or for a large family or banquet meal is considered important enough to evoke a discussion of some length.

ORDINARY BREAKFAST MEALS

Breakfast for people of southern Chinese heritage is a rice *congee,* a noodle soup, or *dim sum* (see the following). Those who eat *congee,* a porridge or gruel usually made from short grain rice and lots of water, often top it with chopped peanuts, diced preserved eggs, pickled vegetables, or cut-up leftover tidbits from a previous meal.

In the north, breakfast is often one or more deep-fried crullers made from a wet wheat paste. These are served with an individual bowl of warm soy milk and a small dish of sugar. Diners take the cruller in hand or with chopsticks, dip it into the soy milk, then lightly into the sugar, and then they bite off the end. This process is repeated until the entire cruller is consumed. Also served at a northern breakfast can be small dishes with peanuts, pickled vegetables, and or other salty snack-type foods. Steamed bread, plain or stuffed, is popular with or without other foods; it is most commonly given to children. In the north, people might eat noodles for their early morning meal with many of the same pickled foods or leftovers that are used in *congee*. Some northerners do eat a rice gruel and they call it *juk*.

DIM SUM

Dim sum meals are also known *dian cai, dien tsin* or *yum cha*. The last one means to "drink tea," the first two to "dot the heart"; they are popular in Guangzhou (Canton) where the idea originated. This eating time can be from early morning into the afternoon, on weekends, and whenever possible on weekdays. Some people enjoy it as a meal, others as a snack. When consumed, people enjoy it at a tea house or at a restaurant that serves it. Popular since the Sung Dynasty, AD 960 to 1279, traditionally this meal was eaten when men went for *dim sum* to read newspapers, talk, and conduct business. These days, northern and southern Chinese men and women of all ages, many generations from one family, and large tables of friends delight in *dim sum*. This is a meal with of a large variety of snack-type foods or foods served in small portions. Diners might eat two, three, or four small buns, dumplings, or other items wrapped in wheat dough, rice paper, bean curd sheet, or rice with meat and/or vegetables wrapped in a large leaf. Most items are steamed, pan fried, or deep fried. In addition to these small items, people might share a large noodle or a rice dish, a special soup, or other dishes such as clams in a black bean sauce, scallion pancakes, spare ribs, tripe, stuffed shrimp, and steamed bread plain or bread stuffed with pork and sweet sauce, many served in small amounts on small plates. Tea is always served, and large eateries have many kinds to chose from be they green, oolong, or black; floral or plain; from inexpensive to extremely costly. The bill for such a meal is customarily tallied by counting the number of each size plate, adding the cost of the tea, and totaling the amount (Anderson, 1988; Chang, 1977; Wong B., 1998).

In Xian, another area that specializes in dumplings, one can have an entire *dim sum* banquet at any time of the day or night. There are restaurants in and around this ancient capital that specialize in 10, 20, even 120 different dumpling dishes. A dumpling banquet is popular in this region, and each dumpling dish would be served as a separate course. The dumplings in the Xian area vary in size and content more so than in any other area in China. The largest ones are made with several tablespoons of filling. The smallest ones have only enough filling to barely cover a baby's fingernail and come in a slightly sweet soup; they are served at the end of the meal. Dumplings are eaten at various times during or between meals; their contents can be animal, vegetable, or any combination thereof. Fancy banquets, *dim sum* or otherwise, are almost always held in private rooms with a waiter assigned to tend to every need. At them and at all meals, no condiments are on the table and no one should consider adding seasonings to any dish served. To do so is to insult those who prepared them (Zhu, 1998).

USUAL MAIN MEALS

A typical mid-day meal when *dim sum* is not served might be similar to a regular breakfast meal. It might be simply a bowl of soup with noodles or rice, steamed dumplings, or buns, or it could be a dinner-type meal with a vegetable dish, rice or noodles, several meat or meat and vegetable dishes, and a soup or two, such as at typical dinner meals. At the end, a fruit or two are served cut up with the intention that they be shared. This main-type meal can be eaten mid-day or in the evening. At these main meals, meat, fish, and seafood dishes are served with or without vegetables, and vegetable dishes come with or without meats or creatures of the sea. Also one or two soups are served with rice or other plain grain dishes to accompany them. Actually, the most popular beverage at these main meals is soup (Newman & Ludman, 1984; Newman, Sirota, & Lei, 1996).

Every diner has his or her own rice bowl, chopsticks, and sometimes a soup bowl and soup spoon. Soup bowls are not always present when families eat informally because soup can be consumed directly from the rice or noodle bowl. At more formal times, diners will also have a flat plate and a small sauce dish as part of their individual place setting. All the dishes, bowls, and soup spoons are usually made of porcelain and the chopsticks of bamboo. Fancy chopsticks can be made of materials such as rosewood, ivory, or jade, as can fancy serving dishes. *Cai* dishes are served in these porcelain plates or bowls, and affluent people can have, in addition, one or more serving items made of jade, such as a jade rice bowl. Knives never appear at any of these meals or at any snack occasion; they are considered weapons of war (Chang, 1977).

When eating a main meal, chopsticks and not flatware are used. For example, rice is eaten by raising the bowl of rice to the mouth with the left hand and taking the rice with the chopsticks in the right and picking some up or more likely pushing the rice into the mouth. The *cai* food is taken from platters or bowls placed in the center of the table with serving chopsticks and moved to each individual's flat plate with a serving spoon. Each person's chopsticks take it from there to the mouth, stopping or holding it over the rice before so doing. At family meals, instead of using serving utensils, family members may use their own chopsticks and take food directly from the serving dishes from the rice bowl to the mouth. At formal meals or when guests are present, the use of service utensils is appropriate and use of one's own chopsticks to garner food from serving platters or bowls is inappropriate.

Food from the meat/vegetable *cai* dishes is never to be put directly on the rice or noodles because to dirty them is considered disrespectful to the rice or other grain and disrespectful to those whose toil produced them, or the person doing so is considered an ignorant peasant. All *cai* foods are cut bite-sized before cooking, and diners take them with serving utensils and put them on their own flat plate. They then use their own chopsticks to pick up a piece of food and make a slight stop over the rice bowl before eating it. Protocol requires one to pick up a food item that is large, hold it in the chopsticks, and bite off a small amount. One continues to hold the rest of that particular food over the individual rice or noodle bowl and does not put it down on the flat plate or anywhere, for that matter, once the food has gone to the mouth.

At family meals, main-meal rice or noodle bowls are often set and filled at the table before anyone comes to the table. The accompanying *cai* dishes arrive at once or almost at once, and they are usually placed in the center equally spaced from each diner. Some families and almost every restaurant use a round lazy-susan–type center arrangement so that the *cai* dishes are equidistant from all diners and easily accessible at all times. This avoids the need to pass them around and disrupt the flow of eating and conversation.

According to Chinese custom, guests and younger family members who want more grain food should hold up their rice bowl in both hands to indicate that they want additional rice or whatever other grain is served and that they give their grain undivided attention. At family meals, custom decrees that no one should eat any food on the table before the rice or other main grain food is served. At formal meals, no food should be eaten until the host begins or advises others to so do. Other meal behaviors include not eating noodles, rice, or other grain foods piece by piece; never picking the best foods from a dish for oneself unless urged to do so; never taking the last items in a *cai* dish; and never leaving a single grain of rice or a single noodle in one's bowl when the meal is over. To leave a grain of rice, children are told, assures a pock-marked face when an adult. Meals are considered finished and diners can leave when everyone's bowl has no more rice or other grain food in it (Fan, 1998).

BEVERAGE CONSUMPTION

All Chinese people drink tea, but not everyone does so at meals. At banquets, no tea may be served; wine or liquor can be served instead or along with the tea. At family meals, there are many reasons not to serve tea. Economics can be a factor, but more than likely not having tea at meals is because soup is the beverage of choice. However, some people believe that tea has a tendency to wash down the food, that filling up with tea is rude to the cook, and that tea is not good for the stomach when mixed with solid foods; therefore, they do not serve it. This clearly is not the case when tea is served before a formal dinner in very tiny cups of an ounce or so. At those times, guests are urged to drink one, two, or three cups of it. Tea is popular after dinner, and it is always served at *dim sum* meals. At both of these times, it is served in larger teacups that hold about four ounces of liquid.

Tea is always offered to guests when they enter a home or just after they are offered a potent liquor, and it is consumed at morning meals and between meals. When served at meals, tea can be consumed whenever the diner wishes, as long as the cup is filled by another. When invited to a meal, one does not drink wine or liquor even if they are poured until the host invites an individual or all diners to do so. After the first toast, an exception to this practice is to lift a glass and toast the host; after that, other guests can be toasted individually. Never, in any case, does one drink an alcoholic beverage alone at mealtimes.

BANQUET AND FORMAL FAMILY MEALS

On these occasions, dishes are served one at a time in courses, sometimes separated by a thick soup near the beginning of the meal and thinner ones nearer the meal's end. It is not uncommon at meals of more than 10 courses for soups or sweet items to punctuate the meal several times. Desserts, as westerners know them, may be part of a banquet meal, but more commonly meals end with a sweet soup and one or more fruits cut up and shared by all diners. A rice or a noodle dish is served at the end of a formal meal, usually after the fish course and before the sweet soup. It is impolite to eat a lot of it because that signals the host that not enough food was served. A formal meal is the only time leaving more than a grain in a dish is considered acceptable, although it should not be the grain in an individual diner's bowl, just that which is left in the serving dish.

Conversation is the entertainment at meals and eating the serious business at hand. Near the end of a banquet meal, the host may initiate a finger game, playing first with one guest and then another. The loser, not the winner, has to drink some wine or liquor, whichever is served. This game is an indicator of appreciation given to imbibing, conviviality, and hilarity. When the meal is over and the host rises, guests do likewise and do not linger and chat but rather leave the restaurant promptly so as not to delay the host, who must leave after all the guests do.

FESTIVAL FOODS

Everyday foods are second-class citizens at festive occasions. Pork is an everyday meat, so poultry and fish are more commonly served as festival foods. If pork is served, it is enhanced with nuts, special fruits, or other less common foods. Fish, mostly as a whole fish with the head pointed to the honored guest or most senior person at the table, is common and served at or near the end of the meal. The more expensive or the more unusual the food item and the more of them that are served, the more status the meal deserves (Newman, 1996).

Certain holidays are times for foods made especially for those occasions. For example, at Chinese New Year, which is also called Spring Festival, a dish of mixed nuts and sugar-coated seeds is served for good luck. A sweet year is wished for with traditional dumplings made with sugar and many seeds inside. Other sweets are popular at this holiday as is the serving of liquor. It is customary to spend New Year's Eve at home with family. The 3 days after are the time for visiting family and friends and being visited by them. Food gifts are brought when visiting, and Mandarin oranges are popular because the word for them in Chinese is a homonym for the word for gold. Thus, they are always bought as gifts when visiting and they are also given to guests when they come to visit. Popular for visitors at New Years is *nian gao,* a sticky rice cake. It is served to cement friendship as are other sticky sweet dumplings offered plain or served in a soup. When visiting or being visited, food is always shared, be it at New Years or other holidays or at special life-cycle events, such as to celebrate the birth of a baby, or at birthdays, marriages, or funerals (Wong S., 1998).

Very special occasions such as a sixtieth birthday or a marriage are times for shark's fin or bird's nest soup, a sea cucumber dish, and perhaps jellyfish, bear's paw, camel toe, or snake. Also served will be one or more chicken dishes, roast duck, pigeon, special mushrooms such as monkey head, expensive vegetables, and shrimp and crab. At birthdays, long noodles wishing long life are a must; at weddings a dish with lotus root as a hope for a long and stable marriage is popular as are dishes with many seeds such as pomegranates wishing the couple many children. At Mid-Autumn Festival, also known as Moon Festival, mooncakes are served. At Dragon Boat Festival, bundles of rice wrapped in bamboo leaves are commonly eaten. As mentioned previously, festival meals frequently end with a whole fish wishing prosperity for all. The cost of the meal, the rarity of the ingredients, and the number of people served mark the importance of the occasion; the more of each of these, the higher the status of the meal (Newman, 1996).

TABLE MANNERS

In the Chinese culture, table manners are important. Many are different from those in Western culture. They are passed by word and deed from generation to generation. Children learn the traditions, values, and beliefs, and they pass them on to their children. One such tradition is never to use chopsticks or a serving spoon to pick over the food seeking out the best pieces for oneself. When one finds a special delicacy in a dish, proper etiquette requires that it be offered to the eldest person at the table or the eldest one seated near you. Politeness is also to take every food item touched; not to excuse yourself when reaching for food; and if a woman, not to reach excessively but rather wait for a man to offer assistance in the form of fine foods out of arm's reach. These and items already discussed, such as that children learn early not to leave a single grain in their bowl and not to stuff themselves but rather leave the table 70% full, are considered important table etiquette; others that refer to chopsticks follow (Chang, 1977).

CHOPSTICKS AND OTHER TABLE IMPLEMENTS

Chopsticks are made from the wood of mature bamboo and from other woods. For the affluent, they can be of ivory, jade, cloisonné, or other hard materials, and at formal meals chopstick rests made of porcelain or metal are used to set them on. When not using them, chopsticks should be placed on these rests or on the side of the plate. Chopsticks are never licked or bitten, just used to get food from plate to mouth. One or more long pairs of chopsticks are found in kitchen areas to stir foods, pick items up, and provide related culinary needs. A fancy pair may be set before the host to allow selecting special delicacies from dishes to place on the plate of a guest.

In restaurants, it is common to remove the paper that chopsticks often come wrapped in and fold it into four across the width, then in half lengthwise to make a chopstick rest out of it. At formal meals there can be two pairs of fancy chopsticks for the host(s) to use. These and other serving chopsticks and those used for cooking are never put in the mouth. Chopsticks are never to be used for any purpose other than to prepare or consume foods; they are not decorations no matter how beautiful they are.

Chinese chopsticks are round at one end and square at the other. The round end goes into the mouth. Should no serving utensil be provided, the square end can be used for serving self or others from communal bowls or platters. At a meal, chopsticks are never left standing in a dish because of the connotation that they then imitate a stick of incense used to worship ancestors. Chopsticks work best when the ends are even with each other. Thus they are tapped against the table or plate to even them but never tapped against a rice bowl. Tradition says that doing so means one's offspring will be destined for poverty. In a restaurant when one has finished eating, the chopsticks are put across the rice bowl or crossed to advise the waiter. While at home, they are put neatly beside the flat plate or rice bowl to signal being satiated.

CHINESE MEAL IDENTIFIERS

Clearly, rice bowls, other Chinese dishes, chopsticks, particular foods, and other items indicate the meal is Chinese, as can the formal order of presentation at a meal. In addition, some seasonings identify Chinese meals as being Chinese. Among these are flavorings such as soy sauce, garlic, ginger (*Zingiber officiale*), star anise, five-spice powder, sesame seeds, ground white pepper, fagara, chili peppers, tangerine peel, and sweeteners such as brown slab sugar. In addition, other sauces say the food and meal are Chinese, including fermented black beans, lily buds, ginseng, fermented bean curd, various fungi such as shiitake mushrooms, and a myriad of sauces (*jiangs*) from sweet to piquant.

CONCLUSION

Chinese food and meals sustain life and are to be shared. In earlier times they were eaten kneeling or squatting; now they are consumed at tables with a chair for each diner. Each meal has grain foods as the bulk of the meal and meat and vegetable or either alone to accompany the grain; a soup or soups, water, tea and/or wine is the beverage at the meal. Having food and beverage together is *yin shi*, meaning one is well off. Eating is *yin* and drinking is *yang*, and all meals need both of these, as well as both *fan* and *cai* dishes and foods that are *yin* and *yang*. The number of dishes that accompany the *fan* varies on the basis of economics, occasion, and circumstance, but one or more are always present at every meal. Chang (1977) indicates that eating food that is not cooked and not eating grain foods are regarded as non-Chinese.

Eating a Chinese meal is a ritual, a social event, and an opportunity for pure enjoyment. Rules do not exist as to what must be served at a particular meal or when. Similar meals are served in the morning, mid-day, or in the evening. The main and continual concern is with the quality of the food. About this and food in general, constant talk of where to find the best food and incessantly comparing and contrasting how to best prepare it are important topics to be discussed as often as possible. The Chinese appreciate meals with harmonious flavors, meals seasoned in the kitchen and not at the table, and meals in which diversity is consciously and continually provided. Meals are social bonds where children learn from their elders, usually eat with their right hand, listen to adult conversation rather than generate it, and grow up to teach their children to do the same. Meals, certain foods, and cooking styles are conscious reaffirmation of Chinese identity, sometimes region of origin. The Chinese use food and meals for communication, expression of status, and as social markers. Meals are spoken about often, planned for frequently, and always enjoyed. About meals, Wen (1975) said it best, "Nothing is more important than eating."

REFERENCES

Anderson, E.N. (1988). *The food of China*. New Haven, CT: Yale University Press.

Chang, K.C. (Ed.). (1977). *Food in Chinese culture*. New Haven, CT: Yale University Press.

Fan, K. (1998). Chinese food rites and rituals. In J.M. Powers (Ed.), *From Cathay to Canada: Chinese food in transition* (pp. 39–42). Willowdale, Ontario, Canada: Ontario Historical Society.

Hom, K. (1989). *Fragrant harbor taste*, New York: Simon & Schuster.

Koo, L.C.L. (1982). *Nourishment of life: Health in a Chinese society*. Hong Kong: Commercial Press.

Lai, T.C. (1984). *At the Chinese table*. Oxford, England: Oxford University Press.

Lamb, C. (1935). *The Chinese festive board*. Shanghai, China: Henri Vetch.

Latourette, K.S. (1964). *The Chinese, their history and culture*. New York: Macmillan Publishing USA.

Newman, J.M. (1981). Bakery products of China. *Cereal Foods World, 26* (8), 395–398.

Newman, J.M. (1984). Chinese food habits: Regional differences. *Foodtalk, 6* (4), 5–8.

Newman, J.M. (1985). Chinese food: Heart, habit, and history. *Current research in history: Sources, topics, methods, and proceedings*. Boston: Schlesinger Library of Radcliffe College and the Culinary Historians of Boston.

Newman, J.M. (1996). Chinese food, unusual delicacies. *Chinese-American Food Society Quarterly, 18* (2), 15–17.

Newman, J.M. (1997). China: Transformations of its cuisine, a prelude to understanding its people. *Journal of the American College of Nutrition, 16* (2), 103–104.

Newman, J.M., & Linke, R. (1982). Chinese immigrant food habits: A study of the nature and direction of change. *Royal Society of Health Journal (London), 106* (2), 268–271.

Newman, J.M., & Ludman, E.K. (1984). Yin and yang in the health-related food practices of three Chinese populations. *Journal of Nutrition Education, 16* (1), 3–5.

Newman, J.M., Ludman, E.K., & Lynn, L. (1988), Chinese food and life-cycle events: A survey in several countries. *Chinese-American Forum, 4*, 16–18.

Newman, J.M., Sirota, L.H., & Lei, X.Y. (1996). Chinese food habit perspectives. *Journal of the Association for the Study of Food and Society 1*, 31–38.

Simoons, F.J. (1991). *Food in China: A cultural and historical inquiry*. Boca Raton, FL: CRC Press.

Tiger, L., & Wolf, R. (1985). *China's food*. New York: Friendly Press.

Wen, C.P. (1975). Food and nutrition in the People's Republic of China. In J. Quinn (Ed.), *Chinese medicine as we saw it*. (NIH Publication No. 75–684). Washington, DC: United States Department of Health, Education, and Welfare.

Wong, B. (1998). Dim sum: heart's delight. In J.M. Powers (Ed.), *From Cathay to Canada: Chinese food in transition* (pp. 15–22). Willowdale, Ontario, Canada: Ontario Historical Society.

Wong, S. (1998). Chinese food in contemporary society. In J.M. Powers (Ed.), *From Cathay to Canada: Chinese food in transition* (pp. 9–14). Willowdale, Ontario, Canada: Ontario Historical Society.

Zhu, H.P. (1998). The cuisine of Northern China. In J.M. Powers (Ed.), *From Cathay to Canada: Chinese food in transition* (pp. 43–48). Willowdale, Ontario, Canada: Ontario Historical Society.

RESOURCES FOR RECIPES

Chang, W.W., Chang, I.B., Kutscher, H.W., & Kutscher, A.H. (1970). *An encyclopedia of Chinese food and cooking*, New York: Crown Publishers.

Chao, B.Y. (1945). *How to cook and eat in Chinese*. Garden City, NY: John Day.

Chen, H. (1994). *Helen Chen's Chinese home cooking*. New York: William Morrow & Co., Inc.

Chen, P.K., Chen, T.C., & Tseng, R.L.Y. (1983). *Everything you want to know about Chinese cooking*. Woodbury, NY: Barrons Educational Series.

Hahn, E. (Ed.) (1968). *The cooking of China*. New York: Time-Life Books.

Hom, K. (1990). *The taste of China*. New York: Simon & Schuster.

Huang, S.H. (1973). *Chinese cuisine*. Taipei, Taiwan: Wei-Chuan Cultural Foundation.

Kuo, I. (1977). *The key to Chinese cooking*. New York: Alfred A. Knopf.

Leung, M. (1976). *The classic Chinese cookbook*. New York: Harper & Row.

Lo, I.Y.F. (1982). *The dim sum book*. New York: Crown Publishers.

Lo, K.H.C. (1979). *The encyclopedia of Chinese cooking*. New York: A & W Publishers.

Newman, J.M. (Ed.) (1994 to date). *Flavor and fortune*. Individual issues of this quarterly discuss various topics and have recipes. Information about the contents is available @P.O. Box 91, Kings Park, NY 11754.

Passmore, J., & Reid, D.P. (1982). *The complete Chinese cookbook*. Dee Why West, NSW, Australia: Lansdowne Press.

Sia, M. (1975). *Mary Sia's Chinese cookbook* (3rd. paperback ed.). Honolulu: University of Hawaii Press.

Tropp, B. (1982). *The modern art of Chinese cooking*. New York: William Morrow & Co., Inc.

Yan, M. (1981). *The Yan can cookbook*. Toronto: Doubleday Canada Ltd.

SUGGESTED READINGS

Bary, W., Chan, W., & Watson, B. (Compilers) (1960). *Sources of Chinese tradition*. New York: Columbia University Press.

Chambers, K. (1988). *The traveler's guide to Asian customs and manners*. New York: Meadowbrook Publishers.

Hsu, L., & Tung, T. (1977). Nutritional concepts and dietary practices in China. *Progress in Food and Nutrition Science, 2,* 483.

Kittler, P.G., & Sucher, K.P. (1998). *Food and culture in America* (2nd ed.). Belmont, CA: Wadsworth Publishing.

Lai, T.C. (1978). *Chinese food for thought*. Hong Kong: Wing Tai Cheung.

Latsch, M.L. (1988). *Traditional Chinese festivals*. Singapore: Graham Brasch.

Lin, H.J., & Lin, T.F. (1969). *Chinese gastronomy*. New York: Hastings House.

Mah, V. (1998). Chinese food traditions. In J.M. Powers (Ed.), *From Cathay to Canada: Chinese food in transition* (pp. 1–8). Willowdale, Ontario, Canada: Ontario Historical Society.

Newman, J.M. (1993). *Melting pot: An annotated guide to food and nutrition information for ethnic groups in America* (2nd ed.). New York: Garland.

Powers, J.M. (Ed.) (1998). *From Cathay to Canada: Chinese Cuisine in Transition*. Willowdale, Ontario, Canada: Ontario Historical Society.

Sinoda, D. (1977). The history of Chinese food and diet. *Progress in Food and Nutrition Science, 2,* 499.

Stepanchuk, C., & Wang, C. (1991). *Mooncakes and hungry ghosts*. San Francisco: China Books and Periodicals.

Tannahill, R. (1973). *Food in history*. New York: Stein and Day.

Townshend, J.R. (Compiler) (1981). *The People's Republic of China handbook* (2nd ed.). New York: China Council of the Asia Society and the Council on International and Public Affairs.

Appendix 10–A

Sample Chinese Menus

BREAKFAST

Southern
   Congee (a hot rice porridge)
   Salted peanuts
   Pickled vegetables (above two items can be mixed into the congee)
   Tea
     or
   Dim sum/yum cha meal of several small items with or without a rice or noodle dish

Northern
   Yao tiao (a fried wheat cruller)
   Soy milk
   Sugar
   Boiled peanuts
   Pickled vegetables
   (yao tiao is dipped in sugar and soy milk and consumed biting a piece off, then redipping before
     the next bite)
   Tea (optional)

NOON (OR LESSER) MEAL

Southern
   Dim sum/yum cha meal of several small items with a noodle or rice dish
     or
   Hundred year eggs with pickled ginger
   Mixed noodle or rice soup or mixed meat and vegetables on noodles or rice
   Soup or tea

Northern
   Tea before or to start the meal
   Dumplings with dipping sauce or steamed bread
   One or more mixed stir-fried dishes
   Noodle soup with or without meat or vegetables

DINNER (OR MAIN) MEAL
   Regional differences are reflected in dish selection and seasonings more than in other differences. At main meals, two or more main dishes not counting rice or noodles and soup are served with the number usually equaling the number of diners. Meals rarely end with dessert, although one or two fresh fruits might be shared at the meal's end. At family meals, most if not all dishes arrive at once or are on the table before the diners sit down. At banquets, the cold platter may be on the table before seating; hot hors d'oeuvres come together or one at a time; all other courses are served one at a time.

For Two
 Stir-fried pork with mustard greens
 Steamed spinach with garlic and preserved eggs
 Steamed rice (in south) or plain or spicy noodles (in north)
 Tomato eggflower soup
For Four
 Steamed vegetable dumplings and dipping sauce
 Egg, kohlrabi, and crabmeat saute
 Sweet and pungent chestnuts with celery cabbage
 Steamed sea bass
 Steamed rice (in south) or plain or spicy noodles (in north)
 Silk Squash and watercress soup
For Six
 Golden shrimp patties
 Chicken livers with quail eggs
 Bean curd with minced pork
 Eggplant and mushroom casserole
 Red-cooked pork knuckle
 Stir-fried mixed Chinese vegetables
 Plain rice (in south) or plain or spicy noodles (in north)
 Squid in hot and sour soup
 Steamed loquat pudding
For Eight
 Red-cooked snails
 Smoked or soy sauce chicken
 Pork and oysters in black bean sauce
 Long beans or mustard greens with dry shrimp
 Stir-fried spiced cucumbers and water chestnuts
 Minced beef in lettuce leaf
 Green onion omelette
 Stuffed eggplant
 Plain rice (in south) or spicy noodles (in north)
 Scallop and rice soup
 Eight precious pudding
Banquet Meal (for Ten)
 Five cold or hot appetizers (or both)
 (i.e., for cold: jellied beef, jellyfish, smoked fish chunks, preserved eggs, sugared walnuts; for hot: stuffed wings, velvet chicken, wrapped shrimp, paper-wrapper chicken, caul fat shrimp roll)
 Stuffed wintermelon soup
 Braised shark's fins with shredded chicken
 Peking duck with Chinese pancakes
 Steamed chicken with ham and bamboo shoots
 Abalone with Shanghai cabbage hearts
 Lion's head (meatballs on cabbage)
 Lotus root cake
 Braised soft-shelled crabs with fish maw in crab sauce
 Phoenix tail shrimp on marinated Chinese broccoli
 Steamed whole fish
 Sweet almond soup
 Litchi snowball
 Tea

Chapter 11

Japanese Meals

Shigeru Otsuka

INTRODUCTION

The most striking feature of Japanese cuisine, and the one that has had the greatest impact on its development, is the taboo on eating animal meat that lasted about 1100 years, from the late seventh to the early nineteenth century. In 675 the emperor of Japan suddenly issued a proclamation forbidding the use of traps, cages, and mechanically triggered spears and the consumption of the flesh of cattle, horses, dogs, monkeys, and chicken.

This emperor is believed to have based his edict against eating meat on a certain Buddhist scripture. Neither China nor any of the countries of the Korean Peninsula, through which Buddhism had been introduced to Japan, ever imposed such a complete prohibition. Needless to say, because the seventh-century Japanese had a strong liking for meat, a single imperial edict could not eradicate meat eating. Later emperors issued two more edicts along the same lines. About 200 years after the first edict, animal meat had finally disappeared from official banquets and gradually from the tables of the common people.

That was the period in which Japanese cuisine as we know it today was taking shape, but meat played no part in subsequent refinements.

One of the natural consequences of the prohibition on meat eating was that people sought to satisfy their palates with fish and other seafoods. In fact, fish came to be regarded as the tastiest of foods. The modern Japanese word for fish, *sakana*, originally meant "cuisine for *sake* (Japanese rice wine)"—in other words, delicacies to be served at banquets.

Especially highly prized was freshwater fish, probably because the people of the landlocked capital, Kyoto, who like all Japanese were fond of *namasu*, raw fish marinated in vinegar, found it much easier to obtain fish that is fresh enough to make *namasu* from freshwater. Nowadays, with modern cold chain facilities, almost all fresh seafoods are available and enjoyed either cooked or uncooked.

Animal meat is the same the year round, but the catches and taste of most fish and shellfish change with the seasons. The prohibition on eating meat led the Japanese to depend on seafood and vegetables as their major nonstaple food items. Both change from season to season, and this made the Japanese sensitive to the seasonal nature of food and led them to take pains to display food in ways that reflected the seasons. This inclination remains strong in the present culinary culture of Japan. Foods in season or early season are cherished as the best treat. Also, meals constitute a veritable pageant of the seasons: vegetables shaped like flowers of the season, thin noodles arranged to resemble a waterfall in summer, dishes garnished with chrysan-themums and colored leaves in autumn.

A traditional ceremony connected with food that is unique to Japan is the so-called carving-knife ceremony, or *hocho-shiki*. In this ceremony the chef, wearing traditional costume, chops, slices, and arranges a fish (often a carp) using only a long carving knife and large chopsticks, never touching the fish directly, as an offer-ing to either gods or nobles. At present, the ceremony is performed in solemn si-lence before a shrine. This liking for ceremony, together with the other facets of Japanese cuisine discussed previously, indicates the extent to which elements hav-ing nothing to do with taste and nutrition have been added to Japanese cuisine and invested with great value in the effort to make food prepared from a limited range of ingredients as pleasurable and beautiful as possible.

However, this does not mean that Japanese cuisine is not tasty. Meat, oil, dairy products, and spices are delicious in themselves and also make other ingredients taste better. The Japanese, lacking these foodstuffs, created two all-purpose season-ings, Japanese *shoyu* (soy sauce) and *miso* (fermented bean paste). The prototypes of both seasonings came from China or the Korean Peninsula in ancient times but were refined into their present forms in Japan. Soy sauce, in particular, is used in almost all Japanese dishes and is the basis of the Japanese flavor.

Another flavor enhancer widely used in Japanese cooking, *dashi*, was devised in ancient times to add body to the taste of delicately flavored ingredients. It is no coincidence that the two main types of *dashi* are made from dried bonito (*katsobushi*) and dried kelp (*konbu*), both from marine products. European equiva-lents of *dashi* have a base of meat broth or dairy products, substances derived from land animals.

Early in the twentieth century a Japanese chemist identified monosodium glutamate as the active ingredient in the *umami*, the flavor-enhancing taste, of dried kelp. Monosodium glutamate, together with sodium salts of inosinic and guanylic acids, known as the *umami* substances in dried bonito and dried shiitake mush-rooms, respectively, is used widely in cooking and food processing in Japan.

Some of the typical, traditional, but still common, foods and the methods for preparing them are discussed here.

RICE AND SUSHI

The rice eaten in Japan is the short-grained variety that, unlike the long-grained variety, has a moderately soft and sticky texture when cooked in the Japanese man-ner: the rice is soaked in water for 15 to 20 minutes, heated with a suitable amount of water (usually about 1.5 times the amount of the grain) moderately at the begin-ning then vigorously until boiling stops. Then the heat is reduced and the rice is kept warm for about 15 minutes without removing the lid.

Sushi

References to sushi appear in early eighth-century laws and regulations, al-though then it was completely different from what it is today. Initially, sushi re-ferred to pickled fish that was made by fermenting salted fish in cooked rice with a weight placed on top. After about 6 months, the rice had dissolved beyond recogni-tion and only the fish was eaten. Today this process—spreading alternate layers of rice and fish in a large wooden tub and weighting them down with a heavy stone—brings to mind pickles rather than modern sushi.

It was not, however, an indigenous invention. The Japanese learned it from the Chinese, who had acquired it from South Asia, where it was already used in pre-Christian times.

In other Asian countries, including some parts of Japan, even today fish is still pickled by this long, drawn-out method. In Japan sushi developed in a different direction. The Japanese long ago began trying to speed up the production of sushi, promoting fermentation by adding *koji*, a yeastlike mold that is also used as a starter in making sake and shoyu. With this method, the rice retained its shape and texture with a sweet and sour taste.

When a method was established for producing fine rice vinegar, sushi makers began seasoning the rice ahead of time with vinegar and *mirin*, a kind of sweet cooking sake. Many kinds of modern sushi were invented by applying this method.

Nigirizushi, or squeezed sushi, the kind most familiar today, is made by laying *sashimi* (a slice of raw fish) on the palm of one's hand, putting on a desirable amount of grated *wasabi* (Japanese horseradish), and placing a small amount of seasoned rice on it. After squeezing and shaping, the *nigirizushi* is served upside down (the *sashimi* on top) in pairs; it is dipped in soy sauce before being eaten. Many seafoods are used and among those most favored are *maguro* (tuna), *hirame* (flounder), *tai* (sea bream), *kohada* (medium-sized gizzard shad), *ika* (squid), *tako* (octopus) legs, *ebi* (prawns), *akagai* (ark shell), *aoyagi* (trough shell), *hotategai* (scallops), *ikura* (salted salmon roe), *uni* (sea urchin roe), and *tamagoyaki* (omelette). *Gari*, thin-sliced vinegared ginger, is used on the side.

Makizushi, or rolled sushi, is made by spreading the seasoned rice on a sheet of *nori* (dried laver sheet), placing the ingredients on top of the rice, and then rolling the whole thing. The roll is cut into thick slices and served. Any vegetable, fruit, and fish can be used as ingredients. The traditional ingredients are seasoned *koyadofu*, freeze-dried tofu, seasoned *kampyo* (dried gourd shavings), *tamagoyaki*, and some green leaves.

FISH: *SASHIMI* AND *TEMPURA*

Sashimi (Tsukuri)

The cuisine that makes raw fish the epitome of epicurean delight is rather unusual. In Japan, the custom of eating fish raw seems to derive more from a genuine love of fresh fish than from the practices of prehistoric times when humans did not yet know how to use fire. For the upper classes living in the inland capital, dining on raw fish was a way of enjoying and displaying the economic and political power that enabled them to obtain fresh fish. Were this not the case, it seems inconceivable that raw fish would continue to be a central feature of Japanese cuisine today despite the development of so many cooking methods.

Ancient records tell of a mythological emperor who dined on *namasu* made with clams obtained near the mouth of Tokyo Bay. The dish pleased him so much that he appointed the man who prepared it to be head chef at court. This story tells us that raw fish has been a great delicacy since ancient times. *Namasu* is made by marinating fish or shellfish in vinegar, which acts as a preservative, removes the fishy odor, and imparts a delightfully refreshing flavor. Mackerel and other fish are still served this way.

After methods for producing good shoyu (soy sauce) were established in the early Edo period (1603–1868), people stopped marinating *sashimi* in vinegar and began eating raw fish slices dipped in shoyu and using *wasabi* as a spice. *Tsukuri*, another term for sashimi, reputedly came into use because of the warriors' distaste for the word "*sashimi*," which literally means "to pierce the body," suggesting defeat in battle.

Almost any fish, shellfish, prawn, or squid is used to make *sashimi* as long as it is fresh enough to consume raw. Materials in season are most delightful. Besides this, the beautiful shape of slices and arrangement of the slices are required to make it look fresh and delicious. To emphasize this, some *sashimi* is served in the complete shape of the fish, head to tail.

Tempura

Tempura is a general name for a deep-fried batter-covered dish. The Japanese had their first contact with Westerners in the mid-sixteenth century, when Portuguese trading ships began venturing to the coasts of Japan. Along with other elements of Western civilization, the Portuguese brought with them such dietary customs as eating meat and using oils. One of the European cooking techniques that was particularly welcomed was coating fish with a thin batter and deep-frying it. Called *tempura*, the Japanese approximation of its Portuguese name, this dish has become a permanent feature of Japanese cuisine.

To make *tempura* batter, an egg is mixed in a bowl, and 1 cup of chilled water is added to it and mixed; then 1 cup of light flour is added and mixed lightly.

Foods are dipped in the batter and gently placed into heated oil and fried until the batter surface becomes brown. *Tempura* is served while hot. *Tempura* is good with salt, but is usually eaten after being dipped in *tentsuyu*, soy sauce–based sauce, with *daikon-oroshi*, grated radish. Originally only seafoods were used to make *tempura*, but today vegetables, sea plants, and even fruits like persimmons are enjoyed as *tempura*. Commonly used foods for *tempura* are *ebi*, *kisu* (a small white fish), *ika*, *shishito* (a long green pepper), pumpkin, sweet potato, eggplant, *renkon* (lotus root), onion, *shiitake* mushrooms, and *nori*.

TOFU AND OTHER SOYBEAN PRODUCTS

Soybeans are used to make a wide variety of foods, including tofu, *aburaage* (deep-fried tofu), *yuba* (soybean milk skin), *natto* (fermented soybeans), *ganmodoki* (deep-fried mixed tofu), kinako (roasted soybean flour), and the traditional seasonings, shoyu and *miso*.

The products made from soybeans all have flavors considered to be among the most typically Japanese. *Miso* soup with tofu, *hiyayakko* (chilled tofu served with shoyu), *dengaku*-tofu (grilled tofu spread with *miso*), and other popular dishes consist entirely of soybean products. Japanese cuisine would not exist without shoyu and *miso*, its key seasonings.

Prominent differences exist between the West and other regions, where butter, cheese, and many other milk products are essentials of the daily diet—what might be called the "milk culture sphere"—and Asian countries that have diets that depend on a wide range of soybean products—the "soybean culture sphere."

Tofu originated in Asia. Chinese legend attributes its invention to the grandson of the Former Han dynasty (202 BC TO AD 8). In fact, it was probably first made in the Táng dynasty (618–907) and became widespread during the Sung dynasty (960–1279). Japanese monks who visited China around the twelfth century probably brought back knowledge of how to make tofu, which seems to have become popular during the Muromachi period (1338–1573).

Tofu is made from soy milk obtained by boiling and crushing soybeans, then, after adding water, filtering the pulpy liquid. To the soy milk, a coagulant called *nigari* (a by-product obtained from desalinated seawater) or modern coagulating chemical is added to solidify the protein, and the mixture is poured into molds to settle or filtered through cotton cloth.

Although it consists of curd, tofu keeps its shape well enough to be handled with fingers or chopsticks. Many Japanese recipes using tofu have been invented over the centuries. Tofu can be used in soups or cooked in just about every way imaginable: boiled, grilled, fried, deep-fried, or steamed. It is interesting to see that in Western countries tofu is used as a salad, an as ingredient of a drink, and to make a "tofu dessert."

The earliest dish using tofu was *dengaku*-tofu. This is made by cutting tofu into thin oblong pieces and grilling them on a double-pronged skewer. The appearance suggested a medieval dance called *dengaku mai*, from which the name was derived. *Dengaku*-tofu was the forerunner of *oden*, a popular hot pot meal, which is mentioned later.

Japanese dishes can be characterized by the extensive use of tofu. Tofu is eaten as *hiyayakko*, *yudofu* (boiled tofu with soy sauce), and *miso* soup and other soups; it is also enjoyed in *sukiyaki* and many other hot pots and *nimono* (boiled dishes seasoned with shoyu and sugar).

Aburaage, *koyadofu*, and *gammodoki* are also used in nimono, soup, and many other dishes.

WHEAT AND NOODLES

Because the hot, humid summers and cold, dry winters of Japan are not conducive to growing wheat, this grain is less common in Japan than in the West. Although wheat is, in fact, grown in many areas, the flour obtained from Japanese wheat does not contain enough gluten to be suitable for making bread. This may account for the great heights to which noodles have been developed in Japan.

Noodles come in three basic varieties: those made by stretching the dough, such as Chinese noodles and Japanese *somen*; those made by rolling the dough flat and cutting it, such as Japanese *udon*; and those, like spaghetti, made by forcing the dough through a hole. Although the end products look similar, the different methods of making them say something about each culture. The Japanese excel at cutting and consider it an important part of cooking.

Somen is very fine noodles, as fine as vermicelli, which is believed to have been transmitted to Japan from China in the early fourteenth century. Unlike Chinese noodles, however, *somen* is made by using pulleys to stretch the dough a little at a time until it is about the thickness of a pencil. It is then caught over horizontal poles at the top and bottom, stretched out in one swoop, and dried. The fine noodles hung to dry give the appearance of a bamboo curtain, creating a distinctive ambience in areas where *somen* is produced. *Somen* is usually served cold, usually in a bowl of cold water with some condiments like grated ginger and *nori* flakes. Always a welcome dish in summer when appetites get weak and something refreshing hits the spot, *somen* is still one of the best lunches and the most popular of summer gifts. The thin *somen*, floating in ice water in a glass bowl, is really nice to look at and to eat in the summer. A slightly thicker version of *somen* called *hiyamugi* is also a favorite of the Japanese.

Udon, the most popular noodle of Japan, appeared at the end of the fourteenth century. These thickish noodles are made by rolling dough flat and cutting it and are characteristic of their square cross sections. Although *udon* is sometimes served cold like *somen*, it is more commonly eaten in a hot soup with various toppings that give each version its name, such as *tempura udon* with prawns and vegetable *tempura* on top, *kitsune udon* with heavily seasoned *aburaage*, or *tsukimi udon* with a raw egg broken on the top while the broth is hot (in "*tsukimi*," or moon-viewing, the yolk is likened to a full moon).

Every town has restaurants serving *udon*, which is reasonably priced and popular. The word *udon* is said to derive from *konton*, meaning chaos, the name of a food that came from China around the ninth century.

Soba is not made from wheat flour but from flour obtained by grinding the seed of the buckwheat. Because the buckwheat flour does not contain enough protein and lacks stickiness to form noodles, usually a small amount (about 20%) of wheat flour is added to make *soba* in the same way as *udon*, only a little thinner. Different from *udon*, the most popular manner of serving soba is a cold dish called *zarusoba*. The *soba* is served in a shallow box with a bamboo rack on the bottom (or the false bottom). The noodles are picked up with the chopsticks and partially dipped in *sobatsuyu* (shoyu-based broth) before eating. Chopped spring onions and grated *wasabi* are mixed into the broth. *Soba* is also served hot in the soup like *udon*.

It is always difficult to eat noodles in the hot soup elegantly, even with chopsticks. In Japan, noodles are picked up with the chopsticks and sucked into the mouth, which, of course, is quite noisy. In the West, this is considered rude, and foreigners in Japan often frown on the many Japanese who slurp noodles. But this way of eating noodles is part of the Japanese culture, and, by imitating this, the noodles might taste different and better.

JAPANESE SEASONINGS AND COOKING

Two fermented condiments, shoyu (soy sauce) and *miso*, are the keys to Japanese cooking. Three decades or so ago, when Japanese cuisine was still relatively unknown in the rest of the world, some Japanese scholars used to argue that Japanese food had completely lost out in the preceding century of international competition and that the fault lay with shoyu. The theory ran that the Japanese shoyu tasted too good, so Japanese cooks became lazy about bringing out the individual aromas and flavors of the basic ingredients or inventing new and specific methods that are most suitable to individual materials, being satisfied simply to season everything with shoyu. This paradox amply testifies to the excellence of shoyu, as well as to the position it holds in Japanese cooking. The argument escalated to the point where people were claiming that Japanese cuisine would dominate the global food market if only the flavor of shoyu could gain acceptance. Now that the situation has, of course, changed dramatically and shoyu and Japanese cuisines seem to delight people outside Japan, we might say that the prediction is coming true.

To be sure, shoyu is an indispensable element of Japanese cooking, used in boiling, grilling, and frying meat, fish, chicken, vegetables, tofu—everything. Moreover, it is distinctive among the world's many sauces in that it is used for seasoning both in cooking and at the table. Shoyu defines "the flavor of Japan," is the common denominator of the taste of home cooking, and is "the all-purpose seasoning."

Even when European food and methods of cooking were introduced, shoyu played an important role in their widespread use among the Japanese people. Especially revolutionary was the coming of the Portuguese and Spanish in the fourteenth century and the contacts made with American and Western eating habits and foods in the last half of the nineteenth century. With the eagerness of the true eclectic, the Japanese welcomed new and exotic ideas along with the new food from across the sea.

However, one special aspect of this enrichment of the Japanese cuisine is especially noteworthy. This is the fact that no matter how eagerly the Japanese took in Western foods, they hesitated to accept Western habits of eating. Instead they incorporated the foods into their own customs and created Japanese-style Western food. This was done chiefly through the use of shoyu.

One example is *sukiyaki*. The first time the common people in Japan had seen beef in more than 1000 years was about 100 years ago. And the first time they picked up beef and looked at it, the first time they worked up enough courage to actually eat it, it was not as beefsteak or as roast beef. Instead they sliced it very thin and boiled it with vegetables and tofu. That was the first *sukiyaki*. Thin slicing was the same method used to prepare raw fish for more than 1000 years. Cooking it at the table seasoned with shoyu was none other than the method that had long been used to prepare fish and poultry dishes.

Going back a little earlier, *tempura* was introduced into Japan in the seventeenth century. The deep-frying method used for *tempura* was quite a new thing to the Japanese of the time. As mentioned previously, the word "tempura" is said to be a corruption of either a Spanish or a Portuguese word. *Tempura* became truly Japanese only after the perfection of a shoyu-based sauce, "*tentsuyu*," to go with it. Of course, *tempura* was developed using traditional Japanese materials: fish, shrimps, prawns, shellfish, seaweed, and vegetables.

When the Japanese come on some new food, their first impulse was to sprinkle it with shoyu or cook it in shoyu. Onions, Chinese cabbage, spinach, Irish potatoes, cabbage, and asparagus all either came to Japan or have come to be widely eaten within the last century, and all of them have come to be appreciated when seasoned and cooked with shoyu and *miso*.

Basically, Japanese cooking is not different from any other kind of cooking in that it consists mainly of raw dishes, broiled dishes, boiled dishes, and deep-fried dishes. However, the distinctive "tastiness," aroma, and color of shoyu are used to give a salty taste in some cases, a special delicious aroma in other situations, and sometimes a few drops of shoyu are used in what the Japanese call *hidden flavor*.

The use of shoyu in Japanese cooking can be outlined as follows:

1. *Sashimi*: Slices of raw fish are dipped in shoyu (or *tamari*, a previous shoyu) and grated *wasabi* before eating. The shoyu is served in a small dish beside the *sashimi* dish.
2. *Sushi* (*Nigirizushi*): *Sushi* is held upside down by the fingers or the chopsticks,

and only the fish is dipped lightly in shoyu or shoyu is spread using a brush. Rice is not dipped because it is already delicately seasoned.

3. Broiled fish: Fish is grilled with an "indirect strong blaze" to obtain a crisp outside and a tender and moist inside. Before eating, shoyu is often sprinkled to add deliciousness to grilled fish dishes. *Teriyaki*, or grilled fish with luster, is a fancy way of grilling a fish fillet. The fillet, usually once grilled, is basted with *tare*, or basting shoyu-based sauce, then grilled. *Mirin*, a sweet sake, contained in the sauce gives an attractive luster and flavor to the grilled surface.

4. Boiled fish: Fish is boiled in a pan with the flat bottom for 15 to 20 minutes over a low flame in a mixture of shoyu, sugar, and water. Care should be taken to remove the fish from the mixture before it begins to fall to pieces. By cooking with shoyu, the smell of fish is removed or masked.

5. Vegetables and meat: Selected vegetables, potatoes, pumpkins, tofu, *aburaage*, beef, pork, chicken, and many other ingredients are boiled in water containing sugar and a bit of salt until tender. A touch of shoyu is added after boiling to add an appetizing flavor.

6. Soup for noodles: Boiled noodles are sometimes dipped in a shoyu-based soup and eaten cold. At other times the noodles are heated and added to the thinner shoyu-based soup with *tempura*, *aburaage*, egg, or other ingredients on top.

7. Shoyu-based sauces: The sauces are made for dipping and grilling different foods. These sauces are applied to the food before, during, and after grilling. More recently, ready-made sauces have been put on the market. Usually the food is marinated in this sauce for some time before it is grilled. Depending on the food, the sauce may also be brushed on the food lightly after it has been grilled to some degree. This sauce can also be used for the boiled fish dishes mentioned previously; it is also good as barbecue sauce.

8. Mixed sauces: When shoyu and vinegar are mixed, their respective tastes complement each other and blend into a delicious sauce (*nihaizu*). This sauce can be used to season salad and boiled vegetables. Sugar is often added to this sauce (*sanbaizu*). A dip made by mixing juices of citrus fruits such as *yuzu* and *sudachi* with shoyu is used for dipping hot pot dishes and boiled foods. This dip is also used to marinate steamed and deep-fried foods. A dip made from grated ginger and shoyu is used with many kinds of foods.

9. Soup: Japanese soup is different from European and American soups in that it is not made from the broth or stock. Typical Japanese soup uses shoyu or *miso*. Vegetables, fish, and meat are added to give body to these soups. The addition of *dashi* (a stock obtained by boiling dried bonito, dried mushrooms, or dried *konbu* [a sea plant]) is indispensable. Chemical seasonings such as monosodium glutamate are also used to add body to the soup.

DASHI (SOUP STOCK)

Dashi is an all-purpose soup stock that is usually made from *katsuobushi* (dried bonito) and *konbu*. Probably no flavor in the world has been the subject of such intricately entangled cultural and scientific debate. In Japan, *dashi* is used to add *umami* to almost all boiled dishes and soups just as bouillon is used in Western cooking to mellow saltiness and give body to the dish. Lacking meat, the Japanese derived *umami* from dried fish, sea plants, and vegetables.

The flavor of *dashi* has been important to the Japanese for well over a millennium, and learning the techniques for skillfully bringing out the *umami* of *dashi* has been an indispensable part of a cook's training. The most common technique for extracting *umami* from *dashi* materials, but without extracting undesirable flavor, is to boil *dashi* materials such as *katsuobushi* flakes and *konbu* sheets for only a few minutes, then immediately strain them to obtain soup stock.

NABEMONO (HOT POT MEALS)

Nabemono, that is "the thing boiled in one pot on the table," play an important role in making winter dining pleasurable in Japan, whether at home with the family or at a dinner party in a restaurant. *Nabemono* are also the quintessence of the Japanese art of imparting a sense of the seasons through the way food is served.

We have a great variety of *nabemono* in Japan: there are *chiri-nabe*, which combines white fish such as cod and sea bream with tofu, Chinese cabbage, edible chrysanthemum leaves, mushrooms, and other vegetables in a *konbu* stock and is eaten with a dip of mixed seasoning as mentioned previously; and *mizudaki*, which is similar to *chiri-nabe* but uses chicken instead of fish.

Udon-suki consists of fish and vegetables cooked with *udon* heated in the broth to top off the meal; *yose-nabe* is made up of chicken, fish, and vegetables simmered in a seasoned stock.

Oden is a distinctive combination of root vegetables, tofu, *konnyaku* (jelly of konnyaku, or "devil's tongue"), *konbu*, various fish meal products such as *chikuwa* and *kamaboko*, octopus leg, and so forth cooked in a seasoned broth.

Shabushabu is the Japanese version of the Mongolian hot pot. A very thin slice of beef is put in boiling water for a second and dipped in a *tare* (thick *miso*-based sauce with sesame) before eating.

Sukiyaki is a beef-and-vegetable dish seasoned with a shoyu-based sauce or simply with shoyu and sugar. Sukiyaki was the commoner's response to the problem of how to cook beef when people were first encouraged to eat it in the late nineteenth century.

These one-pot meals have countless variations that are cooked at the table with everyone helping themselves to whatever they want. This simple style of eating, which some people might consider rude, is among the greatest of treats in Japan.

MODERN JAPANESE EATING PATTERNS

Since World War II, the Japanese diet has undergone a rapid transformation. Faced with a desperate food shortage during and immediately after the war, the Japanese were in no position to worry about dietary traditions or cultural conditions. When they had the time and money to do so, they began rebuilding their dietary culture. Many of the traditional patterns were revived, but in some respects they began moving in an entirely different direction. People who were dissatisfied with Japan's traditional diet began to turn to foreign cuisine in a burst of "internationalization," the word still often used with some joy, sigh, or worry.

A number of factors have contributed to this phenomenon: (1) The period in which we were extricating ourselves from the poverty and deprivation that followed the war coincided with the American Occupation, so we were influenced by the American or Western diet. (2) Rice was in short supply at the time, and relief from abroad came in the form of wheat, sorghum, and other grains that required grinding into flour before they could be prepared for eating. We had to get used to food made from the flour of these grains, which was a historical change because it gave us a wider range of staple foods besides rice. Bread was rationed as a substitute, which helped to put it onto more tables at mealtime. Concomitant with this was the spread of Western foods, such as butter, cheese, ham, and sausage, and Western seasonings such as ketchup, mayonnaise, Worcestershire sauce, and salad dressing, as well as Western-style recipes that do not call for shoyu and *miso*, the traditional all-purpose seasonings for Japanese cooking.

In addition, when fast food and family restaurant chains spread across the country, local products and ways of preparing traditional food fell into decline. The revolutionary changes in the distribution system enabled consumers to obtain a wide variety of processed foods, both Japanese and Western, inexpensively, greatly simplifying home cooking and incorporating dishes from other lands. These developments were totally different from traditional methods of food production, distribution, and consumption.

An example is the school lunch program, which is shown in Table 11–1. After the war, Japan had a serious food shortage. In almost every urban elementary school some children brought no lunch. There was a serious concern about malnutrition in the children. Some teachers got the Ministry of Education involved with

Table 11–1 A Week Menu of School Lunches

Menu	Ingredients	Energy (kcal.)/Protein (g)
Feb. 1. (Monday)		649/32.0
Bread, milk	Chocolate-flavored butter	
Hamburger steak	Beef, onion, eggs	
Spinach aemono	Spinach, tuna (marinated), sweet corn	
Soup	Grilled pork, wakame, bean sprouts, carrots, Welsh onion	
Feb. 2. (Tuesday)		707/24.3
Rice, milk		
Deep-fry, breaded	Squid	
Aemono	Chinese cabbage, bean sprouts, sesame	
Pork soup	Pork, taro, carrots, aburaage, burdock root, Welsh onion	
Feb. 3. (Wednesday)		679/32.3
Bread, milk	Roasted soybeans	
Deep-fried sardines	Boned sardines, eggs, flour	
Sauté	Qing geng cài, Chinese cabbage	
Soup	Tofu, leek, eggs, carrots, shiitake	
Feb. 4. (Thursday)		526/21.4
Bread, milk		
Udon	Beef, kamaboko, carrot, shiitake, Welsh onion	
Aemono	Kidney beans, sesame	
Feb. 5. (Friday)		580/24.2
Rice, milk, banana		
Sushi	Sausage, omelette, pickled daikon, nori (seaweed)	
Thick soup	Potato, carrot, shiitake, konnyaku jelly, starch	
Mangdarin oranges		

Note: Tatsuno City Society for the Research of School Feeding, Hyogo, 1999. For Oyake, Ibo, Honda, Kamioka, Tatsuno, Issai-higasi, and Issai-nishi Elementary Schools
Courtesy of Ms. Eiko Yamamura, City Elementary School, Tatsuno City, Hyogo-ken, Japan.

the problem. Eventually, a free lunch program was begun in the elementary schools, using wheat from the United States provided under postwar aid programs, UNICEF, LARA, and so forth. In this postwar period, bread was served instead of rice, and the menus consisted of Western-style dishes, such as omelettes, hamburgers, sandwiches, and spaghetti. The children who grew up on these lunches developed a preference for Western foods. Today they are adults, and one factor in the rapid westernization and internationalization of the Japanese diet seems to be the unmistakable effect school lunches have had on eating patterns.

A while ago the "Japanese-style diet" was recommended by scientists. Its intent was not simply to promote the traditional Japanese diet but to encourage healthier eating by following the traditional diet, incorporating into it elements of Western and Chinese food.

On the other hand, Japan has the world's longest average life expectancy. Naturally, the greatest factor contributing to this longevity is the great number of elderly people in the population (around 8% are more than 70 years old) relative to the small number of children. People older than 70 grew up on the prewar diet of plain fare, with lots of grains but few meat and milk products, fats and greasy foods, and convenience foods. They ate the traditional diet, which is very different from what we eat today. Some observers take the view that the simple diet these people ate in childhood has enabled them to reach such an advanced age and that there is no guarantee that today's youngsters will live so long.

Today the Japanese are considered to have one of the healthiest diets in the world. This unquestionably relates to our longevity, but it does not necessarily mean that we are wise eaters. It means nothing more than that our preferences at the moment happen to match what the experts say are correct eating habits. For the Japanese, delicacies once consisted of only seafood, rice, and vegetables; meat eating occurred in Japan only about 130 years ago, and it is still climbing. The present level, which we prefer, happens to coincide with what is ideal (Appendix 11–A).

CONCLUSION

The Japanese foods and dishes mentioned previously are traditional ones that have been appreciated for more than 1000 years and reflect the times when people's foods were eaten completely according to place and season. It is one of the joys of living in this land to observe the fruits and vegetables and fish in the local shops as they parade their many colors in unending seasonal change. Even the taste of the same variety of vegetable or fish differs according to the season. Most fish are fattest and taste best before the season for spawning. This season differs according to species, so it is possible to enjoy fish at their very best all year long. The same holds true for fruit and vegetables. Even though this land may be lacking in foods that are available throughout the year, such as meat, poultry, and milk, still the tables of the people living in Japan are decorated with something new and different all through the year.

This is the kind of environment that brought forth Japanese cuisine. In Japan, the people highly value the natural coloring, shape, and aroma of foods, and they add only a bare minimum of flavoring to their foods, preferring to maintain as much as possible the natural goodness, texture, and feeling of the materials themselves. The skill of a cook is, in fact, judged by how well he or she succeeds in bringing out the natural touch and flavor of the materials.

Appendix 11–A

Menu of a Week at a Japanese Home

(September 1998)

Menu	Ingredients	Remarks
Sunday		
Breakfast		
Fried eggs with bacon	Eggs, bacon	
Salad	Boiled spinach, lettuce, tomato	
Toast	Bread	
Milk and coffee		
Lunch		
Fried noodles	Noodles (ramen), ham, onion	
	Pimento	Chopped
Soup	Scallops, qing geng cài	
Fruit		
Supper		
Aemono*	Wakame, small shrimps	*Dressed food
Tempura	Shrimps, pumpkin, onion, squid	
Pickles	Cucumber, daikon	
Miso soup	Potato, aburaage, onion	
Boiled Rice		
Monday		
Breakfast		
Fried ham and eggs	Eggs, ham, pimento	
Fruit salad	Celery, apple, kiwi fruit, tomato	
Toast		
Milk and coffee		
Lunch		
Ramen	Chinese noodles, shrimp, shiitake, vegetables	
Itamemono*	Beef, burdock root	*Fried dish
Broccoli	Broccoli, sesame sauce	
Boiled rice		
Supper		
Teriyaki	Yellow tail	A fillet
Simmered vegetables	Pork, shiitake,* greens	*A mushroom
Miso soup	Taro, aburaage	
Boiled rice		
Tuesday		
Breakfast		
Boiled eggs	Eggs	
Simmered legumes	Young kidney beans, carrots	
Zosui*	Rice, scallops, vegetables	*Porridge of rice
Fruit		
Lunch		
Curry and rice	Chicken, potato, vegetables	
Marinated eggs	Eggs, vegetables	
Yogurt		
Supper		
Sashimi	Sea bream, tuna, young yellow tail	
Simmered aburaage*	Aburaage, taro	*Deep-fried tofu
Sweet-and-sour vegetables	Carrots, turnips	
Boiled rice		
Wednesday		
Breakfast		
Natto	Natto, green onion, nori	
Fried simmered hijiki*	Hijiki,* pork	*A seaweed
Miso soup	Green leafy vegetables	
Boiled rice		
Yogurt	Yogurt with fruit	

189

Menu	Ingredients	Remarks
Lunch in a box		
Omelette	Crab meat, eggs	*Fried and simmered
Itameni*	Potato, minced meat	
Boiled eggplant	Eggplant dressed with mustard sauce	
Boiled rice		
Supper		
Steak	Beef, green peas, carrots	
Salad	Lotus roots, green vegetables	
Soup	Greens, shimeji mushrooms	
Boiled rice		
Thursday		
Breakfast		
Fried eggs	Eggs	
Fried corned beef	Corn beef, cabbage	
Toast		
Milk and tea		
Fruit		
Lunch		
Ramen	Chinese noodles, chopped beef, vegetables	
Simmered vegetables	Carrots, onions, peas	
Supper		
Happosai	Tofu, pork, Welsh onion	*Salsola komarovii
Simmered vegetables	Okahijiki,* dried daikon	Chinese flavor
Soup	Celery	
Boiled rice		
Friday		
Breakfast		
Simmered tofu	Grilled tofu, ginger	
Miso soup	Potatoes, young peas	
Boiled greens	Dressed with pickled plum meat	
Boiled rice		
Fruit		
Lunch in a box		
Hamburger	Minced beef, bun	
Boiled eggs	Eggs, broccoli, mayonnaise	
Salad	Tomato, cucumber, celery	
Milk		
Supper		
Marinade	Steamed salmon with wine, onion	
Salad	Pumpkin, young kidney beans, lettuce	
Sauté	Mushrooms, bacon, butter	
Spaghetti peperoncino		
Saturday		
Breakfast		
Salad	Pumpkin, onion, celery, sausage	Toast
Soup, bread	Sausage, vegetables	
Milk and coffee		
Fruit		
Lunch		
Donburi	Rice, baby clams, trefoils	
Broccoli	Dressed with mustard	
Supper		
Deep-fried chicken	Chicken	
Simmered turnips	Turnips, aburaage	
Miso soup	Taro, young kidney beans	
Boiled rice		

Note: If not mentioned, most ingredients are chopped and seasoned with shoyu, sugar (or mirin, the fermented sweet sake), and dashi (umami flavors such as katsuo-busi, dried konbu, or monosodium glutamate).

Chapter 12

North European Meals: Observations from Denmark, Finland, Norway, and Sweden

Ritva Prättälä

INTRODUCTION

The Nordic countries, Denmark, Finland, Norway, and Sweden,[1] are geographically close and have many political and cultural ties. Until 1814 Norway was part of Denmark and after that an autonomic part of Sweden. Norway became an independent state in 1905. Before the independence Finland was first part of Sweden (–1809), then an autonomic part of Russia (–1917). Swedes, Danes, and Norwegians can communicate with each other by using their native languages, whereas Finnish belongs to another language family and is, therefore, incomprehensible to the others. Most Finns, however, learn some Swedish at school, and 6% of the Finnish population has Swedish as native language.

The physical environment—climate and soil—is most suitable for agriculture in Denmark and southern Sweden. In middle and northern Sweden and Finland, forestry has also been important in primary production. In Norway, the area suitable for agriculture, especially for grain production, is small, but the long coastline provides good opportunities for fishing.

Agriculture, fishery, and forestry still have central roles in the Nordic economy, although most people today earn their living by working in trade and service. In Finland, the change from primary production to industrial work and thereafter to services happened later, mainly after World War II. Denmark, Finland, Sweden, and Norway are today modern welfare states. The standard of living is high; for example, more than 10% of adults have a university degree and more than 30% of the gross national product is allocated to social expenditures.

The physical, cultural, and socioeconomic environments set the framework within which people choose what, where, when, and with whom they eat and how they combine and prepare single food items into dishes, meals, and meal patterns. This chapter will describe similarities and differences in workday meal patterns in Denmark, Finland, Norway, and Sweden. The description starts with the traditional meal patterns. Second, current meals and meal patterns are described. Third, variations and changes in meal patterns, as well as factors contributing to these changes, are discussed.

---

[1]The Nordic countries are Denmark, Finland, Iceland, Norway, and Sweden. Because of the scarcity of research data (because the author does not understand Icelandic), Iceland is excluded from this description.

This chapter reviews studies carried out within nutritional, medical science, social sciences, and ethnology. In the different fields of sciences, the concepts "meal" and "meal pattern" are not used identically. In this chapter meal pattern refers broadly to the number and timing of eating occasions per day and to the type of preparation (cooked, hot meal vs. a noncooked eating occasion). The concept "meal" includes breakfast, lunch, and dinner. Meal composition refers to the foods and dishes that are eaten at a single meal. The chapter concentrates on workday eating in which meals eaten in restaurants—with the exception of lunch canteens and "workplace restaurants"—are not described.

TRADITIONAL MEAL PATTERNS

The traditional Nordic meal patterns were determined by rural work schedules. Workdays were long, and the work was physically demanding and required a high energy intake. Therefore, three or even four hot meals were eaten daily. The first hot meal was eaten before going to work in the fields or forests, the second one in the middle of the workday, and the last meal in the evening. In addition to the three hot meals, two or three lighter meals or snacks were eaten.

In Finland, the three hot meals included bread, porridge or gruel, potatoes, root vegetables, and salted fish. In the summer, dairy products and occasionally meat, game, and wild berries were added to the diet. Boiled potatoes, brown or white sauce, Baltic herring, and rye porridge were usual breakfast items (Prättälä & Helminen, 1990). Thus, the traditional breakfast did not differ from lunch or dinner.

Long workdays and heavy manual labor were also characteristic for early industrial work. Fjellström (1990) has carried out a thorough study of meal patterns in a Swedish industrial community in the 1880s, and she described the typical meal pattern of an industrial worker. During the 10–12-hour long workday, there were three breaks. The meal pattern of the whole family was affected by the work schedule. Before going to work, the worker had coffee and a slice of bread. The first break at work was breakfast at 8–9 AM. Breakfast was a prepared hot meal that included, for example, Baltic herring or pork fat with potatoes, bread, milk, and coffee. The following break was at noon, when the mid-day meal was eaten. This meal consisted of a meat/fish dish, potatoes, bread, and a drink. In addition, porridge, gruel, or sour milk products were served as dessert. At the afternoon break, which took place at 4–5 PM, coffee and sandwiches were eaten. The last meal of the day was eaten between 8 and 9 PM after the workday. This meal, which included porridge and milk, bread, and a piece of fish, meat, or cheese, was slightly lighter than the dinner.

An example of the traditional Norwegian meal pattern can be found in a study of farmers' food habits carried out by Eriksen (1994). She describes meal patterns in rural Norway in the beginning of this century. As in Finland and Sweden, workdays started early, about 6 AM, and the first hot meal, breakfast, was eaten between 8 and 9 AM. Breakfast consisted of porridge or soup, bread, butter, cheese, and milk products. The main meal was either the meal eaten in the middle of the day (lunch) or the one eaten after the workday (dinner). Typical items served at lunch and dinner were porridges, potatoes, bread, herring, soups, and stews made of meat, cheeses, and milk and sour milk. If the lunch had included meat or fish, the dinner may have been based on porridge and milk dishes. Later, before going to bed, some had cold porridge with milk.

CURRENT MEALS AND MEAL PATTERNS

Meal Composition

Today the composition of the first meal of the day is similar in all Nordic countries. It is based on bread, and the only hot item included commonly is a drink, predominantly coffee. Other common items are bread spreads, cheese, milk or sour milk products, and fruit juice. In Norway and Denmark, the second meal of the day is also based on bread and sandwiches (Becker, 1997; Becker & Andersson, 1995; Haraldsdóttir, Holm, Hojmark Jenson, & Moller, 1987; Kleemola, Roos, & Prättälä, 1997; Prättälä, G. Pelto, P. Pelto, Ahola, & Räsänen, 1993; Rothenberg, Bosaeus, & Steen, 1994; Wold, 1985).

The main meal in Finland and Sweden is either lunch (meal at noon) or dinner (meal in the afternoon or early evening). In Norway and Denmark, the main meal is dinner. The basic composition of the main meal is relatively similar in all countries (Exhibit 12–1). Typical components of Nordic meals are hot meat dishes and boiled potatoes. Meat is more common than fish, but in Norway the difference between the two is less than elsewhere. Although rice and pasta are often used as substitutes for potatoes, the Nordic people seem to emphasize the role of potatoes as a component of a proper meal. Vegetable salads are common, whereas desserts, with the exception of fresh fruit, rarely belong to weekday meals. Bread is included, especially in Finland and Sweden. The most common beverages are water, milk, and sour milk. (Ekström, 1990; Haraldsdóttir et al., 1987; Kleemola et al., 1997; Prättälä et al., 1993; Rothenberg et al., 1994; Wandel, Bugge, & Skoglund Ramm, 1995; Wold, 1985).

In Finland and Sweden, bread, fresh vegetable salads, and milk are essential components of lunches served at schools, workplace canteens, or lunch restaurants (Jäntti, 1992; Kruse, Jacobsson, Torelm, & Becker, 1990).

Exhibit 12–1 Examples of Usual Main Meals

Denmark	Frikadeller, kogte kartofler, brun sovs Meat balls, boiled potatoes, brown gravy (*Source:* Haraldsdóttir et al. [1988]; common main dishes, side dishes and sauces in Denmark in 1985.)
Finland	Lihamakaronilaatikko, tomaattikastike Minced meat and macaroni casserole, tomato ketchup (*Source:* Jäntti [1992]; 8-week menu from a Finnish school, lunch on August 15, 1992, lunch includes main dish, salad, bread, bread spread [margarine] and drink [milk, water].)
Norway	Fiskekaker, kogte poteter og gulrøtter, hvit sau Fish patties, boiled potatoes and carrots, white sauce (*Source:* Wold [1985]; common dinner dishes and side dishes in 1981–82.)
Sweden	Stekt strömming, potatismos Fried Baltic herring, mashed potatoes (*Source:* Kruse et al. [1990]; a list of lunch dishes [dagens rätt = dish of the day] from 17 restaurants in 1987–89, lunch includes main dish, bread, salad, and drink [water, milk].)

Finland

Finns daily eat three meals and one to three snacks. The conventional Finnish meal pattern consists of breakfast (aamiainen, aamupala = morning meal/snack, 6–10 AM), lunch (lounas, 11–12 AM), and dinner (päivällinen = day meal, 4–6 PM) (Roos & Prättälä, 1997) (Table 12–1). However, following this norm does not seem easy. Only approximately 40% of adult Finns report that they eat the three conventional meals (breakfast, hot lunch, and hot dinner). More Finns eat one hot meal daily than two hot meals (Helakorpi, Uutela, Prättällä, Berg, & Puska, 1997; Mäkelä et al., 1999; Mäkipää, 1998; Roos & Prättälä, 1997).

The distribution of energy intake over the course of the day shows three energy peaks, at breakfast, lunch, and dinner time. Breakfast contributes 19% of the daily energy intake. Lunch and dinner contribute altogether 44%, and other eating occasions contribute 37%. The energy peak at lunch time is higher than the peak for dinner (Kleemola et al., 1997; Roos & Prättälä, 1997).

The Finnish breakfast is not a cooked meal. Most Finns (87% of men and 79% of women), however, eat breakfast daily (Kleemola et al., 1997). Lunch is usually the main meal of the day. The proportion of Finns eating lunch regularly is higher than the proportion eating dinner. In the early 1990s hot lunch was eaten by 75% of Finnish men and 72% of Finnish women and hot dinner by 68% of the men and 57% of the women (Mäkipää, 1998).

Finns who started school after 1947 have learned the norm of having a cooked lunch. According to a law passed in 1943, elementary schools had to arrange a free lunch for all pupils by 1948. Gradually the free school lunch system was introduced to all levels of basic education (Jäntti, 1992). University students do not get a free lunch, but their meals have been subsidized by about 30% since 1979 (Tainio, 1992). The system is regulated by statutes; for example, the school lunch should provide one third of the daily energy need and should be served between 10:45 AM and 12:30 PM (Jäntti, 1992).

It is thus no surprise that the generation accustomed to free school lunches has in adult age promoted hot lunches at work. Recommendations for workplace lunches were developed in the 1970s, and meal provision at work was included in trade union agreements in both the public and private sectors. The peak year of

Table 12–1 Conventional Meal Pattern in Denmark, Finland, Norway, and Sweden. Summary of Studies Carried Out in the 1980s and 1990s

| Country | First Eating Occasion | | | Second Eating Occasion | | | Third Eating Occasion | | |
	Name, Time	Type	%[a]	Name, Time	Type	%	Name, Time	Type	%
Denmark	Morgenmad	Bread	90	Frokost 11–12 AM	Bread	80	Aftensmad 5–8 PM	Hot	90
Finland	Aamupala aamiainen 6–10 AM	Bread	80–90	Lounas (aamiainen) 11–12 AM	Hot	70–75	Päivällinen 4–6 PM	Hot	60–70
Norway	Frokost 7–9 AM	Bread	75–95	Lunsj matpakke 11–12 AM	Bread	80–85	Middag 3–5 PM	Hot	80
Sweden	Morgonmål frukost 6–11 AM	Bread	90–95	Lunch mitt på dagen mål 11 AM–1 PM	Hot	35–70	Middag kvällsmål 4–6 PM	Hot	35–70

[a] Proportion of respondents who eat the meal regularly.

Finnish workplace catering was 1987. The economic depression in the 1990s decreased the number of Finns working outside the home and diminished support given to workplace catering by employers and the state. In 1982, 50% of state employees regularly ate a hot workplace lunch. Ten years later the corresponding figure was 30% (Tainio, 1992).

In 1991 the Finnish catering services served about 1.8 million daily meals, which equals 0.4 meals per day per person. The majority of these meals were school meals (22.1%), whereas the share of restaurants and hotels was 14.9%, workplace canteens/restaurants 11.5%, hospitals 11.3%, and cafeterias 10.6% (National Nutrition Council, 1992; Tainio & Tarasti, 1995).

Workplace meals have influenced the Finnish meal pattern. In 1984, middle-aged women described reasons for skipping dinner in an interview study. Of the 102 women, 73 ate lunch regularly and 40 had lunch at the canteen at work. The typical explanation for skipping dinner was that all family members ate their main meal at the workplace or in school (Prättälä et al., 1993). On the other hand, Finnish women seem to appreciate common family meals, although the weekday time schedules often make family meals impossible. If common meals cannot be eaten on weekdays, this is compensated by weekend meals. (Mäkelä, 1996).

Sweden

A fixed meal pattern seems very important for Swedes. Sweden is one of the few countries that has included meal patterns in official dietary recommendations. According to the recommendations, Swedes should eat three main meals and two to three snacks daily. The recommended distribution of energy at different meals is breakfast (morgonmål/frukost = morning meal/early food) 20% to 25%, lunch (lunch/mitt på dagen mål = meal at noon) 25% to 35%, and dinner (middag/kvällsmål = mid-day/meal/evening meal) 25% to 35% (Svenska näringsrekommendationer, 1997).

Swedes follow their recommendations in regard to the number of eating occasions, and the mean number of daily eating occasions is about five (Becker, 1992). Typical times of the three main meals are 6–11 AM for breakfast, 11 AM–1 PM for lunch, and 4–6 PM for dinner (Rothenberg et al., 1994). Of these three meals, breakfast is closest to a snack, whereas lunch and dinner are cooked meals (Table 12–1).

The majority of Swedes eat breakfast regularly. Studies carried out among different population groups show that the proportion of Swedes skipping breakfast is 5% to 10% (Becker, 1990, 1997; Becker, Enghardt, & Robertson, 1994).

Lunch and dinner are cooked meals, but all Swedes do not eat both lunch and dinner daily. The majority of Swedes seem to have only one hot meal; based on existing data the figures vary from 35% to 70% (Becker, 1990, 1997; Becker et al., 1994). According to a recent Nordic telephone survey (Mäkelä et al., 1999), 59% of the Swedish respondents ate one hot meal and 34% ate two.

The main meal may be either lunch or dinner. A study carried out in southern Sweden in 1990 suggested that lunch was eaten more often than dinner. Two later studies, one among teenagers and the other among the elderly, showed that dinner was eaten more often and was more important in terms of energy intake (Rothenberg et al., 1994; Samuelson, Bratteby, Enghardt, & Hedgren, 1997).

Like the Finns, the Swedes got used to having cooked lunches at school. Swedish children have received a free school meal since 1946. Free lunch is served in elementary and secondary schools and also in most high schools. Contrary to the

Finnish school meal system, the Swedish one is not regulated by law. The state support for school meals was ended in 1967 and municipalities have since funded the system (Strandell, 1991).

In 1994 about 80% of secondary school pupils (10–16 years old) ate school lunch almost every day. The proportion of pupils eating school lunch was slightly higher than the proportion of pupils eating prepared food for dinner (Becker & Andersson, 1995). According to a more recent study carried out in two regions in Sweden, school lunch is skipped more often than breakfast or dinner. The majority of the pupils ate the school lunch three to four times per week. When the pupils did not eat the school lunch, they chose something else in the school cafeteria (Samuelson et al., 1997).

Catering services have had a major influence on meal patterns of Swedish adults. In 1987 catering kitchens served 4.6 million meals daily (0.55 meals/person/day). The meals are served by different kitchens: 25% school kitchens, 18% staff canteens/restaurants, 22% other restaurants, and 20% hospitals, kindergartens, senior citizens homes, and the like (Anon, 1990).

The proportion of Swedes having a hot lunch outside home may have decreased since 1987. Today more Swedes are unemployed than in the 1980s, and the prices of meals in workplace cafeterias have increased because of changes in taxation. A survey carried out before and after the taxation changes in a sample of private and public enterprises showed that before 1991, 48% of the respondents regularly ate a hot lunch in the workplace canteen and 28% brought food from home. In 1993 about 30% of the respondents ate regularly in the canteen and 44% brought food from home (Lennernäs, Becker, & Hagman, 1994).

Despite the opportunity to have a hot lunch at school or at the workplace, Swedish families appreciate hot meals eaten together with the whole family after work. In a study carried out in middle Sweden among families with children less than 18 years of age (Ekström, 1990), 90% of the participating families had a prepared dinner; at three of four dinners the whole family was present. Women were responsible for cooking whether they were working outside the home or not.

Norway

Norwegians eat about five times on weekdays (Johansson, Solvoll, Bjorneboe, & Drevon, 1997). The conventional Norwegian meal pattern consists of two bread-based meals, one hot meal (dinner), and one snack: breakfast (frokost = early food, 7–9 AM), lunch (matpakke/lunsj = lunch pack/lunch, 11–12 AM), dinner (middag = midday meal, 3–5 PM), and supper (kveldsmat = evening food, 7–9 PM) (Johansson et al., 1997; Wold, 1985) (Table 12–1). In regard to hot meals the Norwegians follow the norm because most (79%) eat only one hot meal daily (Mäkelä et al., 1999). The meal after dinner is most commonly skipped (Wold, 1985).

The Norwegian lunch is not hot in contrast to the Finnish and Swedish ones, and the school meal systems differ accordingly. Since the 1930s, the Norwegian school meal has been based on sandwiches brought from home complemented with milk, which has been offered at a reduced price at school. In 1991, 96% to 98% of elementary school pupils brought lunchpacks to school (Halvorsen, Drevon, & Bjorneboe, 1995).

Norwegian nutrition experts presented the Oslo breakfast in the 1920s, the idea of a healthy sandwich meal eaten at school. The Oslo breakfast, which consisted of bread, milk, cheese, cod liver oil, vegetables, and fruits (Oshaug, 1994), was soon introduced in larger municipalities, but the expense became a problem. Because the efforts to get a free school lunch did not succeed, Norwegian parents were educated to prepare a healthy packed lunch for their children. Norwegians who went to school in the 1920s and 1930s remember films and educational campaigns that institutionalized the concept of the Oslo breakfast (Eriksen, 1994).

In Norway, dinner is the main meal of the day. At dinner the Norwegian families sit down together. Traditional Norwegian dinner dishes with boiled potatoes and vegetables are considered to be more proper and healthy than modern convenience foods such as pizzas or hamburgers (Eriksen, 1994; Fürst, 1985; Wandel, Bugge, & Skoglund Ramm, 1995; Wold 1985).

Denmark

The number of daily eating occasions in Denmark is also about five (Haraldsóttir et al., 1987). The conventional Danish eating pattern consists of three main meals: breakfast (morgenmad = morning food), lunch (frokost = early food/breakfast, 11 AM –12 AM), and dinner (aftensmad = afternoon/evening food, 5 PM–8 PM). In addition, Danes eat two to three snacks (Table 12–1).

In 1985, 90% of adult Danes had breakfast daily, and the corresponding figures for lunch and dinner were 81% and 94% (Haraldsdóttir et al, 1987). Since 1985 the frequency of eating meals has decreased. In 1995, 84% of adults had breakfast daily, 51% had lunch, and 86% had dinner (Fagt & Groth, unpublished data, 1999). Although the conventional meal pattern in Denmark consists of two cold meals and one hot, less than half (45%) of Danes report following the conventional meal pattern every day. Dinner is the hot meal of the day (Haraldsdóttir et al., 1987; Mäkelä et al., 1999).

The Danish school meal system is similar to the Norwegian, and the Oslo breakfast has been the model for a healthy lunch. In Denmark the typical packed lunch consists of sandwiches, milk, and vegetables or fruits. The children bring some food from home, but the majority of schools have a food shop. Shopkeeping is often part of the curriculum, and the pupils keep the shops. The foods for sale are recommended healthy lunch items (Benn, 1993). More than 80% of Danish schoolchildren regularly eat school lunch, but they are rarely served a hot lunch (Due, Holstein, Ito, & Groth, 1991).

Cooked dinner is important to Danes. Dinner contributes 35% of the total energy intake, breakfast about 18%, and lunch 24% (Haraldsdóttir et al., 1987). Dinner is eaten together with the whole family. The Danes prefer homemade food, but it is not always possible in everyday life. The use of ready-to-eat meals is justified by lack of time, not by the taste or healthfulness of these meals. Dinner time is not always fixed and may vary daily according to the schedules of the family members. (Land, 1998).

CHANGES AND VARIATION IN MEAL PATTERNS

Although many empirical studies have described meals in Denmark, Finland, Norway, and Sweden, drawing conclusions about the changes and variation in meal patterns is difficult. The reviewed studies have been published in five languages.

Because the meal is a sociocultural category, the definitions even vary among people speaking the same language. As a result, the studies have not used the same concepts.

The majority of the meal studies have been cross-sectional interview or questionnaire studies covering various time periods. They have often been carried out among small or limited groups and are not based on representative national samples. Only one of the cross-sectional studies has been carried out in Denmark, Finland, Norway, and Sweden by using comparable samples and methods (Mäkelä et al., 1998). A few studies have focused on changes within the individual countries. Meal pattern changes have, however, been discussed in most of the cross-sectional studies by researchers and study subjects.

Despite the variety in approaches, target groups, and methods of the studies, some common characteristics of meal pattern changes and variation can be identified. The first obvious observation is the decrease in the number of daily meals. The traditional Nordic meal pattern consisted of three or even four hot meals daily, but today the number of hot meals is one or two. Cooked meals have been replaced by lighter eating events, but the eating schedule has persisted. None of the cited studies suggested that the pattern of three hot meals was changing into "grazing." The Nordic people seem reluctant to change meal patterns even if maintaining them is difficult. There is plenty of evidence that cooked sit-down meals eaten together with the family are highly valued.

The second uniform observation is the disappearance of the cooked breakfast. The traditional breakfast consisting of boiled potatoes, porridges, bread, milk, and some meat or fish was eaten between 8 and 10 AM. However, the first eating event, a simple snack, had occurred 2 to 3 hours earlier. Remnants of the traditional breakfast can still be recognized in the use of the word breakfast (aamiainen, frukost, frokost) in the Nordic countries. Breakfast can be either the first eating occasion of the day or the first meal of the day. In Denmark, the word *frokost* refers to the sandwich meal eaten at noon (i.e., lunch). In Sweden, some blue collar workers who begin their work day at 6 or 7 AM have a hot meal they call breakfast at work at 9–11 AM. White collar workers who wake up later call their bread-based morning meal breakfast and their mid-day meal lunch (Blom, Lennernäs, Hambraeus, & Anderson, 1991; Fjellström, 1990). In Finland, some still have porridge or other cooked food early in the morning, and they consider their breakfast to be a meal (Mäkelä, 1996; Prättälä, 1998).

Third, the proportions and roles of meal components have changed. The traditional components, meat, fish, potatoes, bread, and milk, still have a place in the modern Nordic meal, and preparing the meal from raw materials is the norm. However, the ideal does not fit in with everyday life. Pizzas, hamburgers, or other convenience foods sell well, and boiled potatoes are often replaced by rice and pasta. Milk or sour milk is still a common beverage at meals, whereas milk dishes are no longer important meal components. The traditional desserts were porridges or gruels that either contained berries or milk or these were added. Today desserts, with the exception of fresh fruits, do not belong to weekday meals. Green salads did not have a place in the traditional Nordic meal, but fresh vegetables are a central part of the modern meal.

The fourth observation is the difference in the type of lunch. In Norway and Denmark the common lunch is based on sandwiches, whereas in Finland and Sweden the hot lunch is the still the norm. The difference in the Nordic lunches is related to the organization of school and workplace meals. In Finland and Sweden all children receive a free hot meal at school, and many adults can buy a meal at the workplace at a reduced price. In Norway and Denmark, this is not the case.

The decrease in the number of hot meals, the disappearance of cooked breakfast, and the changes in the composition of the hot meal are common in all four Nordic countries. These changes can probably be explained by general socioeconomic development and changes in living conditions during this century. Nordic people no longer earn their living in heavy manual labor in the countryside, and the traditional housewife has faded. Today both men and women work outside the home and have less time to prepare meals. Their work is not physically demanding, and they need less energy. At the same time, the availability of fresh vegetables and ready-to-eat meals has increased, and people have more money to buy food.

The differences in the nature of lunch are more difficult to explain. Why do Norwegians and Danes eat a bread-based lunch, whereas Finns and Swedes prefer a hot meal? The observation that the Nordic people have learned different habits already at school age is not a satisfactory explanation. Some other factors have contributed to the variation in school and workplace meal systems.

School meal systems were developed in the Nordic countries in the 1930s and 1940s, a time characterized by institutionalization of food and nutrition policy. The League of Nations and the International Labour Organization had put nutrition high on the political agenda and recommended establishment of national nutrition committees. National nutrition committees were nominated in all four countries, but their activities and influences varied. The variation was connected to the relationship between health and agricultural policies and to the specific traits of the welfare ideology the individual countries adapted. (Kjaernes, 1997).

The Danish welfare policy has been liberalistic; private enterprise and individualistic values have had a more important role in Denmark than in the other Nordic countries (Kosonen, 1998). Furthermore, the national food and nutrition committee has been less influential in Denmark (Kjaernes, 1997). Collectively organized school meals based on regulation and control probably did not fit the individualistic values.

The welfare ideology has been especially strong in Sweden. It has been typical for Swedes to develop centralized institutions aiming at guaranteeing public services, such as free health care, basic education, and school meals (Kosonen, 1998). The work of the Swedish nutrition committee resulted in several activities—among them free school meals—to improve the nutritional situation in the country (Kjaernes, 1997).

In Finland the welfare ideology was not as strong as in Sweden. But the great losses caused by the war and the national efforts to rebuild the country influenced the position of women. The Finnish women were important in the labor force during the war years when the men were at the front. Since the war years, the proportion of women having full-time jobs outside the home has been greater in Finland than in the other Nordic countries (Kosonen, 1998). The Finnish postwar ideology, on the other hand, encouraged families to have many children. Because the Finnish women were needed both as labor force and as mothers, school meal systems and other forms of public services became necessary.

The Norwegian welfare policy has in many respects been similar to the Swedish one. However, other factors may have reduced the pressures to serve free meals to school children. In Norway, the women stayed at home as housewives later than in the other Nordic countries (Kosonen, 1998). Therefore, there was not so much social pressure to serve lunches at schools because the mothers were at home.

Meal patterns are related to the general social policy and socioeconomic conditions of the country. Accordingly, future developments of meal patterns depend on broader social changes of the Nordic region. It can be expected that differences in the Nordic meal patterns may diminish in the future: the Swedish and Finnish school and workplace meal systems have run into economic difficulties, and a greater proportion of the adult population is unemployed. This might lead to a shift from the hot lunch to a bread-based meal. On the other hand, there are few signs of the disappearance of the meal. Some unemployed people prepare more meals at home, and, at least in Finland, charity organizations have started to serve meals for poor and unemployed people. Furthermore, the Nordic people greatly appreciate meals prepared and eaten together with the family.

REFERENCES

Anon. (1990). *Vår Föda 42* (9–10), 573–578.

Becker, W. (1990). Kostvanorna i södra Sverige—Resultat från en provundersökning (Dietary habits in Southern Sweden—Results of a pilot study. In Swedish. English summary). *Vår Föda, 42*(6), 322–333.

Becker, W. (1992). Befolkningens kostvanor och näringsintag (Food habits and nutrient intake in Sweden, 1989). *Vår Föda, 44*(8), 349–362.

Becker, W. (1997). Allför många ungdomar slarvar med frukosten. (Too many teenagers have an unproper breakfast. In Swedish). *Vår Föda, 49*(7), 14–18.

Becker, W., & Andersson, B. (1995). Högstadieelevers matvanor 1994. En av tio hoppar över frukosten. (Food and meal habits of upper level compulsory school pupils. In Swedish. English summary.). *Vår Föda, 47*(7), 9–17.

Becker, W., Enghardt, H., & Robertson, A.K. (1994). Kostundersökningar i Sverige 1950–1990 (100 pp). (Dietary surveys in Sweden. In Swedish). Uppsala: Livsmedelsverkets förlag.

Benn, J. (1993). Kost i skolen i Danmark. Fra fattige borns bespisning til sundhedspaedagogiske projekter (School lunches. From the provision of free meals for school children to school health education projects). *Scandinavian Journal of Nutrition, 37*, 30–34.

Blom, K., Lennernäs, A.-C., Hambraeus, L., & Anderson, A. (1991). Mat i skog och koja—en intervjuundersökning av 40 skogsarbetande män. (Food in forests and cabins—an investigation of 40 woodsmen. In Swedish. English summary). *Vår Föda, 43*(1), 55–65.

Due, P., Holstein, B., Ito, H., & Groth, M.V. (1991). Spisevaner og sundhedsadfaerd blandt 11–15 årige. *Ugeskr Laeger, 153,* 984–988.

Fagt, S. & Groth, M.V. (1999). Unpublished raw data. Institute of Food Research and Nutrition, Veterinary and Food Administration, Denmark.

Ekström, M. (1990). *Kost, klass och kön. (Food preparation, class and gender.* In Swedish. English summary). (267 p.) Doctoral dissertation. Umeå Studies in Sociology No. 98.

Eriksen, S.H. (1994). *En kultursosiologisk studie av matvaner blant bönder. (A socio-cultural study of farmers' food habits.* In Norwegian. English summary). (335 p.). Rapport 1. Trondheim: Senter for bygdeforskning.

Fjellström, C. (1990). *Drömmen om det goda livet. Livskvalitet och matvanor i ett uppväxande industrisamhälle: Stock sågverk 1870–1980 (The dream of the good life. The quality of life and food habits in an expanding industrial society: Stocka Sawmill 1870–1980).* (394 p.). Stockholm: Acta Ethnologica Umensia. Almqvist & Wiksell International.

Fürst, E. (1985). *Mat: arbeid og kultur. Vår matkultur: konflikt mellom det tradisjonelle og det moderne. (Food: Work and culture. Food culture: tradition and modernity in conflict.* English summary). (131 p.) Lysaker: SIFO.

Halvorsen, M., Drevon, C.A., & Bjorneboe, G.-E. (1995). Mat i skolen. (Food in school. English abstract). *Scandinavian Journal of Nutrition, 39,* 151–155.

Haraldsdóttir, J., Holm, L., Hojmark Jensen, J., & Moller, A. (1987). Danskernes kostvaner 1985. 2, Hvem spiser hvad? (375 p.) Levnedsmiddelstyrelsen, Soborg: Ernaeringsenheden.

Helakorpi, S., Uutela, A., Prättälä, R., Berg, M.-A., & Puska. P. (1997). Suomalaisen aikuisväestön terveyskäyttäytyminen. Kevät 1997. Health behaviour among Finnish adult population. *Publications of the National Public Health Institute, B10,* 206 p.

Jäntti, A. (1992). Ravitsemussuositukset peruskoulussa ja lukiossa. (Dietary recommendations in basic and high schools. In Finnish.) In *Joukkoruokailun ravitsemussuositukset* (pp. 50–64). Helsinki: VAPK Kustannus.

Johansson, L., Solvoll, K., Bjorneboe, G.-E., & Drevon, C.A. (1997). Dietary habits among Norwegian men and women. *Scandinavian Journal of Nutrition, 41*, 63–70.

Kjaernes, U. (1997). Framveksten av ernaeringspolitik i Norden. (Origins of nutrition policies in the Nordic countries. In Norwegian). In U. Kjaernes (Ed.), *Utfordringer i ernaeringspolitikken. Raport fra Nordisk ministerråds konferanse, Hässelby Slott, September 1996* (pp. 73–87). Nord: Tema 619.

Kleemola, P., Roos, E., & Prättälä, R. (1997). Päivän tärkein ateria—Aterioiden koostumus vuonna 1992. (The most important meal of the day—Composition of meals in 1992. In Finnish.). *Suomen Lääkärilehti, 52*(9), 1027–1032.

Kosonen, P. (1998). *Pohjoismaiset mallit murroksessa (The Nordic models in transition.* In Finnish). (420 pp). Tampere: Vastapaino.

Kruse, B., Jacobsson, E., Torelm, I., & Becker, W. (1990). Fett och energi i lunchmåltider. (Fat and energy of lunch meals. In Swedish). *Vår Föda, 42*(8), 448–456.

Land, B. (1998). *Consumers' dietary patterns and desires for change* (69 pp). MAPP. Working paper No. 31.

Lennernäs, M., Becker, W., & Hagman, U. (1994). *Matvanor före och efter beskattningen av lunchsubventionerna* (73 pp). Livsmedelsverket. Rapport 4.

Mäkelä, J. (1996). Kunnon ateria. Pääkaupunkiseudun perheellisten naisten käsityksiä. (A proper meal: exploring the views of women with families. In Finnish. English abstract.). *Sosiologia, 33*(1), 12–22.

Mäkelä, J., Kjaernes, U., Pipping Ekström, M., L'orange Fürst, E., Gronow, J., & Holm, L. (1999). Nordic meals. Methodological notes on a comparative survey. *Appetite, 32*, 73–79.

Mäkipää, E. (1998). *Suomalainen ateriajärjestys 1980–1995. Sosiaalisten tekijöiden yhteys ateriajärjestyksen muutokseen ja vaihteluun (The Finnish meal pattern 1980–1995. Association of social factors with change and variation of meal pattern.* In Finnish.). (69 pp). Master's thesis, University of Helsinki.

National Nutrition Council. (1992). *Nutrition policy in Finland. Country paper prepared for the FAO/WHO international conference on nutrition in Rome 1992* (64 pp). Helsinki: Helsinki University Printing House.

Oshaug, A. (1994). Nutrition security in Norway? A situation analysis. *Scandinavian Journal of Nutrition, 38* (Suppl. 28), 26–27.

Prättälä, P. (1998). *Puun ja kuoren välissä. Metsurit ja kirvesmiehet puhuvat terveellisistä elintavoista. (Between a rock and a hard place. Loggers and carpenters talking about healthy lifestyles.* Publications of the LEL Employment Pension Fund. In Finnish. English abstract). (108 pp). LEL Työeläkekassan tutkimuksia 32.

Prättälä, R., & Helminen, P. (1990). Finnish meal patterns. In J.C. Somogyi & E.H. Koskinen (Eds.), Nutritional adaptation to new life-styles. *Bibl Nutr Dieta, 45*, 80–91.

Prättälä, R., Pelto, G., Pelto, P., Ahola, M., & Räsänen, L. (1993). Continuity and change in meal patterns: the case of urban Finland. *Ecology of Food and Nutrition, 31*, 87–100.

Roos, E., & Prättälä, R. (1997). Meal pattern and nutrient intake among adult Finns. *Appetite, 29*, 11–24.

Rothenberg, E., Bosaeus, I., & Steen, B. (1994). Food habits, food beliefs and socio-economic factors in an elderly population. *Scandinavian Journal of Nutrition, 38*, 159–165.

Samuelson, G., Bratteby, L.-E., Enghardt, H., & Hedgren, M. (1997). Mer sött än grönt (Food habits of Swedish Adolescents. In Swedish. English summary). *Vår Föda, 49*(1), 3–9.

Strandell, A. (1991). Praktiska aspekter på måltiden: Skolmåltiden (Practical aspects on the meal: The school meal). In M.L Telegin (Ed.), Måltiden och måltidsordningen. Forskningsrådsnämnden. *Rapport, 91*(3), 87–92.

Svenska näringsrekommendationer. (1997). *Vår Föda, 49*(2), 7–14.

Tainio, R. (1992). Korkeakouluopiskeliuoiden roukailu. (Meals of university students. In Finnish). In *Joukkoruokailun rauitsemussousitukset* (pp. 72–79). Helsinki: Vapk Kustammus

Tainio, R., & Tarasti, K. (1995) Suomalaisen työpaikkaruokailun kehitys. (Development of Finnish workplace catering. In Finnish). In K. Hasunen, P. Helminen, S. Lusa, R. Prättälä, R. Tainio, & V. Vaaranen. (1995). *Yksin vai yhdessä* (pp. 9–18). Työterveyslaitos: Työpaikkaruokailu murrosvaiheessa.

Wandel, M., Bugge, A., & Skoglund Ramm, J. (1995). *Matvaner i endring og stabilitet. En studie av måltidsvaner og matforbruk i ulike forbrukergrupper. (Change and stability in food habits.* English summary). (219 pp). SIFO. Rapport No. 4–1995. Lysaker.

Wold, B.K. (1985). *Mat: arbeid och kultur. Husholdsforbruk og husholdsarbeid—En oversiktsrapport (Food: Work and culture. Household consumption and household work—A survey report.* English summary). (221 pp). Lysaker: SIFO.

Chapter 13

British Meals and Food Choice

D.W. Marshall

I still think the roast beef of old England served with meaty gravy, crisp Yorkshire pudding and crunchy roast potatoes is not only one of the worlds greatest meals, it is something the British do better than anyone else.

—Delia Smith [cookery writer], 1995, p. 142

The meal is alive and well in British and other cultures.

—Wood, 1995, p. 124

Cookery programs are proving popular with British viewers. On one show, *Ready Steady Cook*, the (now) celebrity chefs have 20 minutes to prepare a "meal" using whatever ingredients the contestants elect to bring into the studio. As the countdown begins, the two teams, each made up of a chef and a contestant (usually selected for their limited culinary skills), race to create a meal. The chefs' artistic ingenuity is tested to the full as he, always a man, allocates food to different courses from the assembled collection of ingredients (some of which are certainly not considered part of the daily British fare). Accepting that this is more about entertainment than domestic cooking—it takes place in a well-equipped studio kitchen without the distractions of telephones and children—it is interesting to observe the way in which the meal is literally constructed. The final result is often far removed from the traditional Sunday roast beef dinner mentioned previously, the ubiquitous fish and chips, or the high tea of salmon sandwiches and cake that many associate with the British table. The question addressed in this chapter, accepting the argument that there is essentially no such thing as national food but rather a complex history of colonialism and trade links and cultural exchange (Bell & Valentine, 1997), is what makes a meal distinctly British?

To answer the question one has to begin by asking exactly what do we mean by a "meal"? Somewhat surprisingly few attempts have been made to define or classify the ubiquitous "meal" (Meiselman, 1994). Although individuals have little difficulty in identifying a meal when they see one, most find it hard to articulate a clear definition (Chiva, 1997; Prättälä, Mäkelä, & Roos, 1994). This difficulty, in part, reflects the often taken for granted and implicit nature of the eating experience. Sometimes it is only when the rules are broken or transgressed that they become apparent.

Acknowledgments: Many thanks to Giles Quick and Rick Bell for their help in producing Figure 13–2 and Les Gofton for helpful comments on earlier drafts of this chapter.

A cursory examination shows that meal definitions have been restricted to the temporal aspects of the eating occasions (Gatenby, Anderson, Wlaker, Southón, & Mela, 1995), the assortment of food items, the quality and quantity of food (Skinner Salvetti, Exell, Penfield, & Costello, 1985), the nutritional composition (number of calories, carbohydrate, fat, protein), and the number of people present (de Castro, 1991). Lalonde (1992) takes a broader perspective and draws on temporal, structural, procedural, and social aspects of the meal:

> First, we can understand the meal in its most basic sense as a timely repast. Second, we can objectify the meal in order to discern its structure. Here the meal is perceived as the sum of its parts. Third, the meal may be comprehended as a purposive action, one which follows a 'script' so as to achieve an intended effect. Finally, a meal may be thought of as a social event that creates meaning for the participants (1992, p. 70).

Mäkelä et al. (1995) looked at meals and snacks in Finland and Kentucky and identified a number of distinct differences between these two types of eating occasions (Exhibit 13–1).

More recently Mäkelä et al. (1998)[1] have proposed a framework for analysis of modern food habits that avoids any predefined meal concepts. This "eating system" comprises:

> the eating pattern (rhythm and number of eating events, the alternations of hot and cold eating events), the meal format (the composition of the main course, the sequence of the whole meal), and the social organisation of eating (where and with whom people are eating, who did the cooking) (1998, p. 73).

Ultimately, meals are socially constructed entities that are recognized as such and carry meaning for those present. For a fuller discussion see Chapter 1.

This chapter is divided into four sections organized loosely around the framework proposed by Mäkelä et al. We now turn to look at eating patterns focusing on the daily pattern of meals, the meal structure, and the social organization of eating in British households. The final section considers the "proper meal," which has

---

Exhibit 13–1 Meals versus Snacks

	Meal	Snack
Ingredients	Food groups, variety	Simple, few ingredients
Quantity	More	Less
Quality	Healthy	Unhealthy
Food Preparation	Cooked	Uncooked
Situation	Sitting down	Grabbing food
Sociability	Company	Alone
Planning	Planned	Unplanned

---

[1] This work is part of an ongoing research project on modern food habits in which Nordic researchers are looking at the extent to which regular meals have been substituted by irregular eating patterns.

come to dominate the discussion on British meals, before proffering a speculative summary classification of British meals.

EATING PATTERNS

The "rhythm" of eating is captured in the daily, weekly, and annual distribution of eating occasions or meal patterns. The number of meals per day, their timing, and location all serve to structure our food choice. As "hinges of the day,"[2] they signify the passing of time, signal the start and the end of the working day, and mark off the working week from the weekend. In eighteenth century Britain two meals per day were common, but the introduction of longer parliamentary hours in the Victorian period brought with it a lengthening of the working day and the need for a "formal and substantial luncheon at around 12:30 to 1:00 PM" (Oddy & Burnett, 1992; see also Palmer, 1952; Visser, 1991). At the other end of the social scale, Rowntree's study (1901) revealed a pattern of four meals per day comprising breakfast in the morning, "dinner" at mid-day, "tea" in the evening, and "supper" last thing at night.

Until the middle of the twentieth century, the pattern of three (hot) meals per day, cooked breakfast between 7 and 9 AM, a mid-day meal between 12 and 2 PM, and an evening meal between 4 and 6 PM, was still common in Britain (Burnett, 1995). Although many men came home from work to eat their main meal at mid-day (Warren, 1958), the growing number of women in full-time employment and the emergence of canteens at the workplace contributed to a decline in this practice (Goodman & Redcliffe, 1991). By the end of the seventies the main meal had moved to later in the day, and dinner had become the principal family meal when households would sit down together (Burnett, 1995, Douglas & Nicod, 1974). Meals became polarized toward the beginning (breakfast) and end (dinner) of the working day. The pattern of meals was not unlike that of the leisured society of eighteenth-century Britain (Oddy & Burnett, 1992). British households ate fewer "lunches" and "high teas" at home but ate more mid-afternoon, evening, and late night snacks (Taylor Nelson Sofres,[3] 1998). Add to this the growth in eating out (Warde & Martens, 1999), and one begins to see the gradual erosion of traditional meal times.

Further distinctions can be drawn between the "convenience"-orientated weekday meals and more traditional weekend meals. This idea of a weekly repertoire of meals was captured by the suggestion of a "typical British weekly dinner menu" that consisted of sausages on Monday, sandwiches on Tuesday, lamb or pork chops on Wednesday, shepherd's pie on Thursday, fish and chips on Friday, eat out or pasta on Saturday, and poultry/roast on Sunday (Taylor Nelson Sofres, 1998). These "calendrical" aspects are amply illustrated in Figure 13–1 from Douglas (1996), which depicts how food choice is organized around days and time.

The type of meal eaten is constrained by where we eat, and there appears to be a link between food choice and location. Our own research showed that under-

---

[2] Professor Alexander Fenton, "The Food of the Scots," Lecture 4 "The Hinges of the Day," The Rhind Lectures 1996–97, Museum of Scotland, Chambers Street, Edinburgh. The term has been attributed to Palmer (1952).

[3] This is the United Kingdom's largest continuous household diary panel, representative of the U.K. population, with 2400 households and more than 11,000 individuals recording all food and drink consumed in the home over a 2-week period.

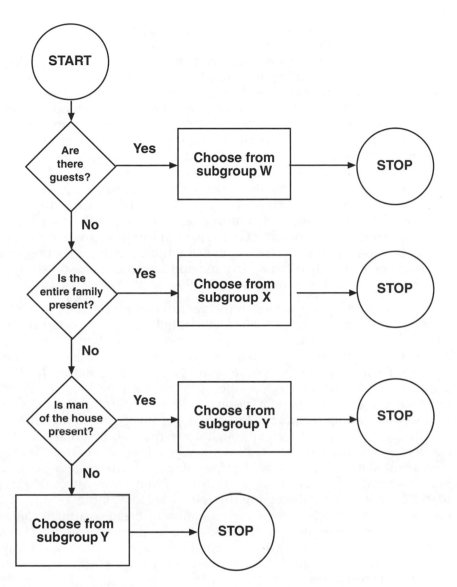

Figure 13–1 Days and Times: Homemaker's Decision Tree. *Source:* Reprinted with permission from M. Douglas, *Thought Styles*, p. 66, © 1996, Sage Publications.

graduate students, when asked to select foods for a dinner, lunch, or snack meal in 11 different physical locations, were considerate of the meal occasion but constrained in their choice by the physical location (Marshall & Bell, 1996).

Regional and Class Differences

Oddy and Burnett (1992), discussing the changes in meals and mealtimes in Britain wrote:

Over the last two hundred years the trend has been for regional and, to a rather lesser extent, social class differences in food to diminish, and for common food and meal patterns to develop under the influence of commercial marketing and advertising: formerly, there were wide differences between England and Scotland, between northern and southern England, and between rich and poor within any one region. The supposed 'preference' of a labourer in southern England for a diet of white bread, tea and, if possible, bacon in contrast to the barley bread, hasty pudding, milk and potatoes of the North was noticed by Eden: further north, in Scotland, the common foods were oatmeal porridge, potatoes, milk, cheese and kale. These differences probably reflected the different availability of foods such as dairy products rather than a true preference based on choice. In all regions, as incomes rose the dependence on bread and potatoes declined and the consumption of meat increased, so that among the wealthy bread and vegetables became mere accompaniments to the preferred food. Over the long term, rising standards of living gave more choice to the working class, and socio-economic differences in diet tended to narrow, but regional differences, especially between England and Scotland, were still strong in the interwar years and even since 1945 (1992, p. 32–33).

Records from the beginning of this century indicate that although little distinguished the social classes in terms of the number of meals eaten throughout the day, those higher up the social scale enjoyed more elaborate meals with greater quantity and better quality ingredients. The lower middle class dined on haricot, boiled or roast mutton, vegetables (carrots, turnips) and potatoes, followed by pudding, fruit, or biscuits and tea or coffee for their main meal of the day, whereas the working class ate meat (pork or bacon), potatoes and/or bread for dinner (Burnett, 1979). Bread, with a scraping of butter, jam, or margarine formed the mainstay of the British working class diet, although it was more common at breakfast and tea. Potatoes were eaten at dinner, and, as noted by one commentator, "potatoes are not an expensive item. But they are an invariable item. Greens may go, meat may diminish almost to the vanishing point, before potatoes are affected" (Reeves, 1979, p. 98). Burnett (1995), discussing the role of bread at meals in British upper and middle class households, notes the relegation of bread to a mere "adjunct to the meal eaten primarily as toast at breakfast or as thinly cut sandwiches at 'tea' or picnics." He contrasted this with the working class meal where bread played a more fundamental role but added that the main "breadwinner" still expected to sit down to a cooked meal in the evening.

Income and class differences continue in Britain and "better health will remain the prerogative of those who have access to and can afford a good diet" (Leather, 1992, p. 91). Warde's (1997) analysis of the Family Expenditure Survey[4] showed that self-employed professionals spent the most on food and white collar workers the least. Although this partly reflected household size, professionals spent proportionally more on beef, fish, fresh vegetables, and fruit and proportionally least on sausages, cooked meats, fish and chips, fresh milk, canned vegetables, potatoes, and tea compared with the working classes. In contrast, the working-class diet was dis-

---

[4] The Family Expenditure Survey collects family expenditure data on a range of items including food. Diary data are collected from more than 7000 households by the Office of Population Census and Surveys for the Department of Employment.

tinguished by comparatively higher expenditure on those items that the professionals spent proportionally less on with a comparatively lower expenditure on fresh vegetables, processed fruit, wine, meals out, and fresh fruit. However, no evidence suggests that the structure and format of the meals was any different, although, as Warde notes, the professional classes were more likely to eat out.

MEAL FORMATS AND STRUCTURED CHOICES

Meals, rather than being some random collection of foods, are highly structured events that follow a series of implicit rules about what foods can be served, where, when, and to whom. Drawing a linguistic analogy, after Halliday (1961), Douglas (1972) proposed a structural analysis of meals that treated food as a code, marking boundaries of social inclusion and exclusion. Analyzing her own family's meals, she classified eating into "daily menu," "the meal," "the course," "the helping," and "the mouthful." For each meal this was further elaborated into primary and secondary structures, courses, elements, and cooking categories. Douglas (1976) used this to illustrate the extent to which food choice was constrained by rules of appropriateness and described the meal as follows:

> A structured event is a social occasion which is organised according to rules prescribing time, place and sequence of actions. If food is taken as part of a structured event, then we have a meal... the meal by contrast [to snacks] has no self contained food items and is strongly rule-bound as to permitted combinations and sequences (1976, p. 83).

This "grammar of eating" could be seen in the British working-class meal, characterized by a tripartite structure with a stressed main course "A" and two unstressed courses "2B." Progression through the courses was signified by a transition from hot to cold, potato to cereal, dominance of the visual over the sensory, and the option to omit elements. Each individual course, or dish, was structured around a stressed item "a" (centerpiece usually meat or fish) and two unstressed items "2b" (staple and trimmings usually potatoes or bread and vegetables) covered in a gravy (Douglas, 1976; Douglas & Nicod, 1974; Nicod, 1979). Variations on this tripartite structure were evident across other eating occasions, or what she called minor meals, and the structure of the grandest meal was reflected in smallest meal.[5]

The meal format or structure, what Lalonde (1992) referred to as "meal as object," was what rendered the meal recognizable, familiar, and acceptable. This structural analysis could be criticized as being too rigid to deal with the complexity and changing nature of social relations or too focused on the symbolic aspects of eating. It did, however, provide a useful insight into how food choice is constrained by the eating occasion and what is deemed appropriate (Marshall, 1993). Moreover, Douglas never lost sight of the meal as a physical, as well as a social, event (Beardsworth & Keil, 1997).

Others have focused their attention on the structure of individual dishes. Mintz (1992) identified three food "categories" that constitute the basis of a meal: the

---

[5] As Douglas (1972, p. 275), drawing on the analogy between language and eating, states, "The smallest, meanest meal metonymically figures the structure of the grandest, and each unit of the grand meal figures again the whole meal—or the meanest meal."

"core" or starchy[6] center, the "fringe" or flavor-giving foods, and the "legume" or protein-giving foods. According to Mintz, it was possible to use this simple "core, fringe, legume pattern" (CFLP) to compare meals across cultures. However, he claims that this structure is currently under threat from processed sugars and fats that have "gradually begun to replace the complex carbohydrates of the core and, thereby erode the structure of the meal itself" (Mintz, 1994, p. 113). Koctürk (1995) examined the ways in which immigrants adapt their eating practices when they experience a new culture. She identified three major food groups—"staple" (potatoes, wheat, rice), "complements" (meat/fish/eggs, milk/cheese, vegetables, legumes), and "accessories" (fats, nuts, sweets, fruits, drinks). These, respectively, played a central, secondary, and tertiary role in the construction of a meal. Like Mintz, she focused on the staple, (usually a relatively inexpensive readily available carbohydrate-rich food with a mild or neutral taste) as the most important element in the meal, the "sine qua non," the element that identifies a cuisine. Whereas boiled potatoes were the main staples in North European cuisine, rice was the main staple in Asia. This might help to explain why in Britain a "proper meal" (more on this later), has to contain boiled or roast potatoes rather than chips[7] or rice (one is reminded of the preceding quotes from Reeves and Burnett). These staples were then combined with complementary foods, and accessory foods were used to enhance palatability and presentation (Koctürk, 1995).

What appears to distinguish the British meal is the emphasis on meat, although the main distinction between major and minor meals relies on substituting bread for potato as the main staple (Douglas, 1976). In a recent article in the *Times,* a writer called on British consumers to make "greatest use of staples and deploy meat modestly, as a garnish with fruit and vegetables ad libitum" (Tudge, 1998). One wonders about the likely success of such a strategy in a nation in which "meat" is a powerful metaphor (Fiddes, 1991) and mainstay of the meal.

SOCIAL ORGANIZATION OF EATING

The social aspects of the meal have been touched on in earlier sections in the chapter and underlie much of the previous discussion. They also are featured in Chapters 1, 7, and 9. For Teuteberg (1992) meals embody social rules and corresponding ways of behavior, present opportunities for social interaction, and carry significance far beyond the mere absorption of food. Meals, according to Warde and Martens in their discussion of eating out, "presume a social ordering of dishes, rules and rituals of commensality and forms of companionship" (Warde & Martens, 1999, p. 3). Although this may be more marked in the public sphere of dining out, domestic meals are subject to social constraints that determine what is appropriate, how food is distributed, and the order of service. Moreover, who is admitted and who is excluded from the event can tell us something about the relationship between the eaters. Of course meals do not *always* involve sharing food or eating with other people,[8] but they are essentially perceived as social events (Mäkelä, Roos, &

---

[6] This starchy center is usually based on staples such as bread, rice, potatoes, tortillas, couscous, boiled and mashed yams, taro, manioc, and so on.

[7] French fries.

[8] A large proportion of what we call "meals" involves eating alone. Lunch and breakfast, for example, are more likely to be solitary affairs, and single-person households have little option if they choose to dine at home.

Prättälä , 1995, see also de Castro, 1991). Simmel's (1910) *Sociology of the Meal* highlighted the social forms of interaction that were implicit in (formal) eating occasions and, in doing so, drew attention to the aesthetics of eating. Simmel emphasized the form, or what he called the "aesthetic stylization," of the "formal" meal. But, as Gronow (1996) points out, applying Simmel's ideas in other social settings may prove problematic. In particular, he notes the different type of relationships between participants at a formal meal and those at a family meal[9] where not everyone is equal. The practicalities of feeding the family, catering to different tastes, and teaching children how to "eat properly" are likely to take precedence over matters of social interaction. Indeed, until comparatively recently it was generally accepted that children should be "seen and not heard" when eating at the table.

Murcott (1997), who was primarily interested in what the meal can tell us about social relations, claimed that the conviviality of the family meal might be little more than a myth. She found limited evidence of its existence beyond the somewhat idealized middle class family. As she commented, "an idea of family meals remains that, an idea. As such it is potentially redolent of ideology and social prescription. The idea portrayed is real enough, but that is not to say it is a faithful reproduction of some truth" (1997, p. 15). An examination of Rowntree's (1901) study of the poor in Britain, Ellis's (1983) study of violence in the home, or Morrison's (1996) discussion on commensality might further dispel some of the myth surrounding the domestic conviviality of this eating occasion. Long, in Chapter 9, finds little evidence of social, jovial, or festive meals. However, accepting the error of equating eating together with conviviality, little suggests that the British are no longer eating together.

What evidence there is suggests that we are still eating together. The family meal was declared "alive and well" after 86% of households in one study claimed to eat a main meal together twice a week or more (Planning Consumer Markets, 1994). In another study more than two thirds (73%) of British adults (*n* = 932 adults) reported eating together as a household unit every day/almost every day or most days[10] (Mintel, 1993). Last year more than half (62%) of those questioned in a survey claimed to sit down to eat together at least once a day (Sainsbury, 1998). Eating together appears to be more prevalent at dinner (Marshall, 1991), in those older than 55 (but common in those between 25 and 34 years of age), in two-person households, and among lower socioeconomic groups (Mintel, 1993). Unfortunately, no detail is available on who was present at these communal meals. Research among children and families in Leicester, England, found that families with young children and a single breadwinner were more likely to eat meals together as a family every day (Dickinson & Leader, 1998). Warde and Martens (2000)[11] found

---

[9] Defined here as mother, father, and child(ren) eating together at the dinner table.

[10] In the study 57% reported eating together every day/almost every day, with a further 16% claiming to eat together on most days. Further market research evidence on British meals can be found in the Warren study (1958) carried out over two weekly periods in August 1955 and February 1956 with a representative sample of 4557 respondents. Although there was no direct reporting on who was present at the meal, most men (91%) and women (93%) in the sample had an evening meal at home. Most men (60%) came home from work for their midday meal, whereas others took food to work or ate in the canteen (20%) or ate in a restaurant or cafe (10%). Women (80%) were more likely to eat their mid-day meal at home, with a small number eating at work or in the canteen (10%).

[11] In Phase II of their study, face-to-face interviews were conducted with 1001 individuals in three English cities: London, Bristol, and Preston.

every household member present at the main evening meal in 73% of households containing more than two people. So eating together has not disappeared, but as Gronow (1996) notes, while meal customs are still considered important, they are not necessarily binding. This is reflected in the fact that not all families eat together at every meal or every day for that matter. Gronow goes on to suggest that the "family meal" is likely to become an "empty ritual" and limited as a meaningful basis of social interaction. This is a view held by Falk (1994) who argues that meals have become marginalized in favor of nonritual "oral side involvements" (snacks). Although Falk does not claim that snacks are supplanting meals,[12] a finding vindicated by market research (Taylor Nelson, 1993), he does argue for a greater diversity of meal patterns. As for sociability he argues that "eating together" rather than "where we eat" or "what we eat" is important. As he says "the role of the meal as a community-constituting ritual has been marginalised" (1993, p. 25).

PROPER (FAMILY) MEALS

Much of the discourse about British meals has been centered on the concept of the "proper meal." This has been characterized by one course, ideally hot meat (or less commonly fish), accompanied by potatoes and vegetables and served with a gravy. Furthermore, these are essentially family meals cooked by the woman of the household, where everyone is required to attend and be seated at the table (Charles & Kerr, 1988; Murcott, 1995, 1982). Referring to these as "hot cooked dinner," Murcott (1995) highlighted the central importance of cooking to eating properly: "A real 'cooked dinner' is a meal (plateful) that requires more than one cooking technique, separate preparation of the various elements, all needing regular attention over a long enough cooking period" (1995, p. 229).

Only by applying the correct cooking techniques—meat should be roasted (baked) and the vegetables boiled—can one hope to create a "proper" meal. A survey in the northwest England suggested that the notion of the proper meal was widespread, and among their Manchester respondents, "meat and two veg" had become synonymous with the idea of eating "properly" (Charles & Kerr, 1988). Analysis of our own food diary data collected from households in northeast England suggested that this remained a popular meal format (Gerhardy, Hutchins, & Marshall, 1995). The extent to which this is a general format, extending beyond the traditional family household or even across cultures, is unclear, although some evidence exists of its broader appeal. For example, many young single people claimed to cook "proper meals" rather than rely on convenience or snack foods, although we have no indication of what constituted a proper meal for these individuals or the frequency with which these were prepared (Mintel, 1992). For households affected by divorce or marital separation the "proper meal" may be particularly poignant (Burgoyne & Clarke, 1983; DeVault, 1991).

For British consumers the proper meal is epitomized by the traditional "Sunday dinner." This features a main course of roast beef carved from the joint (it can be roast chicken, pork, or lamb) served hot with roasted or boiled potatoes, vegetables,

---

[12] Personal communication March 1998.

and gravy; as described by Delia Smith[13] at the beginning of this chapter. Structurally, this is a more elaborate form of the mid-week meal and finds its origins in the formal dinner of the nineteenth century,[14] which saw the introduction of courses (see Mennell, 1985). British middle classes adopted this structure and "simplified it for domestic use into the four-course pattern of soup, fish, roast and pudding or dessert" (Oddy & Burnett, 1992, p. 34). This in turn evolved into the familiar threefold pattern of the formal European dinner with the "overture, climax, sweet final flourish" that serves as the basic structure of the less formal family meal (Visser, 1991). Of all the meals eaten at home, Sunday dinner is the one most likely to involve the household eating together (Taylor Nelson, 1990; Warren, 1958).

The introduction of new foods and dishes in what has been called the "routinization of the exotic" appears to threaten this proper meal format: "it is now routine that we should partake of foreign and international cuisines; and also perhaps that we need not necessarily or explicitly Anglicize them" (Warde, 1997, p. 61). Despite the popularity of ethnic food, particularly among younger British consumers, most households consume one ethnic dish a week on average (Taylor Nelson Sofres, 1998). What we have much less evidence on is the extent to which certain groups, particularly younger consumers, are redefining proper meals to accommodate these dishes and others (Kemmer, Anderson, & Marshall, 1998).

This alleged disappearance of the traditional (family) meal or the increasing diversity of meals may, as Gofton (1995) warns, owe more to the decline in the number of traditional family units than any fundamental change in the behaviour of the group itself. It is true that we do not have to eat our food as part of a meal, but meals and the idea of eating "proper meals" remain important, as we found in our study of young Scottish couples setting up home together (Kemmer et al., 1998). For many of these young couples the evening meal symbolized their togetherness and, at least in the first 3 months of marriage or cohabitation, was an important part of setting up a home. As Charles and Kerr discovered in their Manchester study (1988), respondents placed great importance on providing a "proper" meal for their children, even though this was not always achievable. Although the "proper meal" remains an enduring and important aspect of British food habits, it remains focused on one eating occasion, the main meal, and attributed to one type of household unit.

A SPECULATIVE CLASSIFICATION OF BRITISH MEALS

Table 13–1 offers a summary categorization of British meals, which extends the basic meal category to include celebratory, main (weekend/mid-week), light meals, and snacks. It includes the cooking status and processing of ingredients as part of the structure (see Marshall, 1995), along with temporal and social aspects of eating.
Celebratory/festive meals: Easter, Christmas, birthdays, weddings, and christen-

---

[13] Delia Smith's (1995) detailed description of how to prepare a traditional Sunday lunch includes information on cooking times, depending on how one wants the meat cooked (rare, medium rare, well done); when to start the cooking process for serving at 1:00 PM; and the correct temperature settings. It is interesting that she calls Chapter Seven "The Sunday Lunch Revival and Other Meat Dishes."

[14] *Service á la Russe* was attributed to Russian Prince Kourakin in the 1830s in Paris and was characterized by a succession of dishes being served to diners.

Table 13–1 A Speculative Temporal, Structural and Social Typology

| Meal Pattern | Frequency | Meal Structure and Format | | | | | Social Aspects |
		Meal Structure	Course Form	Cooking Status	Ingredients	Typical Meals	Participants
Celebratory/ festive meals	Infrequent— several times a year	A + 2B (formal meal structure)	a + 2b	Hot cooked, prepared by main domestic cook (usually female) with some assistance from other family members or occasionally outside caterers	Special ingredients, mainly fresh or unprocessed, high quality, variety of foods; emphasis on meat	Wedding meal, christening meal, birthday meal	Household, extended family plus guests; normally seated at table; increasingly eaten outside home
Main meals (weekend-Sunday)	Regular— usually once per week (or less frequently)	2A	a + 2b	Hot cooked, prepared by main domestic cook (usually female)	Special ingredients, specific dishes and foods, mainly fresh, good quality; emphasis on meat	Weekend meal, "traditional Sunday lunch" (midday)	Household eating together at home, eating same food, occasionally extended family and guests; seated at table
Main meals (weekday)	Every day, or every couple of days	A (+ B)	a + 2b	Hot cooked, prepared by main domestic cook (usually female)	Familiar foods, generally acceptable, fresh chilled or frozen, good value, some convenience foods such as cooking sauces	Midweek meal, "dinner" (early evening)	Household eating together at home, head of household present, eating same food; seated at table occasionally in front of television
Light meals	Daily (two or three every day)	A	a + b	Hot cooked/cold precooked or uncooked; prepared by main domestic cook or eater	Frozen or canned products more common, convenience items, ready meals, limited preparation or take home items	Midweek or weekend meal, "lunch" (midday), "(high) tea" (afternoon), (or evening meal if main meal eaten outside home)	At home, not all household members present, head of household or children absent, or individuals eating alone; may be eating different foods; not necessarily at table; some eaten outside home
Snack	Most days	—	a	No further cooking; prepared by eater	Emphasis on convenience, processed foods, or takeaway	Anytime, "supper" (late evening)	Individual eating alone or with others but not necessarily eating same food; eating on the move, often eaten outside home

ings are relatively infrequent occasions and celebrate important life course events or mark important points in the year. These meals are highly ritualized and represent the most complex structural and social menus. In Britain, to take one example, the Christmas dinner menu is clearly prescribed with the customary soup, turkey with stuffing, roasted and boiled potatoes, accompanied by Brussels sprouts and carrots followed by Christmas pudding. At least on the 25th of December we can be sure what most of the British public are eating, although the reality may fail to live up to the idealized images of this as a celebration of family life (Lupton, 1996). Increasingly, these events may be celebrated outside the home, in hotels, restaurants, or provided by outside caterers temporarily relieving the domestic cook of the responsibility (see Chapter 9 for a discussion of holiday meals).

Main meals (weekend): Main meals are typified by the weekly Sunday lunch described previously. As the culinary highpoint of the week, this family event requires considerably more time, effort, skills, knowledge, and money than other domestic meals. The format remains relatively unchanged and centered on "the roast" and, like the celebratory event, requires considerably more time commitment. One survey found British consumers spent an average of 106 minutes preparing Sunday dinner (Sainsbury's, 1998). The emphasis is on fresh ingredients and processed foods, which, although often used for the trimmings are not really considered appropriate for the central elements of the meal. Of all the meals eaten at home, this is most likely to have all the family members present.

Main meals (weekday): Main meals for weekdays are best characterized by the "hot cooked dinner" identified in Murcott's (1982, 1983) study of family life in South Wales. This evening meal is another occasion everyone in the household is expected to attend, although this is not always achievable. Structurally these meals conform to the traditional "meat-and-two-vegetable" model but unlike Sunday lunch are restricted to one or perhaps two courses. Increasingly, chilled and frozen foods are an acceptable alternative to fresh produce for midweek meals as convenience products make inroads in this eating occasion. The expectation is that these meals are eaten at the table.

Light meals: Light meals[15] are perhaps the most controversial category. On the basis of market research findings (Taylor Nelson, 1993), they are characterized by fewer courses (usually one), fewer ingredients, and less time for food preparation and cooking. When compared with main meals, the main course potatoes may be replaced by bread, vegetables may be omitted from the dish, or cheaper cuts of meat used as a centerpiece. These meals are closer to what Douglas described as "minor" meals and are more likely to contain more convenient foods and added value products as the centerpiece (canned, dried, chilled, or ready prepared food). On these

---

[15] This is a relatively new classification used by the Taylor Nelson Family Food Panel. The definition of a light meal offered is a "'low status' formal event (i.e. not snacking between meals) which can occur at any time during the day and encompasses Breakfast through Evening Meal"(Taylor Nelson, 1993, p. 11). It could be argued that this is in some ways a forced classification because of the nature of the data collection.

[16] The BBC television's "6:00 News," October 13, 1998, ran a story that claimed microwave(able)-ready meals were killing cooking as younger British consumers looked for more convenient options. Pasta and seafood were the most popular choices for a romantic meal. Around the same time a newspaper article announced that Delia Smith, whose quote introduced this chapter, was scheduled to launch a television cookery program and accompanying book entitled *How to Cook* (The Nation that Can't Boil and Egg, 1998).

occasions less time is devoted to preparation and cooking.[16] Analysis of food diaries from Tyneside households found that 94% of all foods recorded at meals involved no more than 10 minutes preparation time and 51% no preparation at all (Marshall, 1991).

Snacks: Snacks usually involve single food items that require no further preparation or cooking; at most they need to be reheated. Moreover, the person who eats the food is also likely to be the one who buys and prepares it. They currently account for 17% of in-home eating occasions and are more prevalent among children and younger adults (Taylor Nelson Sofres, 1998).

It could be argued that each of these eating occasions represents a social convention with its own timing, location, structure, and rules of appropriateness. As one moves from "celebratory/festive meals" through the classification toward "snacks,"[17] eating occasions become more informal, less structured, more frequent, include fewer courses, fewer items/ingredients, are more likely to be eaten alone, and are more likely to be prepared individually. Less time is required for planning, food preparation, cooking, and eating, and this is reflected in the type of products served. Consequently, the rules regarding content are less rigid, and "light meals" or "snacks" are more likely to accommodate new foods. Less formal eating occasions are not tied to specific places, times of the day, week, or year. Unlike "celebratory" meals or, to a lesser extent, "main" meals, "light" meals and snacks may be eaten in front of the television or on the move. These do not require the forward planning of the special meals, nor do they demand the time and effort to select, prepare, cook, or even eat the food.

The suggestion here is that we are witnessing a change in the relative importance of these different meals and a wider range of meal formats than previously suggested by the "meal/snack" dichotomy. Market research shows breakfast (35%) to be the most frequent meal eaten in British households, followed by lunch (25%), evening meal (24%), and tea (16%) (snacks and lunchbox meals were excluded). However, in terms of servings, the evening meal remained the most important meal, accounting for 35% of all servings. Further investigation into the distribution of foods across meals and the classification of those meals as "light" or "main" meals shows dinner and breakfast were considered main meals, whereas lunch or tea could be classified as either main or light meals. This main/light classification requires further investigation, but the evidence suggests that a growing number of nonevening meals were classified as "light" (Taylor Nelson Sofres, 1998).

Food selection can be tracked on a meal basis, but there is relatively little information on how foods are combined into recognizable formats. Figure 13–2, based on a correspondence analysis of summary food panel data,[18] attempts to do this at an aggregate level. It is derived from a two-way contingency table of the frequency with which particular foods were chosen for evening meal, lunch, and tea (breakfast and snacks were excluded). Analysis was carried out using JMP Statistical software (SAS Institute, 1994). The correspondence analysis was used to identify similarities and differences between the three-meal occasions in terms of what was eaten, and it provides a pictorial display of the data matrix in a two-dimensional plot. Through

---

[17] This is the direction some researchers believe that meals and food consumption are moving in response to broader economic, social, and political changes (Falk, 1994; Gofton, 1995; Mintz, 1985).

[18] Summary consumption data were made available from the 1997 Taylor Nelson Family Food Panel.

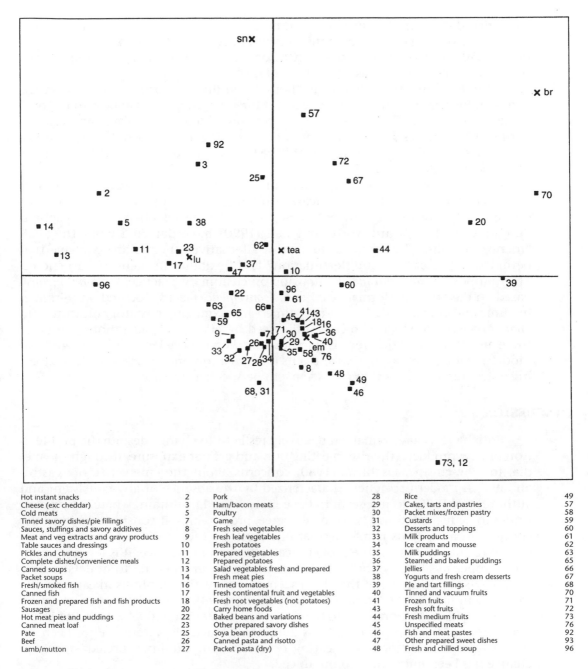

Figure 13–2 Correspondence Analysis Plot of Foods Used at Evening Meals, Lunch, and Tea in Britain. *Note:* br, breakfast; em, evening meal; lu, lunch; sn, snack.

Hot instant snacks	2	Pork	28	Rice	49		
Cheese (exc cheddar)	3	Ham/bacon meats	29	Cakes, tarts and pastries	57		
Cold meats	5	Poultry	30	Packet mixes/frozen pastry	58		
Tinned savory dishes/pie fillings	7	Game	31	Custards	59		
Sauces, stuffings and savory additives	8	Fresh seed vegetables	32	Desserts and toppings	60		
Meat and veg extracts and gravy products	9	Fresh leaf vegetables	33	Milk products	61		
Table sauces and dressings	10	Fresh potatoes	34	Ice cream and mousse	62		
Pickles and chutneys	11	Prepared vegetables	35	Milk puddings	63		
Complete dishes/convenience meals	12	Prepared potatoes	36	Steamed and baked puddings	65		
Canned soups	13	Salad vegetables fresh and prepared	37	Jellies	66		
Packet soups	14	Fresh meat pies	38	Yogurts and fresh cream desserts	67		
Fresh/smoked fish	16	Tinned tomatoes	39	Pie and tart fillings	68		
Canned fish	17	Fresh continental fruit and vegetables	40	Tinned and vacuum fruits	70		
Frozen and prepared fish and fish products	18	Fresh root vegetables (not potatoes)	41	Frozen fruits	71		
Sausages	20	Carry home foods	43	Fresh soft fruits	72		
Hot meat pies and puddings	22	Baked beans and variations	44	Fresh medium fruits	73		
Canned meat loaf	23	Other prepared savory dishes	45	Unspecified meats	76		
Pate	25	Soya bean products	46	Fish and meat pastes	92		
Beef	26	Canned pasta and risotto	47	Other prepared sweet dishes	93		
Lamb/mutton	27	Packet pasta (dry)	48	Fresh and chilled soup	96		

interpretation of the correspondence analysis plot, it is possible to identify those foods most strongly associated with each of the different meals and to speculate on meal formats. One can see from Figure 13–2 that beef, poultry, ham, fish, and vegetables plus potatoes are positioned close together, which lends further support to the existence of the "meat and two veg" platter. However, the close location of pasta/rice dishes and "unspecified" meats to the "evening meal," confirms that these are also being used as main meal dishes. In contrast, products such as canned pasta/risotto and ice cream/mousse are associated with tea, whereas canned meats, cold meats, and soups are associated with lunch.

Although support exists for the "hot cooked dinner" format, the plot reveals considerable variation in what is selected. Moreover, there does appear to be some evidence of an emerging plurality of "proper" British meals. As Warde and Hetherington (1994) suggest, food routine may be becoming more varied. Further support for this was found in a small survey of British undergraduates,[19] in which students were asked to identify the three dishes most often cooked at their parental home. The most popular responses were pasta and sauce (9.7% of responses), chicken with vegetables and potatoes (7.2%), steak with vegetables and potatoes (6.1%), roast dinner (5.3%), fish dish with potato (4.5%), stirfry dishes (4.4%), and spaghetti Bolognese (4.4%). Beardsworth and Keil (1990) have detected what they call "menu[20] pluralism," whereby a number of alternative schemes are available that permit a degree of "menu differentiation" and flexibility in menu (meal) choice. They propose *"traditional menus"* drawing on customary practice, *"rational menus"* based on dietary or performance criteria, *"convenience menus"* focused on minimizing time and effort, *"hedonistic menus"* aimed at maximizing gustatory pleasure, and *"moral menus"* in which choice is influenced by ethical considerations. Whether these are structurally different is questionable. Wood (1995) sees their theoretical model as a long-term aspiration and puts forward a convincing argument for adopting a structural analysis that offers a macro perspective on eating.

DISCUSSION

British meals have remained relatively resilient to change despite the proliferation of new products, the rise in eating out, and greater exposure, through the media, to new cuisine. Fischler's (1980) concerns about the emergence of "gastroanomy" in modern societies, characterized by the absence of any consistent and authoritative rules, do not seem to have materialized in Britain, where eating continues to be regulated and is, to some degree, predictable—if considered within the context of meals. This is manifest, at one level, by the continued importance attached to the ideology of the proper meal, although in practice, it is less common. The "proper meal" is symbolic of British family life and central to understanding how meals are defined, but therein lies a problem. As the emphasis in eating shifts from "communion" to "communication" (Falk, 1994), the social aspects of eating are becoming less important, and trends toward greater individualization and informality render the "proper meal" less dominant. Much less attention has been directed toward other domestic eating occasions, for example, breakfast, where change has been much more pronounced.

However, it is the structural characteristics and the organization and combination of foods into recognizable dishes and courses that provides the cultural charac-

---

[19] This survey was administered to 500 undergraduate students at The University of Edinburgh, Scotland, in 1997.

[20] "Menu" refers here to a set of principles that guide the selection of aliments (any basic item recognized as edible in a nutritional culture) from the available totality (from Beardsworth & Keil, 1997).

ter of eating and highlights the unspoken rules governing our food choice. As Ilmomen (1990) notes, the meal embodies rules of selection, the metaphor (likeness) and metonym (togetherness) that provides the stability in this changing environment. Simply looking at individual foods provides an insufficient understanding of food selection, not only because there is no such thing as a national food, as stated at the beginning of this chapter, but also because other than snacks, most of our meals involve an intricate process of combining foods into recognizable and culturally prescribed meal formats. The distinguishing feature of the meal lies in the combination and presentation of the food. Different combinations, preservation, and cooking processes render different meals in accord with the occasion, and there appears to be considerable flexibility in terms of what is selected (Goode, Curtis & Theophano, 1984). As yet, we have little empirical evidence of what constitutes the emerging plurality of British meals. In a sense it is the structure, rather than the food, that binds us together, and although new foods and variations on old foods (food manufacturers are relatively risk averse) are introduced, meal structures remain relatively intact.

There appears to be something of a shift toward more highly processed food that has been washed, cut, prepared, packaged, and even combined with other foods into ready-to-cook or-eat meals—in short more convenience. Frozen pizza, chilled ready meals, and supermarket take-home-ready-to-eat meals, to name but a few of the options currently available, are finding favor among British consumers on certain occasions. The fact is that food preparation and cooking is now optional, for those who can afford it and who wish to eat in this way. Cooking is becoming a leisure pursuit much to the disgust of some (Independent on Sunday, 1998). In the future our understanding of "proper" may be less concerned about doing the cooking and more with the quality of ingredients, their health characteristics, the nature of the foodstuff (use of pesticides, genetically modified ingredients, use of hormones, or antibiotics, and so forth) as part of the new moral imperative.[21] The fact is that these products do have a place in our meal system and reflect changing attitudes toward what is appropriate. A ready-prepared, "microwaveable," "authentic," ethnic recipe dish that requires no preparation can be cooked and eaten in the same container, avoiding any need to wash pots and pans and is perfectly acceptable while watching a World Cup soccer game (as one British supermarket suggested in its 1998 summer television advertisement campaign).

Warde's (1997) analysis of British eating habits reflects a series of contradictions in food consumption. Yet if one treats "the meal" as a series of subtexts rather than a unitary concept, it is possible to accommodate Warde's modern "antimonies" of convenience and care, health and indulgence, novelty and tradition, and economy and extravagance. There is opportunity to satisfy all these in the diurnal, weekly, and annual pattern of meals. Understanding eating patterns, meal formats, and the changing social nature of eating is the key not only to new product development and product acceptability but also to the adoption of healthful dietary advice, good nutrition, and health promotion. Unfortunately, much of the debate about food choice ignores this contextual influence (Bell & Meiselman, 1995; Meiselman, 1994) as we focus on individual products and consumers. If you still

---

[21] Personal communication with Dr. L. Gofton, May 1998.

need convincing, let us go back to our cookery program. At the end of 20 minutes our television chefs have miraculously transformed the ingredients into something that the viewers now recognize as a meal. How? By adhering to those implicit rules that the audience use every day. We may not always recognize the contents, but the final format is familiar and meaningful.

REFERENCES

Beardsworth, A., & Keil, T. (1990). Putting menu on the agenda. *Sociology, 24* (1), 139–151.

Beardsworth, A., & Keil, T. (1997). *Sociology on the menu: An invitation to the study of food and society*. London: Routledge.

Bell, R., & Meiselman, H. (1995). The role of eating environments in determining food choice. In D.W. Marshall (Ed.), *Food choice and the consumer* (pp. 292–310). Glasgow: Blackie Academic and Professional.

Bell, D., & Valentine, G. (1997). *Consuming geographies: We are where we eat*. London: Routledge.

Burgoyne, J. & Clarke, D. (1983). You are what you eat: Food and family reconstitution. In A. Murcott (Ed.), *The sociology of food and eating* (pp. 152–630). Aldershot, Hants, England: Gower.

Burnett, J. (1979). *Plenty and want: A social history of diet in England from 1815 to the present day*. London: Scolar Press.

Burnett, J. (1995). The decline of a staple food: Bread and the baking industry in Britain, 1890–1990. In A.P. den Harthog (Ed.), *Food technology, science and marketing: European diet in the Twentieth century*. East Linton: Tuckwell Press.

Charles, N., & Kerr, M. (1988). *Women, food and families*. Manchester: Manchester University Press.

Chiva, M. (1997). Cultural aspects of meals and meal frequency. *British Journal of Nutrition, 77* (Suppl. 1), April, S21–S28.

de Castro, J.M. (1991). Social facilitation of the spontaneous meal size of humans occurs on both weekdays and weekends. *Physiology & Behaviour, 49,* 1289–1291.

DeVault, M. (1991). *Feeding the family: The social organisation of sharing as gendered work*. Chicago, IL: University of Chicago Press.

Dickinson, R., & Leader, S. (1998). Ask the family. In *Consuming passions: Food in the age of anxiety* (pp. 122–132). London: Mandolin.

Douglas, M. (1972). Deciphering a meal. *Dædalus, 101*(1), 61–81 (reprinted in *Implicit meanings: Essays in anthropology* (pp. 249–275). 1975. London: Routledge & Kegan Paul.

Douglas, M. (1976). *Culture and food. Russell Sage Foundation Annual Report 1976–77* (pp. 51–58). Reprinted in M. Freilich (Ed.) (1983). *The pleasures of anthropology* (pp. 74–101). New York: Mentor Books.

Douglas, M. (1996). *Thought styles*. Newbury Park, CA: Sage Publications.

Douglas, M., & Nicod, M. (1974). Taking the biscuit: The structure of British meals. *New Society, 19* (30 December), 774.

Ellis, R. (1983). The way to a man's heart: Food in the violent home. In A. Murcott (Ed.), *The sociology of food and eating* (pp. 164–171). Aldershot, Hants, England: Gower.

Falk, P. (1994). *The consuming body*. London: Routledge.

Fiddes, N. (1991). *Meat: A natural symbol*. London: Routledge.

Fischler, C. (1980). Food habits, social change, and the nature/culture dilemma. *Social Science Information* 19(6), 937–953.

Gatenby, S.J., Anderson, A.O., Walker, A.D., Southon, S., & Mela, D.J. (1995). "Meals" and "snacks:" Implications for eating patterns in adults. *Appetite, 24,* 292.

Gerhardy, H., Hutchins, R., & Marshall, D. (1995). Revisiting socio-economic criteria and food choice across meals. *British Food Journal, 97* (8), 24–29.

Gofton, L.R.G. (1995). Convenience and the moral status of consumer practises. In D.W. Marshall (Ed.), *Food choice and the consumer* (pp. 152–181). Glasgow: Blackie Academic and Professional.

Goode, J.G., Curtis, K., & Theophano, J. (1984). In M. Douglas (Ed.), *Food in the social order: Studies in food and festivities in three American communities* (pp. 143–218). New York: Russell Sage Foundation.

Goodman, D. & Redcliffe, M. (1991). *Refashioning nature: Food ecology and culture.* London: Routledge.

Gronow, J. (1996, June 5–9). Regularities of eating. Paper presented at the 5th International Food Choice Conference, St. Louis, MO.

Halliday, M.A.K. (1961). Categories of the theory of grammar. *World Journal of the Linguistic Circle of New York, 17,* 241–291. In Douglas, M. (1972). *Implicit meanings: Essays in anthropology.* London: Routledge & Kegan Paul.

Ilmonen, K. (1990). Food choice in modern society. In J. C. Somogyi & E.H. Koskinen (Eds.), Nutritional adaptation to new lifestyles. *Bibl Nutr Dieta 45,* 30–51.

Independent on Sunday, The nation that can't boil an egg. (1998, May 5). p. 31.

Kemmer, D., Anderson, A., & Marshall, D. (1998). Living together and eating together: Changes in food choices and eating habits during the transition from single to married/cohabiting. *The Sociological Review, 46* (1), 48–72.

Koctürk, T. (1995). Structure and change in food habits. *Scandinavian Journal of Nutrition, 39* (1), 2–4.

Lalonde, M.P. (1992). Deciphering a meal again, or the anthropology of taste. *Social Science Information, 31* (1), 69–86.

Leather, S. (1992). Less money, less choice. In *Your food: Whose choice.* London: HMSO.

Lupton, D. (1996). *Food, the body and the self.* Newbury Park, CA: Sage Publications.

Mäkelä, J., Kjarnes, U., Ekström, M.P., L'orange Fürst, E., Gronow, J., & Holm. L. (1998). Nordic meals. Methodological notes on a comparative study. *Appetite 32,* 73–79.

Mäkelä, J., Roos, E., & Prättälä, R. (1995, April). Ideas of meals in Finland and Kentucky: Results from a collaborative research project. Paper presented at the 4th International Food Choice Conference, Birmingham, AL. In Abstracts of the Fourth Food Choice Conference (1995). *Appetite, 24,* 272.

Marshall, D. (1993). Appropriate meal occasions: Understanding conventions and exploring situational influences on food choice. *The International Review of Retail, Distribution and Consumer Research, 3* (3), 279–301.

Marshall, D. (1995). Eating at home: Meals and food choice. In D. W. Marshall (Ed.), *Food choice and the consumer* (pp. 264–291). Glasgow: Blackie Academic and Professional.

Marshall, D., & Bell, R. (1996). The relative influence of situation and meal occasion on food choice. In T. Worsley & J. Adams (Eds.), *Food choice* (pp. 99–102). London: Eldridge Smith–Gordon.

Marshall, D.W. (1991). *A study of the behavioural variables influencing consumer acceptability of fish and fish products.* Unpublished PhD. Thesis, University of Newcastle-Upon-Tyne.

Meiselman, H. (1994). A measurement scheme for developing institutional products. In H.J. McFie & D.M.H. Thomson (Eds.), *Measurement of food preferences.* London: Blackie.

Mennell, S. (1985). *All manners of food: Eating and taste in England and France from the Middle Ages to present.* Oxford: Basil Blackwell.

Mintel (1992). *Single person households.* Mintel Special Report.

Mintel Leisure Intelligence. (1993). *Cooking and Eating Habits, 4,* 1–29.

Mintz, S. (1992, May 8). A taste of history. *The Higher,* 15–18.

Mintz, S. (1994). Eating and being: What food means. In B. Harriss-White & R. Hoffenburg (Eds.), *Food.* Oxford: Blackwell.

Morrison, M. (1996). Sharing food at home and school: Perspectives on commensality. *Sociological Review, 44* (4), 648–674.

Murcott, A. (1982). On the social significance of the "cooked dinner" in South Wales. *Social Science Information, 21* (4/5), 677–695.

Murcott, A. (1983). It's a pleasure to cook for him: Food mealtime and gender in some South Wales households. In E. Gamarnikow, et al. (Eds.), *The public and the private.* London: Heinneman.

Murcott, A. (1995). Eating at home. In D. Marshall (Ed.), *Food choice and the consumer* (pp. 105–128). Glasgow: Blackie Academic and Professional.

Murcott, A. (1997, January). The lost supper. Perspective, *Times Higher, 31,* p. 15.

Nicod, M. (1979). Gastronomically speaking. In *Nutrition and lifestyles* (pp. 53–65). Conference Proceedings of British Nutrition Foundation. London: Applied Science Publishers.

Oddy, D.J., & Burnett J. (1992). British diet since industrialization: A bibliographic study. In H. J. Teuteberg (Ed.), *European food history: A research review* (pp. 19–44). Leicester: Leicester University Press.

Palmer, A. (1952). Moveable feasts. Oxford: Oxford University Press (Reprinted 1984).

Planning Consumer Markets. (1994). *Consumer Trends, 4,* 64–65.

Prättälä, R., Mäkelä, J., & Roos, G. (1994, November 30–December 4). Lumberjacks and construction workers: Proper meals among working class males in Finland. Paper presented at the 93rd Annual Meeting of the American Anthropological Association, GA.

Reeves, M.P. (1979). *Round about a pound a week.* London: Virago.

Rowntree, B.S. (1901). *Poverty: A study of town life.* London: Macmillan.

Sainsbury's (1998, April). The way we eat now. *The Magazine,* 37–42.

SAS Institute. (1994).

Simmel, G. (1910, October 10). *Soziologie der Mahlzeit.* Berliner Tageblatt. Cited in Gronow, J. (1996, June 5–9). Regularities of eating. Paper presented at the 5th International Food Choice Conference. St. Louis, MO.

Skinner, J.D., Salvetti, N.N., Exell, J.M., Penfield, M.P., & Costello, C.A. (1985). Appalachian adolescents' eating patterns and nutrient intakes. *Journal of the American Dietetic Association, 85,* 1093–1099.

Smith, D. (1995). *Delia Smith's winter collection.* London: BBC Books.

Taylor Nelson (1990, February), What's for breakfast, lunch, tea-time, evening meal: Family Food Panel special report. Taylor Nelson House, 44–46 Upper High Street, Epsom, Surrey KT17 4QS.

Taylor Nelson (1993, November). Light meals: The growth of informal eating occasions: Family Food Panel special report. Taylor Nelson House, 44–46 Upper High Street, Epsom, Surrey KT17 4QS.

Taylor Nelson Sofres (1998). State of the nation: An overview of food consumption trends. Presentation provided by Giles Quick, Director of the Family Food Panel.

Teuteberg, H.J. (1992). Agenda for a comparative European history of diet. In H. J. Teuteberg (Ed.), *European food history: A research review.* Leicester: Leicester University Press.

Tudge, C. (1998, Saturday, March, 28). Farming's future in the balance. *The Times Weekend.*

Visser, M. (1991). *The rituals of dinner: The origins, evolution, eccentricities, and meaning of table manners.* London: Penguin.

Warde, A. (1997). *Food, taste, and consumption.* Newbury Park, CA: Sage Publications.

Warde, A., & Hetherington, M. K. (1994). English households and routine food practises. *Sociological Review, 42* (4), 758–778.

Warde, A., & Martens, L. (2000). *Eating out: Social differentiation, consumption, and pleasure.* Cambridge University Press.

Warren, G.C. (1958). The foods we eat: A survey of meals, their content and chronology by season, day of the week, region, class and age. Survey conducted in Great Britain by the Market Research Division of W. S. Crawford Ltd, Cassell, London.

Wood, R. (1995). *The sociology of the meal.* Edinburgh: Edinburgh University Press.

Part V

Designing and Producing Meals

Chapter 14

Food Service/Catering Restaurant and Institutional Perspectives of the Meal

John S.A. Edwards

Animals feed: man eats: only the man of intellect knows how to eat.
—Brillat-Savarin, 1970

INTRODUCTION

To the caterer (foodservice specialist)[1] meals are equally as difficult to classify. They are generally taken to imply a dish or dishes consumed at any one time (Martin, 1973), although a more classical definition considers them to be the grouping together of various kinds of nourishment taken at a fixed and traditional time (Montagné, 1961).

In the catering industry, the term "menu," first used in 1541 (Martin, 1973), tends to be used interchangeably with the word "meal," where the former is taken to imply a list of dishes available on the day for a meal (Bodenham, 1993) or a list of dishes that are to be served at a meal, (Martin, 1973). Furthermore, a menu is said to specify not only a list of dishes but also the specific order in which they are served in succession at a given meal (Montagné, 1961). This classification is important to the caterer. Routine domestic meals, if not snacks, are likely to consist of a single course (Wilson, 1989). On the other hand, meals eaten outside the home are likely to involve a "starter," "main course," and a "dessert" ("pudding"). In a recent survey,

Exhibit 14–1 Definition of Catering

Catering is defined in the Oxford English Dictionary (1993) as the purveying of food or other requisites, while other authors (for example, Cracknell, Kaufmann, & Nobis, 1985) refer to it as the provision of refreshments in the form of food and drink. In the United States, the term "food service" is favored, although this term has similar connotations in that it is used to describe the provision of food and drink (i.e., meals) for people away from home (Green, Drake, & Sweeney, 1991). "Catering" can perhaps, therefore, be summarized as the provision of food and beverages taken outside the home by people of all ages, in all walks of life, at all times of the day or night, and in every situation (Kinton & Ceserani, 1989).

---

[1] The English (UK) term "caterer" has been used throughout.

Acknowledgments: I am grateful to my colleagues, Joachim Schafheitle and Bill Reeve, for their advice in the preparation of this chapter.

67% of meals eaten in a commercial setting involved more than one course and resulted in more food being consumed than when eating at home. In the same survey, although one fifth claimed to be concerned about weight control, this did not significantly affect the number of courses consumed (Martens, 1997).

The intent then of this chapter is to examine the meal in juxtaposition to the menu from the perspective of the caterer.

A HISTORICAL PERSPECTIVE OF THE MEAL

The type, style, pattern, and timing of meals have changed considerably over the centuries, and, indeed, these patterns have varied between countries, town and country, the nobility, the rich, and the poor. It is not the intent here to provide a detailed chronological account of this development but rather an overview that sets the place of the meal and eating out in history.

According to Brillat-Savarin[2] (1970), meals began with the second age of man; that is, as soon as he stopped living wholly on fruits. Thereafter meat brought the family together and was an occasion at which the spoils of hunting trips could be shared with family and children. Once the human race had spread far and wide, the traveler was able to sustain his journeys by sitting down to what must have been one of these primitive meals. Thus the hospitality industry and the provision of meals out were born.

The development of meals for the traveler continued and in the Christian world would have been provided mainly in monasteries. Thereafter, in Europe, inns and taverns of the early eleventh century would have continued to provide meals out, bread, cheese, and meat for those on the move (Anker & Batta, 1987). Eating out in restaurants, though, is a relatively new phenomenon, and it is estimated that until the middle of the nineteenth century with the growth in rail transport, 90% of the population had never traveled more than 5 miles from their homes. Consequently, they would have had little knowledge and experience of eating out (Tannahill, 1988). Dining out for the middle and working classes was quite unusual, although when single young men needed to live away from home, they would probably have eaten at a pie or cook shop or taken an "ordinary" at a chop-house or public house. Here a slice of meat with vegetables, cheese, and beer would have cost a few pence (Burnett, 1979).

The rich, on the other hand, would have traveled more and eaten out in hotels and inns, although at this period, these generally had a poor reputation and would be used more as a necessity rather than a pleasure (Burnett, 1979). In England, particularly London, "Clubs" would also have been available where the rich could obtain breakfast and other meals.

The middle class in towns might well have taken their midday meal in a "chop-house," but it was not until about 1830 that restaurants became established offering mainly French food. In France, though, restaurants had been established by 1765.

The origin of restaurants is generally attributed to *Monsieur Boulanger*, a soup vendor in the Rue Bailleul, Paris, who named his soups *restaurants* (translated restoratives). In 1765, *M. Boulanger* wanted to increase his menu but was unable to do so because he was not an innkeeper or a member of the corporation of *traiteurs* who were able to sell whole pieces of meat. He therefore offered a dish of sheep's feet in

---

[2] Jean-Athelme Brillat-Savarin, French magistrate, politician, and gastronome, 1755–1826.

a white sauce, thus starting the first restaurant (Montagné, 1961).

However, the first restaurant of note in France, *Beauvilliers*, was established in 1783, but it was not until the French Revolution (1789–1792), which overthrew the aristocracy and abolished "privileges," that eating out in restaurants become popular. Chefs previously employed in the great houses were now available and could establish restaurants or work in hotels, thereby raising standards.

MEAL NAMES AND TIMES

The names and timing of meals have changed over the years and even today vary not only from one country to another but within countries.

In the middle ages (circa 1500) it seems likely that two meals per day would have been the norm. The first "formal" meal of the day would have been dinner, served at around 10:00 AM in the summer and 9:00 AM in the winter. The other "formal" meal was supper, served at around 5:00 PM in the summer and 4:00 PM in the winter. It does, however, seem unlikely that dinner would have been the first "meal" of the day, and foods such as bread and beer would probably have been consumed beforehand.

As the century progressed, meals tended to be served later and later, with dinner slipping to 2:00 PM and supper around 6:00 PM. As a result, breakfast became more and more important until the eighteenth century when it had become firmly established. A similar situation also occurred in France and as *"déjeuner"* (dinner) began to be served later in the day, the first meal became *"petit déjeuner"* (literally translated, small dinner).

However, as dinner was served later, some form of sustenance became increasingly important during the period after breakfast and before dinner. One such meal could also explain the derivation of the word luncheon. During this period the term "nuncheon" had come into existence. This was derived from the Anglo-Saxon word *nōn* meaning noon or midday and *shench* meaning drink. In addition to a drink, probably ale, bread and ale were also provided in the ratio of one loaf of bread and one jug of ale to each worker for "nonsenchis." Gradually the term changed to nuncheons, with the final "s" being dropped in the seventeenth century (Wilson, 1994).

THE EATING OUT OCCASION[3]

Situations in which the caterer has to provide meals are practically limitless. People eat meals out for a number of reasons, and many attempts have been made to classify these occasions. Campbell-Smith (1967), for example, identified 43 separate reasons why people eat out, although Cullen (1994) provides a much simpler distinction of social eating and convenience eating. The former, he suggests, is a means to an end, which must also fulfill a social function if it is to be successful. The latter consists of meals and snacks that enable more time and effort to be devoted to other activities. In practice, social eating is further divided into two categories: the formal social event, which is part of a planned routine and entails dressing up. These meals only adjust slowly to changing circumstances, such as income, and are

---

[3] Eating out refers to food generally eaten outside the home for whatever reasons and provided by a caterer.

probably associated with an older age group. Informal social functions are not connected to any specific activity and dressing up is not involved.

The eating out conundrum though is further compounded for the caterer who must often supply a take-away service or the provision of meals in the home. Including these broader scenarios in the overall thrust of this chapter, eating meals out could be classified into three occasions:

- *Eating out for pleasure*—as a social occasion (both formal and informal), for example, a celebration, convenience, variety, status, culture and tradition, and mood.
- *Eating out at work or for business*—as part of a normal working day or when entertaining business contacts at work.
- *Eating out through necessity*—situations where food is consumed to provide sustenance, a refueling process, and where individuals, given a choice, would perhaps choose not to be, for example, in institutions such as prisons, hospitals, and schools.

However, in all of these situations, the meal should not be considered in isolation but as part of the total eating experience. Whatever the eating out occasion, a number of factors, both tangible and intangible, influence the decision as to where and when to eat. These subsequently affect the enjoyment of the meal and the entire occasion. The influencing factors in this eating out or meal "experience" are summarized in Figure 14–1. These "additional factors" are referred to by scientists working in the field of sensory evaluation as environmental or situational variables that are considered to be significant for food acceptability and consumption (Meiselman, 1996; Meiselman, Hirsch, & Popper, 1988). Social scientists often classify these circumstances as enabling and constraining factors, which include aspects such as the economic ability to be able to afford to eat out (economic access), appropriate social skills (social access), and the levels of provision (Wood, 1990).

THE MEAL PROVISION

In each eating out situation, the caterer can approach the provision of the meal from a number of perspectives (Kotler & Armstrong. 1991):

- The product concept
- The selling concept
- The marketing concept
- The societal marketing concept

The *product concept* assumes that consumers will be receptive to and respond favorably to a "perfect" product. The caterer, using premium quality ingredients and highly trained staff, produces meals that adhere to strict classical guidelines and are perfectly balanced from the gastronomical perspectives such as color, flavor, composition, and presentation. However, the caterer in producing the "perfect" meal may well have lost sight of his customer, whose actual requirements might be somewhat different. He has failed to take into account the meal expectations of the customer (Cardello, 1996).

It could be argued that the move toward healthy eating is part, or an extension, of the product concept. Meals are often planned and produced to conform to nutritional recommendations or healthy eating guidelines in that they contain the cor-

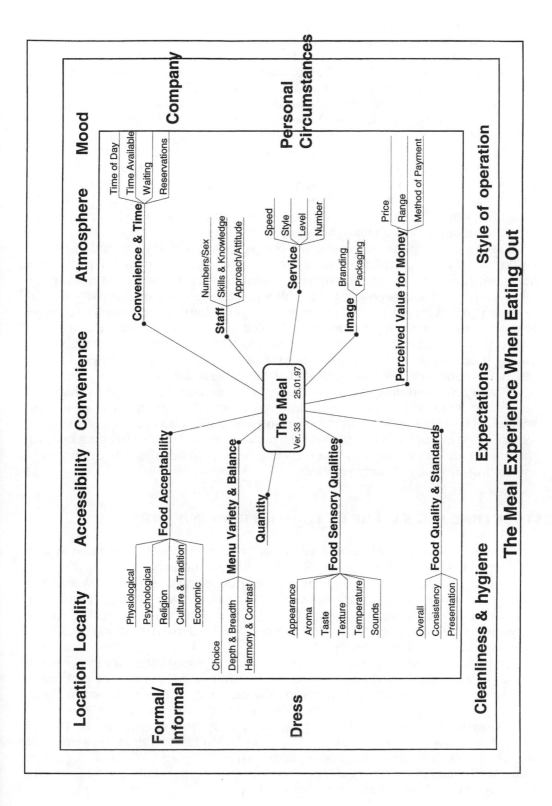

Figure 14-1 The Meal Experience When Eating Out

rect proportion of nutrients such as fat, protein, and carbohydrate. What has been overlooked is that the consumer may not actually like the taste or other characteristics of the finished meal.

The *selling concept* assumes that consumers may not know what they want. They may have certain food likes or dislikes but have no particular views as to what they want to eat. A restaurateur may go some way toward fulfilling or meeting these criteria in the meals provided, but if the sales volume needed for financial success is to be achieved, it may also be required to sell the meals actively or aggressively.

The *marketing concept* focuses on the caterer establishing the precise needs, wants, and expectations of the consumer or the particular market segment he or she wishes to satisfy. This includes not only the composition of the meal but also other factors such as the surroundings under which it is consumed and the price that the consumer would be prepared to pay (the eating out experience). The caterer must then design and produce meals that meet these criteria.

Fast food exemplifies the marketing concept. The meal, the method of presentation, and the entire experience have been very carefully researched. The resulting meal and eating out experience appeals to and satisfies the particular market segment for which it has been designed. Although some promotional activity may be necessary, this is usually limited and directed toward alerting consumers to changes or modifications made in response to perceived changes in the target market.

In practice, most caterers must conform to the marketing concept when producing meals, particularly in the commercial sector, because unless they are able to satisfy the consumer, they are unlikely to be a viable business.

The *societal marketing concept* is similar to the marketing concept in that it attempts to identify the markets' needs and wants but in satisfying those aspirations is more receptive to society and the environment. Meals provided by fast food outlets are again examples of this in that they rapidly respond to the changing needs of the market and, for example, adopt recycling policies to demonstrate or be seen to use environmentally friendly practices.

THE STRUCTURE AND NATURE OF THE CATERING INDUSTRY

The catering industry is composed of a diverse range of outlets operated by staff with mixed skills and levels of ability. It is also an industry that is highly fragmented, and while there are a large number of groups and chain organizations, the main provision of meals is undertaken in small, single, owner-operated and managed units varying in size and style. This is reflected, for example, in the United Kingdom, where 86% of all catering businesses have less than £250,000 annual turnover (Keynote Market Review, 1994).

In most Western economies, the catering industry and the provision of meals eaten out are major contributors to national economic prosperity. The amount of money spent per head on meals eaten out in various countries is shown in Figure 14–2. By far the largest amount of money is spent in the United States (£790), where it is estimated that people eat a mid-day meal away from home, on average, 2.5 times per week, an evening meal 1.5 times, and breakfast less than once per week (Dulen, 1998). Australasia, Singapore, Japan, Hong Kong, and Australia along with Canada all spend in excess of £500 per head on eating out. What is perhaps surprising is the relatively lower amounts spent in Europe.

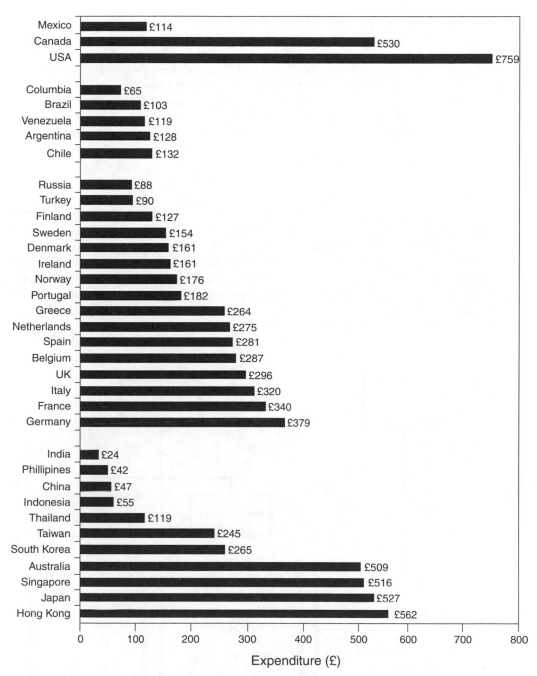

Figure 14–2 Estimated Annual Expenditure per Head on Food Consumed Outside the Home. Source: Data from Foodservice Intelligence, 1998.

It is interesting to question: if people can afford to eat out, do they eat out more because it is cheaper?

Because of its diverse nature, estimates as to the size of the catering industry vary. Worldwide, the hospitality industry, of which the catering industry is a major part, employs 212 million people; one in nine jobs, or 10.7% of the global workforce. It has a global expenditure of $646 billion or 10.9% of world gross domestic product (Henley Centre, 1996). Furthermore, the industry has a significant multiplier effect through the requirement for supporting products and services, for example, the provision of food, beverages, and the supply of services such as laundry, cleaning, and financial services. The resulting multiplier effect is estimated to create approximately 1.3 additional jobs for every one in the hospitality industry and in financial terms £1.75 for every £1.00 spent (Henley Centre, 1996).

The catering industry can be classified in a number of ways, and various attempts have been made. One common and widely accepted classification is the divide between the "commercial" ("profit") sector and the "public" or ("cost") sector. This breakdown is shown schematically in Figure 14–3.

The range and nature of these establishments is diverse, but the factor common to all is that they provide meals for consumption either on the premises or to take away. In addition, the distinctions between the two sectors have become blurred and more difficult to define because they increasingly overlap in their approach to business and the quality and standard of the meals they provide. Furthermore, increased competition in both sectors has encouraged the adoption of newer technologies in areas such as food preparation and cost control to bring about financial savings and to gain a competitive advantage.

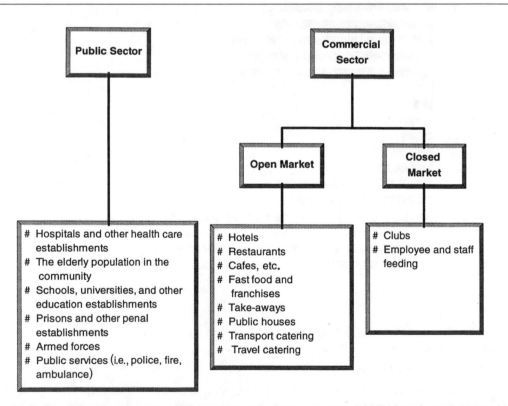

Figure 14–3 Possible Classification of the Catering Industry. *Source:* Adapted with permission from B. Davis and S. Stone, *Food and Beverage Management*, p. 4, © 1992, Butterworth-Heinemann Publishers, a division of Reed Educational & Professional Publishing, Ltd.

The Commercial Sector

The predominant components of the commercial sector are hotels and restaurants, but even here the classification is difficult. The term "restaurant" is invariably used interchangeably with other establishments, such as cafes, snack bars, bistros, wine bars, and public houses. In all cases, though, the primary objective or purpose is to provide meals and other refreshments and to be financially viable. Increasingly, although many hotel restaurants are profitable, they have probably lost their share of the local eating out market (Mintel Marketing Intelligence, 1998). Many hotel restaurants operate at a loss, and profit is generated by the rooms division (British Hospitality Association, 1998); the restaurant is there to provide a service for hotel guests.

The range of "restaurants" is extensive, and most Western countries, in addition to their own cuisine, offer and cater to a range of national and ethnic tastes. The nature of the meals provided also varies from perhaps a simple sandwich, consumed on or off the premises, to a meal consisting of a number of courses. The manner and style in which meals are selected also varies and includes a variety of menus such as table d'hôte, à la carte, or selective menus. The style of service also varies from gueridon and silver service to plated, family, self-, or buffet service.

The Public or Cost Sector

The public or cost sector is composed of outlets where meals are provided at cost or at a subsidized rate and, in general, where the provision of a meal is not the primary function of the organization. It may, however, form an integral part of the overall service of the business and includes establishments such as hospitals, schools, prisons, the armed forces, and the workplace. The image of this sector has, until recently, been one of "second-class meals," where food is perceived as unimaginative and "stodgy." It is also perceived as being prepared and served by staff who lack professional skills, are poorly motivated, and where meals are served in surroundings that do little to enhance the overall meal experience. This situation, however, is changing, and the underlying trend appears to be toward greater sophistication in the market, with companies becoming more customer focused and investing in quality programs (Keynote Report, 1994).

In both the commercial and cost sectors, meals can be provided either by organizations using their own in-house staff or by the use of specialist contract caterers (Food Service Management Company[4]). In the case of the latter, contractors, under the terms of their contract, manage and control the preparation and service of meals (including beverages) and any other services that may be required.

The overall provision of meals in the various sectors within certain countries of Europe is summarized in Table 14–1.

PREPARATION OF THE MEAL

It has often been stated (for example, Glew, Lawson, & Hunt, 1987) that the "ideal meal" is one that has been freshly prepared using high-quality raw ingredi-

---

[4]The term "food service management company" is generally preferred within the contract catering industry (British Hospitality Association, 1995).

Table 14–1 Number of Meals (Millions) Served in Certain Countries of Europe in 1995

Sector	Belgium	France	Germany	Italy	Netherlands	Spain	UK	Total
Commercial								
Hotels	84	630	1,100	773	82	569	563	3,801
Restaurants	112	1,098	1,563	1,643	221	812	422	5,871
Fast Food	97	420	701	86	63	177	567	2,111
Cafes/take aways	138	704	2,248	1,139	596	946	1,108	6,879
Bars/pubs	109	300	765	228	113	298	1,163	2,976
Travel	45	194	195	254	77	282	427	1,474
Leisure	115	352	651	740	165	334	1,102	3,459
Total commercial	700	3,698	7,223	4,863	1,317	3,418	5,352	26,571
Cost								
Staff catering	246	633	2,977	538	225	307	1,327	6,253
Health care	238	899	1,570	708	255	602	758	5,030
Education	93	910	475	458	68	292	897	3,193
Services	81	351	511	339	112	305	149	1,848
Total cost	658	2,793	5,533	2,043	660	1,506	3,131	16,324
Grand total	1,358	6,491	12,756	6,906	1,977	4,924	8,483	42,895

Source: Data from Foodservice Intelligence, 1998.

ents and then served immediately in appropriate surroundings. This is usually referred to as the traditional or cook-and-serve system. However, for a caterer to achieve this requires highly skilled staff with imagination and flair and, because of the fluctuations in demand, large numbers of staff. Inevitably it results in a meal prepared and served in this way being beyond the reach of most individuals eating out for pleasure and certainly for those eating out at work or through necessity.

Overcoming this can, in part at least, be achieved by using less fresh ingredients and relying more on ready prepared and convenience foods. Although this removes much of the "drudgery" and many of the low-skill operations from the kitchen, it only partially helps to reduce the reliance on a skilled labor force and similarly fails to remove the peaks and troughs associated with the production and service of fresh food. The meal needs to be prepared and then "held" at the appropriate temperature, either hot or cold, until required for service. If not served immediately after preparation, deterioration in quality and nutritional profile quickly ensues, and the risks from microbiological contamination are increased.

This situation was partly addressed by the introduction of cook-freeze in the 1960s, cook-chill in the 1970s, and sous vide in the 1980s. The primary purpose of these "technology"-driven systems (albeit "low tech") was to separate production of the meal from its service and in doing so overcome some of the difficulties and to create a number of further advantages. Using cook-chill and cook-freeze techniques, meals or the component parts of meals can be prepared centrally using "factory" production and assembly techniques. They are then either blast frozen or chilled and stored for up to 1 year or 5 days, respectively. When required for service, they are taken to the outlet where they are to be consumed, defrosted if necessary, regenerated, garnished, and served. This process is shown schematically in Figure 14–4.

A number of benefits have been attributed to these systems, including the following:

- The centralization of the majority of the skilled staff into one location where meal production can take place enabling:
  - The number of staff, particularly skilled staff, to be reduced
  - The use of factory-style production techniques, which improves productivity, eliminates some of the bad craft practices and control over waste, and permits greater uniformity of standards across a range of meals and dishes
- 24-hour availability of meals that only require to be regenerated and served at the point of consumption
- Elimination of peaks and troughs in meal preparation, which can be produced centrally in "factory runs" in more social hours
- Greater control over aspects such as menu variety, uniformity of standards, and presentation
- Improved hygiene practices
- Greater use of automation and efficient large-scale equipment
- Less duplication of meal preparation equipment at the point of consumption
- Centralized and therefore more appropriate and effective purchasing, storage and control procedures
- Control over menu variety

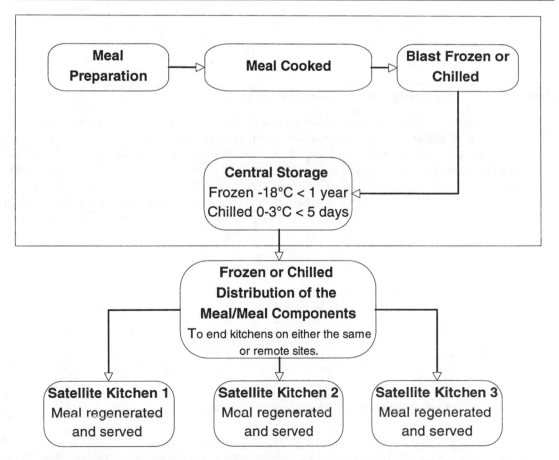

Figure 14–4 Schematic Representation of the Cook-Freeze and Cook-Chill Catering System

However, a number of disadvantages also accrue using these systems, including the following:

- Greater energy use in the preparation, chilling, transportation, and subsequent regeneration.
- Not all dishes suitable for these processes.
- A perception of lower quality when the system is used.
- Some recipes and methods need to be modified.
- High capital costs needed to establish a central production unit.
- More difficult to personalize meals for individual requirements.
- Greater risk of food contamination if careful hygiene practices are not used.

A more recent development in meal preparation is an extension of the cook-chill system, namely sous vide. The term (translated *under empty*) refers to a process adapted from two other processes patented in the United States in the early 1970s and developed in France in collaboration with an American packaging company. The original rationale centered on weight loss in *paté de foie gras*, although in early research (Pralus, 1985) it was suggested that flavor retention and textural qualities were also heightened. In this system, raw ingredients are prepared, blanched, or browned as necessary; the food is then inserted into heat-resistant "plastic" pouches, a vacuum drawn, and the food cooked to a predetermined temperature. This is followed by rapid chilling and, depending on the nature of the cooking process, is treated in a similar manner to cook-chill, and the food is stored in excess of 5 days.

Prevalence of Catering Systems Today

The prevalence of these four primary meal production systems is difficult to determine because few studies have been undertaken. The most recent study of hospitals in the United Kingdom was undertaken between 1994 and 1995 and included hospitals with more than 100 beds (Ervin & Edwards, 1995). Results show that 67.9% of hospitals use a traditional cook and serve system, 24.8% use a cook-chill system, and 4.1% use a cook-freeze system; the remainder use a combination of the systems. Results of a survey in the United States, again in hospitals with more that 100 beds, are shown in Table 14–2.

Table 14–2 Characteristics of Catering Systems in US Hospitals

Hospital Size (No. of Beds)	Conventional		Cook-Chill		Cook-Freeze		Other		Total
	N	%	N	%	N	%	N	%	N
<200	600	95.7	17	2.7	6	1.0	4	0.6	627
200–299	421	92.5	16	3.5	5	1.0	13	2.9	455
300–499	391	86.5	37	8.2	11	2.4	13	2.9	452
500 +	238	73.0	67	20.6	6	1.8	15	4.6	326
Total	1,650	88.7	137	7.4	28	1.5	45	2.5	1,860

Notes: Percentages may not total 100 because of rounding.

Source: Adapted with permission from M.F. Nettles and M.B. Gregoire, Operational Characteristics of Hospital Foodservice Departments with Conventional, Cook-Chill and Cook-Freeze Systems, *Journal of the American Dietetic Association*, Vol. 93, No. 10, pp. 1161–1163, © 1993, American Dietetic Association.

As can be seen, although only 7.4% of those hospitals surveyed have adopted a cook-chill system, approximately one fifth of them are large hospitals with more than 500 beds. Few hospitals have adopted a cook-freeze system (1.5%), and most hospitals (88.7%) continue with their traditional system. It is also interesting to note that the majority of hospitals that have installed cook-freeze systems did so before 1982, whereas the installation of cook-chill systems is spread over the entire period and shows little sign of slowing.

STYLES OF SERVICE FOR THE MEAL

Styles of service of the meal have also changed throughout the ages. Until the early part of the nineteenth century, a meal would have consisted of two, perhaps three, courses (stages or services), *Service à la Français* or French Service, with each course consisting of a number of dishes. The first course (*entrée*) might have consisted of soup, hors d'oeuvres, hot entrées, vegetables, and salads and would have been laid out symmetrically on the table before diners entered the room. In addition to the food, elaborate centerpieces, silver, and other artefacts would also have been incorporated into the table decoration to create maximum impact. Hot dishes may have been kept warm with metal or earthenware containers (*timbales*), although dishes would often be allowed to go cold. Diners helped themselves to food but, because they could not always reach every dish, often had to restrict their meal to foods that were within easy reach. Once diners had eaten sufficient of this course, a second course (*remove*) was then brought in. This consisted of lighter meats but laid out in a similar symmetrical style. The delineation between what was served in the first and second course was vague and depended to a great extent on what was available (Brears, Black, Corbishley, Renfrew, & Stead, 1997). The third course, if not included in the second course, was composed of fruit, sweets, and cheese.

This style of service lasted in England until 1856 when a "new" style, *Service à la Russe* or Russian Service, was introduced. The introduction of this style of service is generally attributed to Felix Urbain-Dubois, a French chef who had worked at the Court of the Czar of Russia (Cracknell & Nobis, 1985). Although the structure of the meal remained largely unaltered, the presentation and service of the meal changed. In place of dishes set out on the table before diners entered the room, each course, now consisting of one dish, was brought into the dining room where it was offered to diners. Diners, using a spoon and fork, then helped themselves. This style of service had a number of obvious advantages, not the least of which was to ensure that food was served at the correct temperature, diners could taste all the dishes that had been prepared, and waste was reduced considerably. The change in service also permitted the introduction of menus such as table d'hôte and à la carte, the introduction of new implements such as carving and other trolleys, and a more cost-effective use of skills. All these contributed to the enjoyment of the meal (Cracknell & Nobis, 1985). Subsequent refinements, with waiters assisting the diner with the service of the food, provided faster service, and the food was transferred to the diner's plate without "destroying" the dish too much.

In the United States, the introduction of the "blue-plate special" during the 1920s provided a further style of service. Instead of the various dishes for a main meal being served onto the plate by a waiter, individual foods were placed into separate compartments of an oval platter in the kitchen. The platter was then brought out into the restaurant and put down in front of the diner. This style of service proved to be extremely popular and led ultimately to the introduction of plated service (Cracknell & Nobis, 1985). In its original form, plated service was restricted to the cheaper end of the market, where fewer skills were needed in the service of the meal. The time taken for the meal occasion was also reduced, and portion control was easier to exercise. More recently, plated service has become extremely popular at the top end of the market with the introduction of *cuisine nouvelle*. Here the plate, normally larger than previously used, forms the centerpiece on which the food can be artistically arranged. Because the waiter's serving duties are largely confined to placing the plate down in front of the diner, the artistic presentation is not disturbed.

Other variations used by caterers in the service of the meal include various styles of buffet and self-service, for example, finger, fork, modern, and classical buffets, and a drive-through service popular with fast food outlets.

THE MALE–FEMALE DIVIDE WHEN EATING OUT

Although considerable literature has been written and research done on gender differences in food habits, little empirical research has been undertaken into eating out (Martens, 1997). Many studies into eating at home still consider the cooked meal to be a male preserve (see Chapter 13). Eating out, therefore, despite commercial cuisine being inherently "masculinist" (Wood, 1990), provides women the opportunity to choose foods they prefer (Wood, 1992).

Men are more likely to eat a starter than women, who in turn are more likely to eat a dessert. When asked what they had chosen for their main course when they last ate out, although a similar number of men and women ordered meat for their main course, 58% of men chose beef compared with 42% of women. On the other hand, 37% of men chose chicken compared with 63% of women. Women also consumed more fish, 58% compared with 42% (Martens, 1997).

In many countries, women eat out as frequently as their male counterparts, although they do not always receive similar treatment. They are often regarded as being fussy, unlikely to be high spenders, and probably poor tippers. In effect, despite being a valuable market segment in their own right, they have been marginalized by the hospitality industry and are treated as either an appendage to a male client or one element within the family (Wood, 1992).

LEGAL CONSIDERATIONS

In many countries, the meal has a number of legal connotations that must be taken into consideration by the caterer in the provision of a meal. In the United Kingdom, for example, there is a legal obligation on the part of the restaurateur not to mislead customers or potential customers with inaccurate descriptions. The menu, therefore, needs to accurately reflect and describe the food or meal being served. Prices of both food and drink must also be displayed, and there is an obligation for those prices not to be misleading. When a dish is described as "homemade," it should be prepared from basic ingredients in a domestic-style kitchen and not food brought in having been prepared under factory-like conditions (Pannett & Boella, 1996).

In the "purchase" of a meal, the menu, in a more formal restaurant, is merely an invitation to treat and has no legal effect. What the customer selects to make up his or her meal then becomes an offer to contract. The server in taking the order, and the items being available, accepts the offer, and a legally binding contract to supply the meal is made. If the meal items are not available, no contract is formed (Pannett & Boella, 1996).

Where the sale of alcoholic beverages takes place, a liquor license is required, and where a meal is served this could be a "restaurant license." However, a license can only be granted and a meal served where the "premises are structurally adapted and bona fide used... providing the customary main meal at midday or in the evening...." Where this applies, meals can only be served to people seated at a table, although in this context, a table could be a counter or other similar fixture (Field, 1988). The term "customary meal" is difficult to define and has also been the subject of legal argument; a customary meal may well be a sandwich or other similar snack. In general, a meal refers to the main or substantial meal of the day.

Food hygiene and food safety are becoming increasingly important in the preparation and service of a meal. The reported incidence of bacterial food poisoning, for example, *Salmonella*, has increased, and public concern was one of the main reasons for the enactment of food legislation, such as the Food Safety Act of 1990 (Sprenger, 1998). It is essential that the caterer be aware of and comply with this legislation and, where necessary, show due diligence in the procedures used.

CHANGES IN MEAL PATTERNS WHEN EATING OUT

A number of changes have taken place when eating out, and Table 14–3 summarizes current and forecast changes in eating out in Europe. As can be seen, growth in the commercial sector increases by 14% up to the year 2000, with fast food estimated to grow even faster (30%). However, little growth is forecast in the cost sector.

Table 14–3 Recent and Forecasted Changes in Eating Out in Europe (Millions of Meals per Year)

Sector	Year and Percentage Change				
	1990 M	1995 M	% Change 1990–1995	2000 M	% Change 1995–2000
Commercial					
Hotels	3,701	3,801	3	4,478	18
Restaurants	5,530	5,871	6	6,699	14
Fast food	1,596	2,111	32	2,737	30
Cafes/take aways	6,565	6,879	5	7,516	9
Bars/pubs	3,372	2,976	−13	3,354	13
Travel	1,360	1,474	8	1,635	11
Leisure	3,074	3,459	13	3,973	15
Total sector	25,198	26,571	5	30,392	14
Cost					
Staff catering	6,485	6,253	−4	6,308	1
Health care	5,098	5,030	−1	5,069	1
Education	3,287	3,193	−3	3,153	−1
Services	1,874	1,848	−2	1,822	−1
Total sector	16,744	16,324	−3	16,352	0
Totals	41,942	42,895	2	46,744	9

Source: Data from Foodservice Intelligence, 1998.

Changes in Eating Out for Pleasure

Increases in the standard of living have led, in most countries, to a rise in the number of people eating meals out for pleasure. In the United Kingdom, eating out as a social occasion has increased considerably, with 32% of the population claiming to eat out at least once per week. Weekly, men tend to eat out more than women, 37% compared with 25%; 61% of those less than 31 years old compared with 33% for those older than 31 years old; 54% single, compared with 39% for married couples; and 57% of those from the North compared with 41% from the South (Roche, 1997). Although English food remains the most popular single national category of food, ethnic or foreign cuisine is now the most popular (Table 14–4). In the United States, fast and convenient appear to be key, with 44% claiming to eat out at fast food restaurants once per week or more (Table 14–5) (Dulen, 1998).

In addition, the British, when eating out, have traditionally been rather reticent to complain if the occasion has not been satisfactory. However, this seems to be changing, and 61% say they complain if they are not satisfied compared with 28% 5 years ago. Older groups at the higher end of the socioeconomic scale are most likely to complain (Roche, 1997). Furthermore, although most people regard eating out as being a pleasurable occasion, it can, nevertheless, be extremely stressful for a number of reasons (Roche, 1997):

- Dealing with the wine waiter
- Complaining—for yourself or on behalf of a member of your party
- Having someone else speak for you to the staff
- Being outfaced by the waiter in front of friends
- Being embarrassed by ignorance or lack of knowledge of dining etiquette
- Being served by unfriendly staff or receiving bad service

Table 14–4 The Popularity of Ethnic and National Cuisines in the United Kingdom

Type of Cuisine	Percentage Choosing
Traditional British	46
Chinese	34
Italian	28
Indian	27
Fast Food	14
Mexican	8
Mediterranean	5
Other	2

Source: Reprinted with permission from Roche, *Eating into the Millenium: The Rennie Report,* © 1997, Roche Products, Ltd.

Table 14–5 The Popularity of Eating Styles in the United States

Where Eating Out	Quick Service Restaurants (%)	Casual/Family (%)	Fine Dining (%)
> Once per week	25	15	1
Once per week	19	20	3
2–3 times per month	18	23	6
Once per month	10	18	14
< Once per month	13	17	43
Never	6	2	16

Source: Reprinted with permission from J. Dulen, Changing Taste, *Restaurants and Institutions,* Vol. 108, No. 3, pp. 58–69, © 1998, Resturants and Institutions.

- Tipping
- Not knowing in advance how much the bill is going to be
- Being made to feel uncomfortable in excessively formal restaurants
- Feeling the need to dress up

Changes in Eating Out at Work or for Business

Changes in working habits have also influenced the provision of meals in the workplace. In the United Kingdom at the turn of the decade, 73% of the workforce claimed to take a meal break every day at midday; by 1997, this had declined to 44%. Over the same period the number of people claiming not to take a meal break at all at midday has risen from 7% to 29% (Compass, 1997).

The time available for a meal break in the United Kingdom has also decreased in the decade; in 1990 39% of the workforce had less than half an hour for their midday meal break, and by 1997 this had risen to 44%. The number of workers reporting working through their meal break has also increased from 17% to 20% during the same period (Compass, 1997).

It is interesting to compare these figures with other European countries (Table 14–6) where, due in part by culture, meal breaks at midday are somewhat longer.

Patterns of alcohol consumption at the midday meal break have also changed, and 83% of the workforce claim never to drink at lunchtime compared with 68% in 1990.

As a result of these changes, the style and pattern of the midday meal has also changed, and more workers bring a packed meal to the workplace than ever before (Table 14–7).

Table 14–6 Time Taken for Midday Meal Breaks in Europe

Country	Length of Midday Meal Break and Percentage of Respondents				
	< ½ hr	½–1 hr	1–2 hr	> 2 hr	Don't know
United Kingdom	39	55	5	—	1
France	27	48	22	1	2
Germany	47	36	9	1	7
Italy	15	28	35	12	10

Source: Adapted with permission from Compass Services, A Comparison of Working Lunchtime Habits in Europe, *European Lunchtime Report,* © 1992, Compass Group, U.K.

Table 14–7 Changes in the Type of Meal Consumed at Work in the United Kingdom

Style of Meal	Percentage of Workers
Bring in packed meal	45
Purchase from sandwich bar	19
Use staff restaurant	18
Purchase from fast food/take away	8
Purchase from supermarket/store	8
Use wine bar/pub	6
Use a restaurant	4

Source: Adapted with permission from Eurest, The Sixth Annual Survey of Lunchtime Habits, *Eurest Lunchtime Report,* © 1997, Compass Group, U.K.

Despite these changes, the provision of subsidized meals remains extremely important in the United Kingdom as shown in Table 14–8.

RESTAURANT AND INSTITUTIONAL PERSPECTIVES OF THE MEAL—THE FUTURE

Although eating out is a relatively new phenomenon, it is one that has grown in popularity over the past 20 years and looks set to continue in the commercial sector but remain relatively stable in the cost sector (see Table 14–3). In the United States, 40% to 45% of all food dollars are spent on eating out, a figure that is likely to increase to 50% to 55% by the year 2005 (Sloan, 1998). The increase in eating out world wide can be attributed to a number of factors, including the following:

- A growing population
- An aging population
- Increased free time, caused in part by
  - A lowering of the age at which people cease full-time work
  - Technology
- A higher disposable income
- A desire to spend this free time and higher disposable income on travel, leisure, and other pursuits outside the home

Increases in the number of people eating out will in itself generate change, and in a fiercely competitive market the need to be aware of changes and developments will become paramount. In response to this and to take advantage of this increased demand, there will undoubtedly be themes, styles, and fads that rise and fall. See for example Table 14–9.

Similarly, the demand for ethnic and specialty restaurants will wax and wane, but what are the underlying criteria that need to be considered? Many of these factors are interrelated and interdependent, but it is perhaps important to be aware of how they might impinge and impact on the eating out experience.

Table 14–8 The Importance of Worker Benefits

| | Workers Rating Very and Fairly Important | | | | | | | | | |
| | UK 1997 | | UK 1990 | | F 1990 | | D 1990 | | I 1990 | |
Worker Benefit	%	P	%	P	%	P	%	P	%	P
Pension plan	88	1	85	1	89	3	80	1	78	=2
Private health insurance	67	2	66	2	96	1	64	4	59	5
Flexible hours	—	—	62	3	85	4	71	=2	78	=2
Subsidized travel	54	4	—	—	—	—	—	—	—	—
Free or subsidized meal	58	3	46	4	74	5	53	6	76	4
Crèche or child care facilities	52	5	40	5	90	2	71	=2	85	1
Company car	32	6	36	6	53	6	38	7	53	6
On-Site leisure facilities	30	7	38	7	49	7	60	5	49	7

Note: P = Position; F = France; D = Germany; I = Italy.

Source: Adapted with permission from Compass Services, A Comparison of Working Lunchtime Habits in Europe, *European Lunchtime Report,* © 1992, Compass Group, U.K.

Table 14–9 Increases in the Number of Dishes To Be Ordered in the Coming Year in the United States

Dish	% of Respondents
Hamburgers	15
Pizza	14
Baked potatoes	12
Side salads	12
Fish	11
Steak	11
Entrée salads	10
French fries	10
Grilled/broiled chicken	10
Soup	8

Source: Reprinted with permission from J. Dulen, Changing Taste, *Restaurants and Institutions,* Vol. 108, No. 3, pp. 58–69, © 1998, Restaurants and Institutions.

Changes in the Availability of "New" Foods

Perhaps one of the biggest changes or developments in the eating out experience is likely to be the introduction of new and genetically modified foods. In the 1960s and the 1970s, the "green revolution" increased yields of crops such as wheat by 100% (Kleiner, 1996). This rapid increase in yield has declined dramatically, but if the growth required to feed an increasing world population is to be sustained, alternative techniques must be sought and found.

Genetic modification of both crops and animals appears to be one solution. Genetic modification has the ability, for example, to increase a crop's yield, produce a "new" crop or species, or reduce resistance to fungal attack (Coghlan, 1998). Other developments are also taking place; for example, developments in humidity and temperature control combined with two "sterilizers," ozone and negative air ions, have enabled the shelf life of fruit to be extended for up to 5 months (Hadfield, 1998).

More Casual and Family Dining

Although there will always be a requirement for "upscale" restaurants that use primarily high-quality ingredients and cook-and-serve production techniques to prepare meals, the move is likely to be away from this style, primarily because of high costs and skilled labor, the latter invariably being in short supply.

To overcome these challenges, the trend may well be more toward the "de-skilling" of food preparation and food service by using pre-prepared foods and processes such as cook-freeze, cook-chill, sous vide, or a hybrid of these. The craft and skill of food preparation, the culinary art, will be "operationalized" in that it will be modified into "food assembly." Operatives who are unskilled or semi-skilled will assemble meals from pre-prepared, pre-portioned ingredients. Combined with the introduction of "newer" technology-driven systems, this will also enable the caterer to offer a greater variety of meals while maintaining uniform standards.

However, as eating out becomes the "norm," meals are likely to be "simpler" in style, service, and the environment in which they are served. This process is relatively well established in the United States, but elsewhere it is likely to increase if the rising numbers are to be satisfied. Increasing emphasis on aspects such as food quality, food hygiene, and control procedures are further likely to lead to the standardization of the meal experience.

As eating at home continues to be less family oriented (single parent families) and casual, this will have parallel consequences. As prices are reduced, eating out will fall within the reach of more families, which in itself will create greater availability of eating out opportunities and longer opening hours. Meal replacement, food from supermarkets to rival that available from restaurants, is readily available in the United States and is a growing market. It is not, though, always regarded as superior to similar restaurant food, with 34% considering it to be "somewhat worse" and 45% about "the same" (see Figure 14–5).

Increase in the Number of Outlets

The catering industry is generally regarded as being an industry that is easy to get into and out of and will continue to attract the entrepreneur and those individuals wanting to work for themselves. Preparing and serving meals is often seen as an attractive second career and requires limited skills and capital to establish a business. Unfortunately though, this relative ease of entry is also the primary cause of failure. Helped by the growth in franchise opportunities, where failures are considerably lower, the number of outlets must be set to increase.

Against this, however, there will be a number of negative forces. A lowering of the alcohol limits for driving and the continued rigid enforcement will necessitate people reconsidering their eating out patterns and either changing their style of dining, consuming less alcohol, or reconsidering the location of the establishment. The latter will also be influenced if the encouragement to be less reliant on private motor vehicles is successful.

Notwithstanding, eating meals at home, until recently, was the standard or norm for most people, and meals were only consumed away from the home on specific occasions. This situation has now changed, and people eat out for a number of reasons, including pleasure, business, or through necessity. World wide the consumption of meals out of the home is set to increase further, and caterers must be

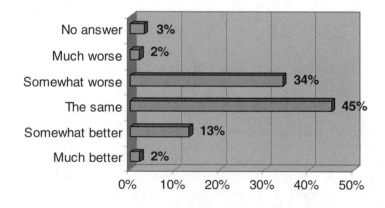

Figure 14–5 How Freshly Prepared Meals from Supermarkets Compare with Freshly Prepared Meals from Restaurants. *Source:* Reprinted with permission from J. Dulen, Changing Taste, *Restaurants and Institutions,* Vol. 108, No. 3, pp. 58–69, © 1998, Restaurants and Institutions.

aware of trends and developments if they are to take advantage of these developments. Furthermore, if home meal replacements are included, eating meals at home
will continue to decline and eating out will in itself become the standard or norm.

REFERENCES

Anker, M., & Batta, V.K. (1987). *Basic restaurant theory and practice*. Harlow, Essex: Longman Scientific and Technical.

Bodenham, D. (1993). *The food dictionary*. Bristol: Redcliffe.

Brears, P., Black, M., Corbishley, G., Renfrew, J., & Stead, J. (1997). *A taste of history. 10,000 years of food in Britain*. London: English Heritage and the British Museum Press.

Brillat-Savarin, J.-A. (1970). *The philosopher in the kitchen* (Translated from *La Physiologie de Goûte*). Harmondsworth, Middlesex: Penguin Books (Original work published 1825).

British Hospitality Association. (1995). *Contract catering survey*. London: Author.

British Hospitality Association. (1998, May). *Hospitality Matters*. London: Author.

Burnett, J. (1979). *Plenty and want. A social history of diet in England from 1815 to the present day*. London: Scolar Press.

Campbell-Smith, G. (1967). *Marketing the meal experience. A fundamental approach* (pp. 75–78). Guilford, Surrey: University of Surrey.

Cardello, A.V. (1996). The role of human senses in food acceptance. In H.L. Meiselman & H.J.H. MacFie (Eds.), *Food choice, food acceptance and consumption* (pp. 1–82). London: Blackie Academic Publications.

Coghlan, A. (1998). You won't find a spot on these plants. *New Scientist. 159*(2142), 18.

Compass. (1997). *The Eurest Lunchtime Report 1997. The sixth annual survey of lunchtime habits*. London: Compass Group UK Division.

Cracknell, H.L., Kaufmann, R.L., & Nobis, G. (1985). *Practical professional catering*. Basingstoke, Hampshire: MacMillan Education.

Cracknell, H.L., & Nobis, G. (1985). *Practical professional gastronomy*. Basingstoke, Hampshire: MacMillan Education.

Cullen, P. (1994). Time, taste and technology: the economic evolution of eating out. *British Food Journal, 96*(10), 4–9.

Dulen, J. (1998). Changing taste. *Restaurants and Institutions, 108*(3), 58–69.

Ervin, J., & Edwards, J.S.A. (1995). *Hospital catering 1995*. Bournemouth University: A Worshipful Company of Cooks Centre for Culinary Research Publication.

Field, D. (1988). *Hotel and catering law* (5th ed.). London: Sweet & Maxwell.

Foodservice Intelligence. (1998). 84 Uxbridge Road, London W13 8RA, England.

Glew, G., Lawson, J., & Hunt., C. (1987). The effect of catering techniques on the nutritional value of food. In R. Cottrell (Ed.), *Nutrition in catering* (pp. 53–74). Camforth, Lancashire: Porthleven Publishing Group.

Green, E.F., Drake, G.G., & Sweeney, F.J. (1991). *Profitable food and beverage management: Planning*. New York: Van Nostrand Reinhold.

Hadfield, P. (1998). The secret of keeping food fresh. *New Scientist, 159*(2143), 20.

Henley Centre. (1996). *Hospitality into the 21st century. A vision for the future*. London: Author.

Keynote Market Review. (1994). *UK catering market* (6th ed.). Hampton, Middlesex: Keynote Publications Ltd.

Keynote Report. (1994). *Contract catering* (7th ed.). Hampton, Middlesex: Keynote Publications Ltd.

Kinton, R., & Ceserani, V. (1989). *The theory of catering* (6th ed.). London: Edward Arnold.

Kleiner, K. (1996). Superwheat to field the world. *New Scientist, 152*(2053), 9.

Kotler, P., & Armstrong, G. (1991). *Principles of marketing* (5th ed., pp. 10–16). Englewood Cliffs, NJ: Prentice-Hall.

Martens, L. (1997). Gender and the eating out experience. *British Food Journal, 99*(1), 20–26.

Martin, R. (1973). *International dictionary of food and cooking*. London: Constable.

Meiselman, H.L. (1996). The contextual basis for food acceptance, food choice and food intake: The food, the situation and the individual. In H.L. Meiselman & H.J.H. MacFie (Eds.). *Food choice, food acceptance and consumption* (pp. 239–263). London: Blackie Academic Publications.

Meiselman, H.L., Hirsch, E.S. & Popper, R.D. (1988). Sensory, hedonic and situational factors in food acceptance and consumption. In D.M. H. Thomson (Ed.). *Food acceptability* (pp. 77–87). London: Elsevier Applied Science.

Mintel Marketing Intelligence. (1998, January). *Hotels*. London: Mintel International Group.

Montagné, P. (1961). *Larousse gastronomique* (p. 614). London: Paul Hamilyn.

Nettles, M.F., & Gregoire, M.B. (1993). Operational characteristics of hospital foodservice departments with conventional, cook-chill and cook-freeze systems. *Journal of the American Dietetic Association, 93*(10), 1161–1163.

Oxford English Dictionary. (1993). Vol. 2. Oxford: Clarendon Press.

Pannett, A., & Boella, M. (1996). *Principles of hospitality law* (4th ed.). London: Cassell.

Pralus, G. (1985). *La cuisine sous vide—Une histoire d'armour—La cuisine de l'An 2000*. Pouilly-sous-Charlieu, France: Briennon.

Roche Products Ltd. (1997). *Eating into the millennium*. Welwyn Garden City, Herfordshire: Author.

Sloan, A.E. (1998, January). Food industry forecast: Consumer trends to 2020 and beyond. *Food Technology, 52*(1), 37–44.

Sprenger, R.A. (1998). *Hygiene for management* (8th ed.). Doncaster: Highfield Publications.

Tannahill, R. (1988). *Food in history*. London: Penguin.

Wilson, C.A. (1994). *Luncheon, nuncheon and other meals. Eating with the Victorians*. (Food and Society Series, Vol. 7.) Stroud, Gloucestershire: A. Sutton Publishers.

Wilson, G. (1989). Family food systems, preventative health and dietary change: A policy to increase the health divide. *Journal of Social Policy, 18*(2), 167–185.

Wood, R.C. (1990). Sociology, gender, food consumption and the hospitality industry. *British Food Journal, 92*(6), 3–5.

Wood, R.C. (1992). Gender trends in dining out. *Nutrition and Food Science, 5*, 18–21.

Chapter 15

Integrating Consumers, Developers, Designers, and Researchers into the Development and Optimization of Meals

Howard R. Moskowitz

INTRODUCTION

Today, more than ever before, product developers are being called on to create products quickly, to assure the marketer that the product will pass hurdles for acceptance, and to guarantee that the product will meet cost constraints. Given the frenetic pace, it is no wonder that newer, more structured methods for concept and product development are being adopted in the food industry. The development of meals is even more complex than the development of foods because team members in the corporation must determine what will comprise the meal, what they must create versus what they must buy "off the shelf," and what concept should underlie the meal.

By means of both principles and a case history, this chapter shows how a commercially oriented development program works from concept to product to package. The details of the case history come from an actual project, with the specific details disguised to maintain confidentiality. The approach, however, reflects the new thinking in business-driven development. After the case history, the chapter will then switch to a more detailed analysis of several topics touched on in the case history.

THE TRADITIONAL APPROACH

Over the past 60 years the traditional approach has been a relaxed development of food items motivated by a consumer market that was growing, whose needs were evolving, but who were prepared to pay more for quality, variety, and convenience. Advertising promoted the benefits of "homemade," there was plenty of time to create meals "from scratch," and consumers were not yet surfeited by the multitude of choices available. Business and daily life were more relaxed, the meal was an occasion for the family to gather together, and the lifestyle was such that "eating-out" was an occasion rather than the behavioral norm.

Within such an environment research and development product developers did not have to ensure that their products would be optimally acceptable, primarily because the increasing consumer demand became a "tide that lifted all boats." Problems in the development of new menu items were hidden by the general prosperity. Increasing population growth in the 1950s and 1960s, stagflation in the 1970s, and a general prosperity without much competition in the early 1980s all reduced innovative development that is often seen in more competitive business environments.

It might take 6 months to a year for a technical developer to create a product, followed by cautious market introduction or (in the case of larger companies) extensive preintroduction testing.

Viewed from today's vantage, the development in days gone by was primitive, nearsighted, and, more often than not, only modestly related to the business issues driving the corporation. For example, as recently as 10 years ago developers generally concentrated primarily (and often only) on the physical stimulus (e.g., a prepared meat product). The developer was handed the product concept by the marketing group and instructed to create a product to match the concept. There was relatively little connection between the marketing group and the product development group other than that the product developer took responsibility for realizing the concept in physical form.

TODAY—TEAM DEVELOPMENT

More recently the notion of "team development" has become popular. The team is composed of a product developer, marketer, market researcher, package designer, and other support people as needed. The team develops or refines the concept and then moves forward to product development. The need still exists for concepts to precede products, but the "splendid isolation" has evaporated. In many corporations today's team identifies trends in the market, sets business objectives, musters the relevant resources, creates the prototypes, does the necessary premarket research (often called "presearch"), creates the packaging, launches the product, and manages the product through its first year or two.

To this end an entire discipline has developed in the area of new products, with different ways of identifying opportunities (e.g., using primary or secondary research, using competitive intelligence), with methods for creating concepts, quantifying the potential of these concepts, and then managing the prelaunch and launch activities. The notion of cross-functional teams is becoming increasingly popular. Journals such as *Product Innovation Management* (a journal sponsored by the Product Development and Management Association [PDMA]) regularly deal with the corporate structures needed to develop and nurture new products. Commercial consulting firms specialize in new product development, bringing together the consultant and the different members of the corporation. A variety of consultancies have been founded that sponsor conferences on "best practices" for new product development. All this activity has emerged only in the past decade and a half as product development has become more responsive to the business environment and to consumer needs.

PART 1: CASE HISTORY ILLUSTRATING HOW A MEAL PRODUCT IS CREATED BY CORPORATE MARKETERS AND DEVELOPERS

The best way to understand the creation and optimization of meals is by means of a case history that approaches life. The case history here concerns lunch products. The original business objective was to identify what components should be present in an "off-the-shelf" children's lunch meal and then to optimize the components of this lunch product for maximum acceptance. The manufacturer had to optimize the concept (so parents would buy the lunch product) and then optimize both the products and the package (so that children would eat the lunch product, and demand it again). This chapter shows how the people in different corporate functions think about the problem, design the study, execute it, and analyze the data.

It should be noted at the outset that the approach espoused in this chapter, statistical design, modeling, and optimization, is not new (Griffin & Stauffer, 1990; Joglekar & May, 1991). It has been used over the past two decades at an increasing rate to develop products, many of which have enjoyed success in the marketplace. The statistics, sensory principles, validation, and market success can be attested to in a variety of industries.

The Business Background

The manufacturer wished to create a line of lunch products for both food service and for supermarkets. The specific objective was to enter the contract feeding market with a line of school lunch products that could be purchased in the cafeteria itself, from local stores, or from vending machines. The manufacturer was informed by the contract feeding company that it was to compete against several other manufacturers for a lucrative, multiyear contract. Consumer acceptance and cost of goods were to be paramount in deciding which manufacturer would win the contract. The manufacturer was a major player in the consumer package goods industry, with little experience in creating complete meals. The request to create the linear of school lunch products would not have been entertained except that through separate analyses of trends it became apparent that consumption patterns were changing. It became clear that more meal occasions were "away from home," and that high-quality convenience meals (but not frozen dinners) would be emerging as popular alternatives to "scratch meals."

Getting Started—Defining the Problem and Specifying the Development Tasks

The first issue to arise is problem definition. In the creation of a packaged meal the marketer usually has a good idea of what the product(s) should do. For the packaged meal the marketing goal was to launch a line of lunch products that children would like, that would contain nutritious food, whose meat items represented components from the company's current production lines, and whose additional components could be purchased "off the shelf." In the mind of the marketing director the overall business objectives in launching the product line were quite clear—"tracking research" dealing with in-market products revealed a major opportunity in the children's convenience market, with no similar "prepackaged" lunch products available.

General marketing issues must be translated into specifics for product developers to create concepts and products. The overall marketing objective simply serves as the general directive for the marketers and the product developers. The specific problems appear below:

1. General product concept for a prepackaged lunch—What general type of product appeals to consumers from the viewpoint of a concept (or a communication)? Concepts are statements about a product rather than the product itself. How should the product be described? By developing a strong product

concept, the marketer makes the developer's job easier because the concept provides a blueprint for the physical product(s) (Cattin & Wittink, 1982; Green & Srinivasan, 1978).

2. Number of components and the selection of particular components—What should the lunch product contain? What are the particular components? How many items should go into the lunch? Does dessert make a difference? Are there differences between two-, three-, and four-component lunches?

3. Number of different products to incorporate in the line—How many different lunch products can the market sustain without becoming overly crowded? To the extent that the manufacturer can market many different lunch products (namely, different combinations of components) the manufacturer will occupy the niche and prevent competitors from entering. On the other hand, more products in the line mean more "facings" in the store. Supermarkets charge for their space in the form of "slotting fees," which are the equivalent of rent, with the shelf considered as a piece of real estate to be rented out. More facings cost more in slotting fees. If the lunch products do not sell, the manufacturer loses the money and may even have to pay a penalty to "de-list" the item from the supermarket. Perennial tension occurs between optimism and realism in product development, especially when items can be created easily and sent out into the market.

4. Optimal formulation of the meat item—What is the physical formulation of the key component, in terms of ingredients and process, to maximize acceptance, to create a product that fits the concept, and/or to minimize cost of goods, and so forth?

5. Package design—What should the package look like to attract mothers and children? Are there general "rules" about the features to show on the package, for both the current items and for future items to be marketed? That is, can the manufacturer identify "hot buttons" on the package (e.g., color, general design) that increase the likelihood that the product will be attractive and inviting in the context of a crowded supermarket and in the context that the item will be located in the refrigerated compartment.

Problem Definition and Stages

Faced with this request, the manufacturer assembled a team composed of marketing (having direct interaction with the contract feeder) and product development (with the responsibility of creating the actual products). The final team was composed of marketing, marketing research, packaging, product development, sensory analysis, and purchasing.

The team specified the following six stages:

• Stage 1: Define the requirements in the request.
• Stage 2: Develop an overarching concept to link together the different products that the line would be composed of.
• Stage 3: Identify the specific components of the packaged meal.
• Stage 4: Identify which particular products in the packaged meal could be purchased "off the shelf" versus which had to be created in the manufacturer's facilities
• Stage 5: Develop and optimize the meat component.

- Stage 6: Develop and optimize the package.

Stage 1—Define Requirements and Capabilities

Because each company and every product development team has its strengths and weaknesses, no development task is really as straightforward as desired. If the company or the team specializes in the particular products for contract feeding, the project requirements may be straightforward (e.g., the ingredients to be procured, the difficulty, the cost). If the company or team does not specialize in the specific product, the next question is whether the task can be properly executed in a reasonable time at a reasonable cost.

The manufacturing business involved processed meat products under the company's own brand, as well as co-packing for other companies. The expertise for developing the meat item was resident in-house. The expertise for developing full packaged meals for contract feeding was not, however, resident in-house. The project manager in charge developed a cross-functional team composed of in-house representatives and external contractors.

The team concluded that the project was feasible, although it required a number of stages of consumer research to identify the appropriate options (which form the substantive content of this chapter). An analysis of the market revealed that besides the meat, the other components of the lunch could be procured at a reasonable price to ensure profitability should the manufacturer win the contract. The packaging group confirmed that the company could create attractive, shelf-stable packages, incorporating the meat product along with other luncheon items. Thus, the early vote was to accept the challenge to create the meal and to compete for the contract.

Stage 2—Develop the Overarching Concept for the Lunch Line

Commercial products need conceptual "blueprints." A blueprint is a product concept telling the developer what the product should contain. The product concept is developed with consumers to ensure consumer acceptance.

Concepts are created in many ways. One method relies on an expert who comes into the process and suggests what the team should create. The expertise may come from years of experience or from the individual's job title (namely, president or vice president of marketing). Other methods rely on ideation and creativity sessions in which the participants sit as a team and come up with ideas about the product. Both of these methods are "top down," relying as they do on the complete

---

Exhibit 15–1 A Full Concept ("Top Down")

Introducing Munchables
Munchables is the full school lunch, full of great taste and nutrition. Have any of six different meat sandwiches, a cheese, a fruit, and a soft drink. You can get Munchables at your school cafeteria, in the refrigerator case, or at the convenience store near you. You now have great taste, great nutrition. And...as a special treat, we've included a gift—sports cards or doll cards.

concept. Exhibit 15–1 presents an example of a "top-down concept." Quite often these concepts must be refined and retested among consumers.

A contrasting method of concept development, "bottom-up" or "conjoint analysis," does not require the participant to create complete concepts. Rather, the participants in a creativity (or ideation) session create parts of concepts ("elements"). After the full set of elements has been created, the elements are then "mixed and matched" by computer, according to an experimental design to create test concepts. The panelists evaluate these test concepts on "interest" and other attributes deemed relevant. From the ratings the researcher can trace the interest in the concept to the presence or absence of the specific elements.

To create the concept, the developer explored the key features of the lunch product, using a modified version of conjoint measurement (IdeaMap; Moskowitz, 1994; Moskowitz & Martin, 1993). The underlying assumption of the approach is that consumers may not be able to articulate what they want in the lunch product, but they will recognize it when they see it. In the up-front ideation phase the team creates a set of "categories" (e.g., statements about packaging, product, use) and within each category creates a set of elements. The elements are the *specifics* and comprise snippets of ideas.

Table 15–1 shows the categories and elements for the lunch product. These elements were generated through "brainstorming sessions." In the session participants listed the different possible categories and then populated each category with many elements. The brainstorming sessions were conducted with two groups of panelists—one group composed of professionals in the food service business and the

Table 15–1 Examples of Categories, Elements, Interest (Utility), and Feasibility

	Interest Utility	R&D Feasibility
When Used		
For meal after sports	13	7
For a fast pick-me-up snack	13	8
For lunch	4	8
For an early supper at school	–2	6
Components of the Meal		
Meat + fruit + beverage	8	5
Meat + beverage + candy	8	5
Meat + starch + beverage	6	5
Meat + starch + candy	5	8
Packaging		
Items packaged in a see-through container	4	7
Items packaged in a space-efficient box	2	6
Items packaged along with a surprise gift	1	8
Items packaged with a separate preparation container for meat/starch	–3	5
Orientation and Benefits		
Contains all the items you need to make lunch	4	6
Oriented both to kids and to teens	2	8
Oriented to teens and adults	–2	10
A product that's fast to make and fun to eat	–4	10
How to Acquire It		
Sold at special counters in school cafeteria	7	7
Comes in different packs, for variety	7	–3
Buy through a vending machine	6	–9
Available at a local variety store, in five packs, for the whole school week	3	7

other group composed of consumers. The two brainstorming sessions generated a total of 138 elements. These elements were culled down to those dealing with product features only (reducing the set from 138 to 43). The remaining elements dealt with advertising positioning. For the illustrative purposes of this chapter, the set of 43 product and purchase-related elements were further culled down, refined, and then put into five categories composed of four elements each (a total of 20 elements).

By themselves the elements in the left side of Table 15–1 represent only the building blocks from which concepts can be created. To create the optimal concepts it is important to bring consumers into the loop. One method espoused by those working in conjoint analysis combines the elements into small, easy-to-understand combinations, and obtains consumer reactions to these combinations. If the combinations vary according to an experimental design, the researcher can subsequently estimate the contribution of each particular concept element to consumer acceptance of the entire combination. If the rating attribute changes from interest in the concept to "appropriate for a given end use," the researcher estimates the contribution of each element to that particular rating attribute. Other groups, such as research and development (R&D), can also be brought into the concept development and evaluation.

The actual interview was composed of two parts. The panel was composed of 53 teenage consumers, representing the likely target population for the meal. In Part 1 the panelists rated purchase interest in 50 different concepts for the lunch created by "mixing and matching" the 20 elements in different combinations according to an experimental design. The experimental design can be composed of any one of a variety of different statistical layouts (Box, Hunter, & Hunter, 1978; Plackett & Burman, 1946). The ratings of consumers are related to the presence or absence of the concept elements by means of "dummy variable regression analysis." At the same time, a small group of product developers and managers from R&D evaluated the same set of 50 concepts, rating each on "feasibility." Both consumers and experts used an anchored 1–9 point scale adapted from the hedonic scale (Peryam & Pilgrim, 1957: 1 = not interested in purchase vs. 9 = definitely interested in purchasing; 1 = not feasible vs. 9 = definitely feasible). This first portion lasted 20 minutes. From this first part of the project the developer synthesized a concept. Table 15–1 shows the part-worth contribution of each concept element from the regression analysis to consumer ratings of "purchase interest" and to the R&D estimates of "feasibility."

Part 2 of the study dealt with the selection of individual items for the meals. In Part 2, run a week later with the same 53 consumer panelists, each panelist assessed 26 different items (by name only). The task was to rate both the acceptability of each item and the fit of that item to the meal as specified by the optimum concept. The creation of the optimum concept will be dealt with shortly.

Table 15–1 (right hand portion) shows the (partial) results from the modeling. The ratings from each panelist were first transformed, so that a rating of 1–6 on the 9-point scale was transformed to "0" (representing "not interested"), and a rating of 7–9 was transformed to "100." This conversion from a metric to a binary value is often done by market researchers who are interested in the percent of panelists showing a specific behavior (e.g., percent interested in a concept) rather than interested in the "magnitude" of the response.

Each concept element generates its own part-worth use corresponding to the coefficient in the regression equation relating interest to the presence or absence of the element. In the case of feasibility as rated by R&D, the coefficient shows the relation between the presence or absence of the element and whether the combination is considered to be feasible. By combining consumer and R&D data into one comprehensive model, the marketing group can create concepts that meet two objectives—interest and feasibility.

From the utilities in Table 15–1 the developer synthesized various concepts by setting objectives for the concept to achieve. Table 15–2 shows three concepts, the best, a middling concept, and the worst, respectively. First, the concept was optimized for consumer acceptance without any additional consideration regarding R&D feasibility. Second, the concept was optimized for consumer acceptance, subject to a minimum level of feasibility (feasibility >24). Below that minimum level management believed that the product could not be developed given the company's current capabilities and investment plans.

The development can take different directions based on the selection of elements in the final concept. Some of these elements may require a great deal of effort to realize, whereas other elements may be less acceptable but also far more feasible. By developing product concepts in a structured manner the marketer and the product developer ensure that the product they are developing will achieve both consumer requirements and developmental feasibility.

Stage 3—Identify Components of the Meal

When a manufacturer creates meals composed of components, all the components should be acceptable; they should work together and should fit the concept.

Table 15–2 Three Concepts (Best to Worst) and the Utilities of Their Components

		Interest Utility	R&D Feasibility Utility
	Best Concept (Synthesized Concept To Be Used Later)		
	Additive Constant	32	27
A4	For a fast pick-me-up snack	13	8
B4	Meat + fruit + beverage	8	5
C1	Items packaged in a see-through container	4	7
D4	Contains all the items you need to make lunch	4	6
E3	Sold at special counters in school cafeteria	7	7
	Sum of utilities + additive constant	68	60
	Middling Concept		
B1	Meat + starch + beverage	6	5
C3	Items packaged along with a surprise gift	1	8
D2	A product that's fun to make and to eat	−4	10
A4	For a fast pick-me-up snack	13	8
E2	Buy through a vending machine	6	−9
	Sum of utilities + additive constant	54	49
	Worst Concept		
A2	For an early supper at school	−2	6
B2	Meat + starch + candy	5	8
C2	Items packaged with a separate preparation container for meat/starch	−3	5
D2	A product that's fun to make and to eat	−4	10
E4	Available in local variety store, five packs, for the whole school week	3	7
	Sum of utilities + additive constant	31	63

Table 15–3 Average Ratings of Liking and "Fit to Concept" for 11 Meats and 5 Beverages that Could Be Used for the Lunch Product

	Liking	Fit Best Concept (Table 15–2)
Meat		
Turkey slice	62	55
Meatballs	62	62
Chicken chili	60	55
Roast beef	58	62
Bologna	54	63
Hot dog	50	67
Knockwurst	43	49
Schnitzel	42	50
Pastrami	39	49
Spiced ham	36	51
Meat/vegetable patty	22	36
Beverage		
Apple juice	52	55
Orange juice	58	59
Cranberry juice	62	61
Lemonade	57	58
Tomato juice	58	59

One week later, after the concept evaluation was run, the same 53 panelists were brought back to the test facility and shown the synthesized "optimal" concept (best concept in Table 15–2). The panelists then evaluated each of 26 items for acceptance and "fit to the best concept." (The questions asked were: "How much do you like this food item?" and "How well does this food item fit this particular concept?".) The data generated a profile of item by attribute shown in part in Table 15–3 for the 11 meats and the 5 juice beverages.

By themselves, the ratings of liking for the 11 different meat items only show the aggregate acceptance. The mean ratings for liking do not immediately provide a set of products for a line of meals. That is, if the manufacturer is asked to create a set of meals, it may be imprudent simply to select the two top-scoring meats. Certainly these two meals perform the best in terms of how consumers rate liking, but the two meats may appeal to the same group of consumers. A more prudent approach identifies two (or more) items so that together these items satisfy the greatest number of consumers. To identify the best set of meat items, the best set of beverage items, etc. requires a TURF analysis (Total Unduplicated Reach and Frequency; Moskowitz, 1994, 1997).

TURF estimates the total proportion of consumers who would be satisfied with at least one of the items. TURF prevents the creation of a line of items in which all items appeal to the same "majority" set of consumers but fail to satisfy other consumers. TURF analysis is well known in market research when the goal is to create a line of items that will enjoy the maximal success by appealing to the most number of consumers. TURF analysis is relevant for meal development discussed here because it allows the developer to identify a set of meal components that, at least by themselves, enjoy the greatest potential for market success. TURF works on a product category basis (e.g., meats alone). TURF is an enumerative procedure, which lays out all single items (e.g., single meats), pairs of items (e.g., pairs of meats), triples of items (e.g., triples of meats), and so forth. For each single item, pair of items, and so forth, TURF estimates the number of consumers who, on the basis of their liking ratings, would like *at least one* of the items. Using this enumeration method the researcher quickly identifies the combination of items that satisfies the greatest number of consumers. It could easily turn out that the combination of a very popular item and a modestly popular item would satisfy more consumers than would another combination of two similar items.

For this meal project there are 11 single meats, 55 pairs of meats (11 x 10/2), and 165 triples of meats (11 x 10 x 9/6). Because TURF uses a simple computational algorithm, it is straightforward to estimate the total number of panelists satisfied with at least one item in any offering of meats. The same routine computation is performed on every combination to arrive at the percent of consumers satisfied with at least one item. Table 15–4 shows the best four pairs of meats and the worst four pairs of meats. The best strategy for the manufacturer is to work on chili (definitely) and on either hot dog, ham slices, or turkey slices as the best backup. Marketing decided to spend most of the effort on chili because it is popular with children and could provide a point of difference from the meats chosen by the other suppliers competing for this contract. The marketing group also planned to use the TURF data to provide the contract feeder with additional information on alternative meats that would go into the meal line.

The reader should keep in mind that the liking ratings and the TURF simply provide the researcher and marketer with those components of the meal that show the greatest promise. Furthermore, the TURF analysis is a computational device, not a theory of food choice. That is, TURF is simply an enumeration method and a heuristic. No underlying theory dictates the selection of different items for the meals. Finally, as presented here, the TURF analysis depends highly on the magnitude of the liking ratings assigned by the panelists. A panelist who scores every meat high will generally be satisfied with any combination of meats simply because any combination is bound to contain a high-scoring meat. A panelist who is more critical, scoring every meat low, will generally not be satisfied with any combination of meats simply because that panelist's criterion for scaling acceptability is quite stringent. (Ways around the scaling problem include normalizing the ratings, or using the rank orders instead of the liking ratings themselves.)

Table 15–4 Pairs of Meats in the Line and the Proportion of Panelists who Like at Least One of the Meats very Much (namely, whose Rating for at Least One Meat Exceeds 65 on the 0–100 Point Liking Scale)

Meat No. 1	Meat No. 2	% Satisfied
Chili	Hot dog	68
Chili	Ham slice	62
Bologna	Hot dog	60
Chili	Turkey slice	58
Pastrami	Spiced ham	43
Knockwurst	Meat/vegetable patty	41
Pastrami	Meat/vegetable patty	30
Spiced ham	Meat/vegetable patty	27

Stage 4—Identify Which Components in the Meal Must Be Made versus Purchased

Quite often developers can purchase many of their components either from other manufacturers or "off the shelf." The product developer employed by this manufacturer realized that the only product that needed to be created was the meat item. The remaining items (e.g., dessert, fruit, candy, napkin, juice) could all be purchased and at a relatively low price if purchased in bulk.

Although Stage 4 seems quite small in the sequence, in reality it presents a key stage for those business situations in which a company can either produce a product or "outsource" the production. In commercial situations requiring a product (namely, meals) composed of several components, the perennial tendency is to want to create everything "in-house." Yet, the economics are such that more often than not the more items that can be purchased off the shelf as components and repackaged, the more likely it is that the company can profitably compete as a business. Certainly in the case of prepackaged meals, the more that a company can "outsource" by purchasing at an advantageous price, the better the economics of the product will be.

Stage 5—Optimize an Expensive Component in the Lunch Product

The company specialized in creating and marketing meat products. One of the products in the TURF analysis was a chili product based on chicken. Stage 5 deals with the optimization of the chili product. The reader should note that a great deal has been written about product optimization (Gacula, 1993; Moskowitz, 1996a), so product optimization will be covered only in a cursory way in this chapter.

Chili products vary on a variety of ingredients, including the type of meat, the beans, and the flavorings. Depending on the specific composition, the chili may taste spicy or mild, appear light or dark, be thick or thin, have a preponderance of meat or a preponderance of beans, and so forth. To win the order, the product developer recognized that he or she would have to create a chili product that teenage consumers would find acceptable. The cost of the chili product varies with the components of the chili. More meat of higher quality can generate an unfeasible cost, whereas less meat and of lower quality will generate a cost-effective product, but one that no one will accept.

When creating a product, the developer may follow one of two paths. The first path consists of creating several prototypes that represent different directions. The developer creates "rifle shots" and submits these to consumer testing to identify which chili prototypes perform acceptably well. From these candidates the developer then selects the one chili to use in the final lunch product. This first path is easy to follow, relatively cost-efficient, and quite popular. Creating and testing a limited number of rifle shots does not yield the requisite knowledge that can be used later on, should the developer need to cost reduce the product or modify it to achieve other, newly emerging, objectives. What is easy up front often later may turn out to be problematic.

The second path is composed of a more disciplined approach and is gaining favor as business becomes increasingly competitive. Rather than creating the few rifle shots, the developer uses experimental design. Experimental design identifies the physical variables under the developer's control (ingredients, process) and creates systematically varied prototypes with different levels of those physical variables. The developer tests these prototypes, develops models relating consumer reactions, cost, and other dependent variables to the independent variables, and then optimizes the product. At the end of the sequence the developer discovers the relation between the physical formulation and the consumer reactions and identifies possible formulations that are highly acceptable and cost-effective for the product. In this case history the developer used experimental design but augmented the design with several best guess rifle shots. This joint strategy gives the best chance for success by combining design and educated guesses (Box et al., 1978).

The product development team at first thought they knew all about chili from previous experience in the product category, but during the course of the development, they realized that they knew relatively little. To accelerate the development of knowledge and maximize the value of the research, the development team selected a "screening design" to identify the key drivers of product acceptance. Screening designs allow the developer to explore a wide variety of different product formulations. Often they are used as the first step in a multistep process for product development. In the case of a contract feeder, the screening design may actually be the key and often the only product research step (other than simple product evaluation) because the ultimate financial pay-out for an optimal product is simply not high enough to warrant extensive research.

Screening designs were developed to identify the impact of many different variables on a single dependent variable (Plackett & Burman, 1946). Through screening designs the developer combines a large number of different variables at two levels (high vs. low, present vs. absent, etc.). The design matrix ensures that the independent variables are present in a balanced array so that each of the two levels of an independent variable is present with each level of every other independent variable. Through regression analysis the developer determines what each independent variable contributes to the consumers' ratings.

For the chili study the product development team identified 10 different formulation variables that they believed would have an impact on both acceptance and cost. The appropriate and most cost-effective screening design was a 12-run Plackett Burman design. The 12-run design provides the minimum number of runs needed to estimate the contribution of each of the 10 different variables. (Note that if the developer wants to increase the confidence in the measurement of effects, he or she would use more runs, such as the 16- or the 20-run design, albeit with the same 10 variables.)

The panel was composed of 42 children and teens, who were positive to a concept about chili as part of a lunch box and who were not averse to a chili made from chicken. Each consumer evaluated the 12 chili prototypes and three in-market competitors that served as benchmarks. (Benchmarks are important because they allow the developer to compare the scores achieved by the experimentally designed prototypes to scores achieved by known in-market products.) The product evaluation required two sessions, each session lasting 2½ hours. The panelists rated each product on a set of liking, sensory, and image/use attributes, using anchored 0–100 scales (Moskowitz, 1985). Panelists find anchored scales easy to use, having been exposed to them in school. The anchors on the extremes of the scales reduce any ambiguity. Panelists also find it easy to intermix sensory and liking ratings and have no trouble distinguishing the attributes.

By themselves, the summarized data (ratings for the products on attributes) teach the developer a great deal. One key measure is the range of acceptance. The wider the range of liking across prototypes the more likely that the ingredients affect acceptance. A second key measure is the range of sensory levels achieved. If the products differ from each other on sensory attributes but achieve similar liking scores, the developer concludes that almost any product in the set will do as long as the product reaches the requisite level of acceptance. In this case the chili prototypes achieve a wide range of liking (22 points). Thus, it is quite likely that the product developer can create acceptable products by varying the ingredients. (In some cases the ingredients and process conditions do not seem to affect acceptance, even though they affect the sensory characteristics. In such cases the developer should probably aim to produce the lowest cost product, subject to achieving reasonable acceptability and subject to the sensory characteristics being compatible with the product concept.)

An idea of the impact of each ingredient on each attribute can be obtained from models relating ingredients to consumer ratings (see Table 15–5). The model is composed of an additive constant and a coefficient. Each coefficient of the regression model shows the effect on the attribute when the particular ingredient (independent variable) changes from state "0" (absent, low level, option No. 1) to state "1" (present, high level, option No. 2). The additive constant shows the expected rating of that attribute when all 10 independent variables are simultaneously set to "0."

Table 15–5 immediately shows the developer which independent variables exert strong effects and which can be ignored. For example, variable J shows a large effect on liking (-7) when it goes from option No. 1 (coded as 0) to option No. 2 (coded as 1). Furthermore, variable J affects liking of aroma (-7), liking of taste (-8), and thickness (-13). From Table 15–5 five variables appear to affect liking (B, C, E, F, and J, respectively) to different degrees and in different directions. The developer now knows on which variables to focus and the best levels to choose. Should the developer wish to stop here, he or she might select the best performing of the 12 prototypes or synthesize the best product for overall liking corresponding to "0" options of B, C, E, and J, and the "1" option of F. The remaining independent variables have little effect on overall liking.

Table 15–5 Additive Effect of Each Variable (A–J) on Each Rating Attribute

	Constant	A	B	C	D	E	F	G	H	I	J
Like overall	52	−1	−7	−6	3	−6	7	−2	2	−1	−7
Like appearance	62	0	−8	−9	−4	−2	−6	0	1	0	−1
Like aroma	35	−3	−3	15	5	−2	3	0	2	6	−7
Like taste	58	2	−1	6	−5	−1	2	−9	−4	3	−8
Like texture	65	−14	−1	−16	1	−2	8	−9	−7	−3	−3
Sensory—dark	57	9	−9	−3	−12	−4	1	−1	6	−9	6
Sensory—thick	57	−1	4	−8	1	−4	0	1	6	−9	−13
Sensory —spicy	42	10	4	-5	−2	6	3	−4	4	4	−4
Sensory—meaty	56	−4	−2	8	1	2	0	−15	9	−6	2
Sensory—tomato	38	−5	12	3	−8	0	−4	11	8	7	−4
Sensory—gritty	50	3	5	−2	−4	4	−6	−9	3	−4	6
Sensory—oily	57	−8	0	−1	−7	−5	−11	6	6	6	0
Image kid vs. adult	67	−7	−2	−10	−3	−6	−2	3	−4	5	−3
Image—different	54	5	−5	0	−5	2	−5	−3	−2	−7	2

Stage 6—Package Design for the Meal

Although not typically thought of as a product development task, package design is becoming increasingly important as foods are consumed away from home. The package is often the first aspect of the product to which the consumer is exposed. If the package is attractive and inviting, the consumer is more likely to purchase and accept the product inside. Conversely, if the package is unattractive, the consumer is less likely to accept the food and may even attribute negative sensory characteristics to the food. Today, more than ever before, packaging is recognized as critical in differentiating one product from another and driving purchase.

Traditionally, package design has been left to the expert, artist designer, who, briefed about the background of the product and the marketing objectives, creates an appropriate package in "splendid isolation." This artistically orientated isolation worked in the past, but in the increasingly competitive arena of fast-moving consumer goods, marketing the package is becoming more important as part of the product. Marketers are demanding that the designer substantiate consumer acceptance of the package and in many cases no longer rely solely on the designer as the expert whose word is sacrosanct.

Package design can be handled in the same way as concepts, namely, by one single package or by experimental design of package features. In this study the marketing manager suggested that the team systematically vary the features of the meal package, combine these features to create prototype packages, and test these packages among consumers. The packages for the full meal were created by experimental design on a computer (which superimposed features to create the package). The experimental design of the features enables the package designer to understand how the different features of the meal package drive acceptance and image.

In this study the researcher and package designer identified four general categories of features (name, color, background, message). Each of these four categories was varied in different ways (options, elements) to yield a total of 17 elements. The elements then became the 17 variables that were then arrayed according to a customized experimental design to generate 40 different test packages. Each test package contained one element from each of the four categories. Panelists evaluated the 40 different package combinations presented on a high-resolution computer screen. Each package was rated on liking, appropriate for children versus adult, and fit to the best concept (shown in Table 15–6).

Designs in which every prototype has one element from a category must be analyzed differently from concepts, whereas a category can be absent from a concept. The regression model for four categories is composed of 13 predictors (variables) rather than 17 predictors. That is, one element (or design feature) from each category, the "reference," is deliberately left out of the equation. The reference can be any one of the elements in the category. The additive constant is the estimated value that would be assigned to a package composed of the four reference elements. The coefficient for an element in the regression model corresponds to the estimated change in the rating if that specific element replaces the reference element in the package. The coefficients are relative values, not absolute values. Table 15–6 shows these coefficients, based upon the 40 packages composed of the 17 elements. The reference element is assigned the weight "0" for each of the three rating attributes. The remaining two coefficients in the category must be interpreted relative to that reference value.

Table 15–6 Coefficients of the Additive Model for the Meal Package, Showing the Contribution of Each Design Feature

	Interest	Child vs. Adult	Fit Concept
Constant	32	44	48
Rectangular Pack Shape (Reference)	0	0	0
Oval pack shape	12	6	4
Square pack shape	10	5	4
Interior arrangement 1 (Reference)	0	0	0
Interior arrangement 2	6	1	3
Interior arrangement 3	5	3	1
Interior arrangement 4	5	1	2
Interior arrangement 5	1	−1	0
Cover 2 (Reference)	0	0	0
Cover 1	12	4	12
Cover 3	10	2	13
Cover 4	12	4	12
Cover 5	11	6	11
Picture 3 (Reference)	0	0	0
Picture 1	5	5	4
Picture 2	6	5	3
Picture 4	7	2	3

The final decision is to select the winning elements. The additive model in Table 15–6 provides a blueprint of the package design features, enabling the team to craft a winning package for the submission. Winning in this context means a combination of high acceptance and an agreement with the strategy. Table 15–7 shows three packages, with Package No. 3 selected as the final package with which to move

Table 15–7 Three Meal Packages, Designed According to Different Objectives

	Interest	Child vs. Adult	Fit Concept
Package 1—Maximize Acceptance			
Constant	32	44	48
Oval Shape	12	6	4
Picture 4	7	2	3
Interior arrangement 2	6	1	3
Cover 1	12	4	12
Sum	69	57	70
Package 2—Maximize Interest and Make It Oriented Toward A Child			
Constant	32	44	48
Rectangular shape	0	0	0
Picture 4	7	2	3
Interior arrangement 5	0	0	0
Cover 3	10	2	13
Sum	49	48	64
Package 3—Maximize Interest As Well As Fit To Concept			
Constant	32	44	48
Oval shape	12	6	4
Picture 4	7	2	3
Interior arrangement 1	0	0	0
Cover 1	12	4	12
Sum	63	56	67

forward. By understanding how the different components of the package design "drive" a variety of consumer attributes, the package designer can now communicate, through the design, a desired "image" or set of image attributes.

General Discussion—Business Applications

The Role of Consumer Research in Commercially Oriented Meal Design

Traditionally the design of prepackaged meals in a commercial environment was an informal task, often done by internal staff (e.g., home economists). Over the past decade, and in the face of increasing competitive pressure, meal design is fast falling into the scope of consumer research. Appreciation of the importance of consumer input manifests itself in two different ways:

1. Stage at which the consumer is brought into the project: Researchers are more willing to solicit consumer inputs earlier in the development phase. Traditionally, consumers were brought into the research when the products or meals were finished. The consumers were asked to rate one or two stimuli in the final development stages. If consumers rejected both stimuli, the developer had to begin again. Now consumers are brought in quite early for both qualitative (discussion) and quantitative (rating/evaluation) tasks.
2. Consumers as guides through a database of alternatives rather than simply as evaluators: Researchers are more willing to assess a large number of stimuli varied in a disciplined fashion. Today, developers and researchers are willing to create large numbers of stimuli, be they concepts, products, or package designs, and work with consumers to identify winning and losing stimuli (with winning/losing referring to acceptance and/or appropriateness for various end uses). Thus, a growing body of data exists from consumer research on reactions to multiple stimuli and a parallel growing body of "rules" exists that developers follow on the basis of consumer data.

Excellent versus Adequate Development

The approach presented here reflects a highly disciplined, potentially expensive system for developing meals and their components. In many commercial situations there is simply neither time nor budget for a disciplined, extensive development cycle. In these more pressured situations the developer often settles for judgments, inexpensive research (e.g., focus groups), or rapid-fire, shoot-from-the-hip evaluation of a limited number of prototypes. Any research, properly conducted, can provide information. The developer must assess the benefits of the research versus the cost. Disciplined, comprehensive research is warranted in those situations in which the researcher does not know much about the product or when being correct can result in a highly profitable product. Adequate research is warranted in those situations in which the researcher knows a lot about the product or when being correct results in a modestly profitable product. Every commercial concern involved in development makes these tradeoffs, either consciously or unconsciously, when assigning corporate resources to a project.

The Knowledge Value of Designed Experiments in Meal Development

In the corporate environment significant emphasis is placed on understanding consumer needs and then translating those needs into actual products. All too often

in the corporate environment little time is available to consult the scientific literature. Little time is available to think about the scientific aspects of meal design
because the pressure to create concepts and prototypes is never-ending and often
all-consuming. There is a growing bank of literature in food science on the properties of particular products but no home for business-oriented literature on meals,
their components, and the interaction of the consumer and the meal. As a consequence, anyone in industry attempting to create meals has to rely on scattered information in the literature and on corporate archives.

Systematic exploration of meal variables, whether at the concept or product (or
even packaging) stage, is critical for the knowledge base. Many developers admit,
after the fact, that the systematic exploration of the features of a meal, from concept
through packaging, is the most important part of their developmental process.
Only designed experiments possess the depth and force the discipline needs to edu-
cate the developer beyond the momentary reaction to the immediate emergencies.
Over time, and as management in companies becomes increasingly sophisticated
and appreciative of research, we may expect to see more "experimentally designed"
studies in the applied environment. This trend may seem obvious or irrelevant to
some who believe that the corporation somehow "knows" what the components of
the meal should be, what the product should taste like, and what the package
should look like. In actuality, nothing could be further from the truth.

PART 2: BASIC RESEARCH FOUNDATIONS UNDERLYING APPLIED MEAL DEVELOPMENT

This section deals with several research issues that are similar to those faced by
the marketer and developer in our case history. The data shown below deal with
empirical study data in greater detail.

The Relative Importance of Components in a Meal to the Acceptability of the Meal

More than 15 years ago, the author and a collaborator (Rogozenski &
Moskowitz, 1982) presented data on the overall acceptability of the meal versus the
acceptability of meal components. The issue was the degree to which the overall
liking of a meal could be predicted from the weighted sum of the individual liking
ratings. To the degree that one could identify the important components of the
meal, the meal developer could justify the effort on some aspects of the meal more
than on others.

The approach presented soldiers with both meals and meal components and
instructed them to rate the liking of each stimulus on a 9-point hedonic scale. The
result from that study was an equation:

Meal liking = 5.68 + (2.07 x entree) + (0.53 x starch) + (0.42 x vegetable) +
(0.25 x salad) + (0.57 x dessert)

This equation means that most of the liking of the meal is a function of the
acceptability of the main item, with the side dishes being far less important. In
other subsequent work (with frozen convenience meals) the author has found similar types of equations and similar orders of magnitude. This means that liking for a
combination of a food item is not simply the arithmetic average of the liking of the

components but rather a weighted average. Turner and Collison (1988) found a similar dominance of the entrée, in their equation:

Whole meal liking = 0.57 + 0.43 (entree) + 0.14 (potato) +0.14 (starter) + 0.21 (sweet)

Time Preference

Twenty-five years ago Joseph Balintfy at the University of Massachusetts introduced the concept of "time preference" for menu planning (Balintfy, Duffy, & Sinha, 1974; Moskowitz, 1980). Balintfy's hypothesis was that customers would select different items in a meal as a function of the degree of acceptability, but that acceptability was a function of at least two variables. One variable is the intrinsic acceptance of the item, whereas another variable is the loss of acceptability as a result of frequency of consumption. That is, according to Balintfy, a food can be highly acceptable, but if it is eaten too frequently, it loses its acceptability. A loss of acceptability is only temporary, however, in the world of time preference. That is, with sufficient time after the last consumption, the consumer would again choose that food item.

Figure 15–1 shows an example of the time-preference curve that might develop for a food item following Balintfy's notion (but simplifying the approach considerably). The panelist's job is simplified considerably from the original approach to time preference. When evaluating single items or an entire meal, the panelist is asked to imagine that the last time that the item (or meal) was consumed was X

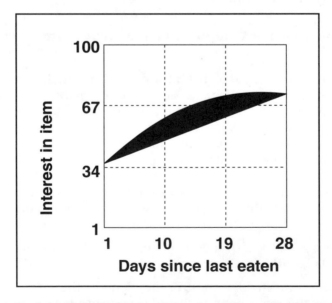

Figure 15–1 Schematic of the Time Preference Curve for an Item. The Curve Shows the Interest in Consuming an Item (e.g., on a 0–100 Point Scale) That Would Be Assigned to the Product Assuming That the Last Time the Panelist Ate the Product Was a Specific Number of Days Ago. The Area Subtended by the Curve Is the Boredom Factor.

days ago, where X can take on different values. X might range from 1 day to a week to a month to 3 months, or even to a year or longer. The panelist is instructed to rate "interest in the item" using a scale. The panelist has no trouble understanding the instructions and comes up with an answer. More often than not, the panelist recognizes (without prompting by the experimenter) that there is an innate time-preference curve and realizes that the more distant in the past the panelist has eaten the product (or meal) the more liked the product is right now.

Balintfy's idea is attractive for commercial product development because it recognizes the notion of "boredom" in meals. Commercially oriented product development for meal items always has to take into account the misrepresentation of consumer acceptance data. Companies offering items that quickly become boring stand to lose a great deal of money. Companies offering items that maintain interest (namely, have a flat time-preference curve) stand to make a great deal of money, especially if the intrinsic acceptance of the product is high and the item can be consumed time after time without losing its acceptance.

Moskowitz and Rogozenski (1975) looked at the time-preference curves for a number of foods, using soldiers as panelists. They found that many meat items exhibited steep time-preference curves, so that acceptance falls off with repeated use. In contrast, some less acceptable but staple items, such as breads, exhibit time-preference curves with lower slopes. The lower slope may be either because the item is a staple (in which case consumers are used to repeated consumptions), or the item may become a staple because it exhibits a flatter time-preference curve.

The concept of time preference is important for designing meal products, especially in light of the fact that today consumers have a plethora of choices. To the extent that the product developer understands what factors make a product "boring" over time (namely, lead to a steep time-preference curve), the developer will be able to avoid these factors. It may well turn out, however, that for the most part the same factors that make a product highly acceptable also lead to steep time-preference curves.

Concept Segmentation and Meal Preference Patterns

The notion of segmentation has been widely accepted among marketers and is beginning to be accepted among basic researchers. Research on sensory preferences shows significant individual differences in liking, especially when it comes to the chemical senses of taste and smell (Moskowitz, 1981; Pangborn, 1970). These differences manifest themselves in acceptance ratings for foods. Preference segmentation transcends actual physical stimuli, however, and applies to concepts as well. Elements in a concept statement may appeal to one person but not to another.

A good example of concept segmentation applied to meal concept comes from a study of frozen food dinners. The study was composed of 200 panelists chosen to participate because they both purchased and prepared frozen meals. The panelists were women, ages 21–55, living in three U.S. markets (New York, Chicago, Los Angeles). The panelists each evaluated 100 concepts about frozen dinners. The concepts were created from a total of 297 elements. Each panelist evaluated a different set of concepts, using the IdeaMap method of conjoint measurement (Moskowitz & Martin, 1993).

Each panelist generated an individual model showing the utilities of all 297 concept elements following the procedure outlined by Moskowitz (1996b). The

Table 15–8 Winning Elements for a Frozen Food Study for Total Panel and for Two Segments

Element	Total	Seg 1	Seg 2
Additive constant	60	91	49
Meals carefully prepared using all natural ingredients	8	−5	12
Contains less fat	8	0	11
Authentic pasta dishes for 99 cents	7	0	10
Everything premium except the price	6	0	8
Meals carefully prepared using all natural ingredients	8	−5	12

Note: Segment No. 1 (Seg 1) appears to be principally interested in a low price. Segment No. 2 (Seg 2) appears to be principally interested in health benefits.

study generated two segments, differing in the concept elements that were most appealing. Table 15–8 shows the strongest (winning) elements from the study for the total panel and for two segments. The total panel shows a mixture of types of elements that win and lose, respectively. No clear pattern exists for the total panel, suggesting no organizing principle governing winning versus losing elements. On the other hand, the segmentation reveals two clearly different groups. Segment No. 1 is interested primarily in a lower price for frozen dinners. Segment No. 1 is basically interested in frozen dinners (with an additive constant of 91, meaning that even if the concept had no elements present but just presented the idea of a frozen dinner, panelists in segment No. 1 would still be interested in buying the product). In contrast, segment No. 2 is interested in health benefits. Their constant, 49, is moderate, meaning that with no elements present in a concept these panelists would only be moderately interested in buying the frozen dinner. On the other hand, with the proper health benefit emphasized for the dinner, the interest of the panelists would increase.

Subjective Aspects of a Meal—Multicompartment Frozen Meals

In a separate study of frozen dinners, the author recently had the opportunity to compare consumer reactions to a number of different tray configurations. As part of the study, and before the actual consumer evaluations of concepts (some of which included the tray), a small group of 12 consumers profiled the different concept elements on a set of nine descriptive semantic scales, ranging from "traditional-contemporary" to "wholesome." The first part of the study generated a matrix of concept element times semantic scale. In the second part of the study another group of 180 panelists rated different combinations of concept elements. A combination or test concept was composed of two to five concept elements, including package design in some cases. The systematic combinations were the test concepts. Every panelist evaluated a totally unique set of combinations. With 245 concept elements in that study there were many tens of millions of combinations that could be tested. The concept evaluation only looked at 16,000 combinations (100 combinations for each of 160 panelists). Those combinations sufficed to estimate the part-worth use of each concept element.

We are concerned here with the semantic scale profiles of nine different trays. The trays were shown as colored pictures that were taken from photographs containing actual foods. These nine tray elements in the concepts provide their own miniexperimental design because there are three different shapes (oval, round, rectangle) and three compartments (one, two, three).

Table 15–9 shows the basic data extracted from the "dimensionalization" part of the study, in which panelists rated the different meal trays on various semantic scales. Each column consists of a specific shape (oval, round, or rectangle) and a specific number of compartments (one, two, or three). The semantic scales went from a low of 1 to a high of 9. The nine semantic scales appear as the first column, which shows the top part of the scale only (namely, contemporary stands for traditional = 1 to contemporary = 9).

The nine dinners are laid out in the form of a complete factorial design (three shapes times three compartments). The factorial design permits the creation of a simple model showing the part-worth contribution of each shape and each number of compartments to the consumer rating. The regression model is composed of five terms—an additive constant, the contributions of two of the three shapes, and the contribution of two of the three compartments. One shape and one compartment are left out of the equation and arbitrarily assigned a value of 0. These are the reference shape and compartment. They must be left out of the equation because they are not independent of the other two shapes and compartments. The additive constant shows the expected value of the combination of the reference shape and compartment number.

With this in mind one can see from Table 15–10 that the number of compartments exerts a greater effect on the ratings. For example a three-compartment dinner tray is perceived as far more traditional (-1.0) than a one-compartment tray (0), far more tasty (1.3), far more convenient (1.3), and far more satisfying (1.2).

Improving Meal Combinations

Meal combinations are composed of individual items. A previous study (Moskowitz & Klarman, 1977) attempted to represent meal items in a geometrical space with the closeness of two items in that space proportional to the degree to which the items were compatible. That approach did not work because of "intransitivities"

Table 15–9 Semantic Profiles of Dinner Trays Varying in Shape and Compartments

Shape	Round	Oval	Rect	Round	Oval	Rect	Round	Oval	Rect
Compartments	1	1	1	2	2	2	3	3	3
Traditional vs. contemporary	5.6	5.7	3.8	5.2	5.3	4.9	3.8	4.1	4.2
Low value vs. high value	5.0	4.3	4.7	4.9	4.6	4.9	5.4	5.2	4.4
Tasty	4.8	5.4	5.1	5.8	5.7	5.0	6.3	6.7	6.2
For one person vs. family	5.4	5.6	5.2	4.9	5.2	5.2	6.1	6.3	6.9
Expensive vs. affordable	6.8	7.2	5.8	7.3	7.3	6.9	6.4	6.9	6.0
For occasions vs. daily	3.2	3.3	5.3	2.8	2.7	3.3	4.0	3.9	3.8
Convenient	6.2	6.1	6.0	6.9	6.8	6.1	7.2	7.9	7.0
Satisfies	4.8	5.6	5.8	6.0	6.1	5.2	6.4	6.8	6.4
Wholesome	5.3	5.8	6.0	6.0	6.3	5.3	6.0	6.2	6.2

Table 15–10 How the Shape of the Tray and the Number of Compartments Drive Subjective Impressions of the Meal Contained Therein (The Constant = Expected Rating of a Round, One-Compartment Tray)

	Constant	Shape Round	Shape Oval	Shape Rect	Comp One	Comp Two	Comp Three
Traditional vs. contemporary	5.1	0.0	0.2	−0.6	0.0	0.1	−1.0
Low value vs. high value	5.0	0.0	−0.4	−0.4	0.0	0.1	0.4
Tasty	5.1	0.0	0.3	−0.2	0.0	0.4	1.3
For one person vs. family	5.2	0.0	0.2	0.3	0.0	−0.3	1.0
Expensive vs. affordable	6.7	0.0	0.3	−0.6	0.0	0.6	−0.2
For occasions vs. daily	3.7	0.0	0.0	0.8	0.0	−1.0	−0.1
Convenient	6.2	0.0	0.2	−0.4	0.0	0.5	1.3
Satisfies	5.2	0.0	0.4	0.1	0.0	0.4	1.2
Wholesome	5.6	0.0	0.3	0.1	0.0	0.2	0.4

(e.g., food A might go well with food B, food B might go well with food C, but food C might not go well with food A). A, B, and C are all from different parts of the meal.

For this reason, the notion of a general single rule for food combinations was abandoned in favor of a mix-and-match system, the TURF analysis discussed previously. This section deals with the TURF approach for a more extensive problem in menu development, again for frozen meals.

The TURF analysis discussed previously has been used in a number of studies to improve the acceptability of commercially available meals. These meals are composed of individual items that may be included in a dinner because of the confluence of dietitian inputs and economic factors. For example, quite often meat loaf is paired with mashed potatoes and diced carrots. Meal combinations can be optimized by identifying new, potentially more acceptable components. For example, in one particular instance, a meal made up of meat loaf, mashed potatoes, and diced carrots appeals to 31% of the total population, based on the TURF analysis. By substituting scalloped potatoes with either green beans, green beans and slivered almonds, or baby carrots, the manufacturer can increase the acceptance of the meal from 31% to 35% of the population.

A Meta-analysis: Number of Components in a Meal and the Complexity of the Main Item

Quite often researchers think of main meal items as belonging to a single, homogeneous category that as a group tend to have the same general properties of being a main item. Recent work suggests, however, that main meal items may differ from each other, depending on the nature of a specific item. This is a meta-analysis of the data, or an analysis that looks for higher order patterns.

One example of this meta-analysis comes from a study of different alternations of an existing set of meals in the market. The study focused on how to increase overall interest in specific frozen meals by changing the number and type of side dishes. The panelists rated interest in single items (main) and interest in the main item paired with one, two, or three additional side dishes. The main item was held constant. The panelists evaluated different sets of meals in a rotated order, so that the meal description would not always have the same main item. The data, shown in Table 15–11, suggest that the simpler main entrée (e.g., a simple meat) requires more side dishes to maximize meal acceptance than do complex main entrées and that a single complex side dish can substitute for two simpler side dishes. These results appear (in part) in Table 15–11, showing the percent of consumers that rate interest in a particular combination of one, two, three, or four combinations of items.

GENERAL DISCUSSION—BASIC RESEARCH ISSUES IN THE WORLD OF BUSINESS APPLICATIONS

Academic Search for Principles Versus the Developer's Expedient Solutions

The commercial world differs from the academic world in the way they approach the solution of problems. Academics look for general rules underlying specifics. The experiments that academics conduct are generally done to discover the underlying principles. The aim of the experiment is to confirm or to deny and in some cases to create the necessary data from which to draw trends.

Applied researchers in the commercial realm also need to have principles to guide development, but usually do not have the resources, the requisite time, or the reward structure to discover underlying principles. A great deal of corporate understanding of the components of a meal, what goes into making an acceptable meal, what makes a product boring, and so forth, comes from the gradual learning gleaned by one individual or by a small team. Applied research tends to be more haphazard, more focused on answering immediate questions. It is unfortunate, however, that in the competitive commercial world in which such knowledge is needed most, little attention is paid to general rules that could make future product development all the more easy. Even analyses such as the relative importance of items in a menu (e.g., entrée vs. starch vs. vegetable) tend to be considered as "academic issues" and are jettisoned in favor of the quick, expedient answer.

Table 15–11 Maximum Percent of Panelists Interested in a Specific Meal, Composed of 1–4 Items

	Items	*Interest*
No. 1	Salisbury steak	29
	Salisbury steak + mashed potatoes	41
	Salisbury steak + mashed potatoes + sweet corn	43
	Salisbury steak + mashed potatoes + sweet corn + strawberry cheesecake	51
No. 2	Fried chicken	39
	Fried chicken + parsley buttered new potatoes	47
	Fried chicken + parsley buttered new potatoes + Corn	56
	Fried chicken + baked potatoes + broccoli + apple crumb	60
No. 3	Chicken chow mein	33
	Chicken chow mein + 2 mini chicken egg rolls	37
	Chicken chow mein + 2 mini chicken egg rolls + fried rice with almonds	43
	Chicken chow mein + oriental stir fry rice pilaf + 2 mini chicken egg rolls + apple plum crisp	48
No. 4	Beef enchiladas	34
	Beef enchiladas + Mexican style rice	38
	Beef enchiladas + Mexican style rice + southwestern picante beans	40
	Beef enchiladas + Santa Fe–style potatoes + corn + apple plum crisp	42

Databasing

Databases are becoming increasingly popular in the business community. Until now most databases that menu developers relied on have come from the records of family-style and quick-serve restaurants. Current technology allows the researcher to identify those particular items that often are selected with each other. Databases using only actual products provide only half the story, however. These "behavioral databases" tell the developer what has happened but do not point to new possibilities. Only attitudinal data (e.g., responses to statements about menus), or sensory data (responses to actual products) can transcend the shackles of "what is" to "what could be." The creation of a database for menu preferences is not particularly difficult, but again it requires a change of the paradigm from "reactive" to "proactive."

The Role of Meta-analysis

In commercial developments of meals, the creation of "rules" is becoming important. Rules allow the developer to go beyond the empirical data and identify general "organizing principles." Rules may be as simple as what products go together or as complex as algorithms to identify optimal combinations of foods that maximize acceptance and minimize cost. Rules are obtained through meta-analysis, consisting of a step back from the empirical data to ask "what is going on." A rule might be that "for complex entrees themselves comprising multiple components, one needs fewer side dishes, and any more side dishes adds less incremental acceptance." As mentioned in the introduction, the reader should not expect ongoing systematic investigation by companies of what should be incorporated in a meal that these companies themselves create and market. There is simply no time and no immediate payback for the research. Academic researchers interested in the study of consumer reactions to meals may make a great contribution to the field by developing these rules through a "meta-analysis" of data. Those researchers can be certain of a warm welcome by corporate researchers of any such rules that they discover.

REFERENCES

Balintfy, J.L., Duffy, W.J., & Sinha, P. (1974). Modeling food preferences over time. *Operations Research, 22,* 711.

Box, G.E.P., Hunter, J., & Hunter, S. (1978). *Statistics for experimenters.* New York: John Wiley & Sons.

Cattin, P., & Wittink, D.R. (1982). Commercial use of conjoint analysis: A survey. *Journal of Marketing, 46,* 44–53.

Gacula, Jr., M.C. (1993). *Design and analysis of sensory optimization.* Trumbull, CT: Food and Nutrition Press.

Green, P.E., & Srinivasan, V. (1978). A general approach to product design optimization via conjoint analysis. *Journal of Marketing, 45,* 17–37.

Griffin, R., & Stauffer, L. (1990). Product optimization in central location testing and subsequent validation and calibration in home use testing. *Journal of Sensory Studies, 5,* 231–240.

Joglekar, A.M., & May, A.T. (1991). Product excellence through experimental design. In E. Graf & I.S. Saguy (Eds.), *Food product development* (pp. 211–230). New York: Van Nostrand Reinhold.

Moskowitz, H.R. (1980). Psychometric evaluation of food preferences. *Journal of Foodservice Systems, 1,* 149–167.

Moskowitz, H.R. (1981). Sensory intensity vs. hedonic functions: Classical psychophysical approaches. *Journal of Food Quality, 5,* 109–138.

Moskowitz, H.R. (1985). *New directions in product testing and sensory analysis of foods.* Trumbull, CT: Food and Nutrition Press Inc.

Moskowitz, H.R. (1994). *Food concepts and products: Just in time development.* Trumbull, CT: Food and Nutrition Press Inc.

Moskowitz, H.R. (1996a). A commercial application of RSM for ready to eat cereal. *Food Quality and Preference, 8,* 191–202.

Moskowitz, H.R. (1996b). Segmenting consumers on the basis of their responses to concept elements: An approach derived from product research. *Canadian Journal of Marketing Research, 15,* 38–54.

Moskowitz, H.R. (1997). Preferences for southwestern salted snack flavors: Individual differences at the conceptual level and the construction of a product line. *Journal of Sensory Studies, 12,* 11–24.

Moskowitz, H.R., & Klarman, L.L.E. (1977). Food compatibilities and menu planning. *Canadian Institute of Food Technology Journal, 10,* 257–264.

Moskowitz, H.R., & Martin, D.G. (1993). How computer aided design and presentation of concepts speeds up the product development process. *Proceedings of the 46th ESOMAR Conference,* 1993, Copenhagen, Denmark, 405–419. Amsterdam: ESOMAR.

Moskowitz, H.R., & Rogozenski, J.G., Jr. (1975). A system for the preference evaluation of cyclic menus. *Technical Report 75–46-OR/SA.* Natick, MA: U.S. Army Natick Laboratories.

Pangborn, R.M. (1970). Individual variations in affective responses to taste stimuli. *Psychonomic Science, 21,* 125–126.

Peryam, D.R., & Pilgrim, F.J. (1957). Hedonic scale method for measuring food preferences. *Food Technology, 11,* 9.

Plackett, R.L., & Burman, J.D. (1946). The design of optimum multifactorial experiments, *Biometrika, 33,* 305–325.

Rogozenski, J.G., Jr., & Moskowitz, H.R. (1982). A system for the preference evaluation of cyclic menus. *Journal Of Foodservice Systems, 2,* 139–161.

Turner, M., & Collinson, S. (1988). Consumer acceptance meals and meal component. *Food Quality and Preference, 1,* 21–24.

Chapter 16

Meal Design: A Dialogue with Four Acclaimed Chefs

J.M. Schafheitle

INTRODUCTION

Just as the artist is faced with a blank canvas on which to create a masterpiece of color, texture, and form, the chef has an empty piece of paper—a menu—on which to compose a nutritionally balanced and gastronomically exciting meal, anticipation of which should satisfy and delight the dining experience of his customers. This chapter seeks to describe and analyze the development of meal design from a historical to a modern context. It highlights the sequence of dishes and their place on the menu. Four internationally acclaimed chefs give their views, in dialogue form, on the elements they as professional chefs consider fundamental to good meal design. Their reflections are based within a framework of questions with menu examples provided.

Possibly the first thing we think of when considering meal design is the menu, a French word that denotes minute details as applied to a bill of fare or sequence of a meal. Its main purpose is to inform the diner of the dishes and courses available (Senn, 1954); it informs the diner accurately about the food offered and provides a blueprint for the kitchen operation to develop (Fuller, 1974). The menu choices offered reflect the availability of food and the chefs' ability in relation to skill, competence, and time available to produce dishes. To the diner the menu choice provides a balance of flavors and textures and is a signal of what is available on a particular day. In western society, meals are composed of a series of courses: starter, main course, and sweet, and although the number of courses may change to reflect the occasion, the basic *sequence* remains the same. The movement through courses should ascend progressively from light, delicate, and more complex flavors to rich, more full-bodied and simple flavors (Olney, 1985), with each course providing a contrast to the one preceding it. The desired result may often be difficult to achieve; essentially the only thing to remember is that the palate should be kept fresh, surprised, and excited throughout the meal and not overwhelmed.

HISTORICAL PERSPECTIVE

The history of Europe has been colorful and turbulent, with kings and emperors fighting for possession of its soil. Food choice was within the realm of the nobility and ruling classes because they were the only people with the means to choose; the peasants usually ate what the land naturally provided or spent their meager wages on basic staples. Eating fashions changed and adapted as a consequence of food availability, but the central focus in the homes of the ruling elite was the cook.

French recipes were the backbone of meal design; French was the language of the kitchen, the menu, and the food culture. French cooks dominated the kitchens and their influence permeated the household.

Meal design evolved over hundreds of years (Aron, 1975) into a style of service that became known as *Service à la Française*. This opulent buffet style contained a number of meal choices and featured all of the food on the table before the commencement of the meal. The food was invariably cold, and where you sat usually dictated what you ate. Despite a mistrust of all things Russian, it was generally acknowledged that *Service à la Russe* had benefits over the French style. The table was laid with a floral centerpiece or candelabra, linen, cutlery, and glasses; the food was then served in a sequence of dishes so that everyone enjoyed the same food at the same time. This was the forerunner of the modern "silver service."

The rise of the middle class (Mennell 1985, Visser 1991) opened up the "etiquette" of eating and dining. Dynamic changes in the fabric of society affected the social structure, aristocracy, and royalty; the formation of nationhood, particularly in Britain and France, brought about changes in the relationship between fashion, custom, table manners, and food dishes. Meal choice no longer remained the sole province of the rich; meal design began to develop in the style we recognize today.

EARLY INFLUENCES ON MEAL DESIGN

The eighteenth and nineteenth centuries were the heyday of opulence and extravagant culinary showpieces in the homes of the wealthy. The move toward "eating out" started with the emergence of hotels and restaurants becoming popular with the wealthy middle class. Dishes were given the names of historical events, famous people of the past and present, geographical areas, and seasons; all were listed in the Répertoire de la Cuisine (Saulnier, 1976); for example, the cooks bible, which detailed the dishes, their components, and garnishes in French, naturally, but left the chef to choose how the food would be cooked.

The personality of the chef also became prominent at this time. He commanded the respect of his kitchen brigade and of a wider audience; books containing his recipes, culinary tips, and methods were widely available and were influential not only in culinary circles but domestic ones also. The striving for perfection became an art and a vocation. Marie Antoine Carême, Urban Dubois, Emile Bernard, Alexis Soyer, Auguste Escoffier, and Edouard Nignon were all French chefs of high repute, who made their mark in the culinary world in their lifetime and left a legacy of literature that still inspires chefs of today (Page & Kingsford, 1971). The food culture of the French, admired and imitated by royalty and the aristocracy of Europe traveling with the French army and to the colonies, was practiced in all civilized western societies. It encompassed food, the combination and sequence of dishes, marriage of food with beverages, manners, and rituals of the table and culminated in a set standard that became the accepted norm that was emulated and held up as the ultimate in fine dining.

This standardization of good to high-quality menu compilation was unchallenged throughout the western world until the early 1970s, when Cuisine Nouvelle, a new style of cooking, began to emerge. This acknowledged and focused on a wider consumer profile, their needs, and wants; dishes were light and delicate, visually pleasing, and full of the natural flavor of the food. Fernand Point was considered the Godfather of Nouvelle Cuisine; Paul Bocuse, the Troisgros brothers, and Roger Verge were followers until they themselves became innovators and leaders (Blake &

Crewe, 1978). Menus were highly descriptive and often detailed all the components of a dish. Promotion of this new understanding was propagated through the dimension of the modern media. The role of the food guide and critic to publicize an establishment's food became an important marketing and promotional tool that chefs took every opportunity to exploit. Television programs took on the mantle of educators, informing the public in an entertaining way how food should be cooked and what combinations are acceptable.

The influence of French gastronomy expressed in meal selection, the art of cooking, and the language of the menu may now be in decline as people rediscover their own regional or national food heritage. Acknowledgment of the merits of other food cultures has resulted in the availability of a great diversity of new dishes and food combinations.

BASIC PRINCIPLES LEADING TO GOOD MEAL DESIGN WHEN PLANNING A MENU

J.A. Brillat-Savarin, the eighteenth century magistrate, philosopher, and gastronome, said *"menu mal fait, dîner perdu"*—a badly composed menu is likely to spoil the best dinner—a point confirmed by Escoffier, the twentieth century chef, culinary leader, and writer who spent many hours over the composition of his menus and wrote of the difficulties and hazards involved in this exacting art (Escoffier, 1965). Menu planning has been defined (Pauli, 1979) as an art and a science that requires an extensive knowledge of food, the basic method of preparation and service, understanding of the nutritive value of foods, and a sensitivity to the aesthetics of foods in combination.

The traditional sequence used to construct a meal (shown in Table 16–1) has been based on the historical French influences in which a formal menu might con-

Table 16–1 Traditional Sequence of Courses

Course No.	English	French	Food Example
1	Cold hors d'oeuvre	Hors-d'oeuvre froid	Oysters
2	Soup	Potage	Consommé
3	Hot appetizers or small entrées	Hors-d'oeuvre chaud or entrées volantes	Beef marrow on toast
4	Egg or farinaceous dish	Oeuf ou farineux	Poached egg
5	Fish	Poisson	Blue trout
6	Hot entrée, garnished	Entrée chaude	Breast of chicken
7	Main course—joint of meat or poultry with garnish of vegetable and potato	Relevé ou pièce de résistance —ou grosses pièces	Saddle of veal
8	Cold entrée	Entrée froide	Ham mousse
9	Flavored water ice	Sorbet	Champagne sorbet
10	Roast with salad, poultry, game, some meats	Rôti et salade	Roast pheasant
11	Hot or cold vegetables	Entremet de légumes	Gratinated asparagus
12	Hot sweet	Entremet chaud	Sweet soufflé
13	Cold sweet	Entremet froid	Vanilla bavarois
14	Cheese savories, hot or cold, cheeseboard	Entremet de fromage	Emmentaler cheese straws
15	Selection of fresh fruit	Dessert	Seasonal fruit selection
16	Coffee	Cafe	

Source: Data from E. Pauli, *Classical Cooking the Modern Way,* © 1979, CBI Publishing Company, Inc.; and W.K.H. Bode, *European Gastronomy,* © 1994, Hodder & Stoughton.

Table 16–2 Modern Shortened Menu

Course No.	Menu Sequence	Food Example
1	Cold hors d'oeuvre	Game pâté
2	Soup	Clear soup with diced vegetables
3	Hot fish course	Poached fillet of sole with creamy mushroom sauce
4	Meat with garnish	Roast rib of beef with Yorkshire pudding
5	Dessert (or cheese)	Peach Melba

Source: Data from E. Pauli, *Classical Cooking the Modern Way,* © 1979, CBI Publishing Company, Inc.: and W.K.H. Bode, *European Gastronomy,* © 1994, Hodder & Stoughton.

sist of 13 to 16 courses, each of which conformed to the classical and universally accepted French style. The number of courses has been reduced considerably through the trend toward simplification of eating patterns and fewer menu choices may be offered; occasions and demand to warrant a large choice of foods and courses at any given time are seldom, and a menu of three to six carefully selected delicacies, expertly prepared and served, are more appreciated by diners today.

A shortened menu version based on the classical French sequence and structure has evolved (Pauli, 1979) and is shown in Table 16–2. This follows the same classical

Exhibit 16–1 General Guidelines for Designing a Meal

Vary	**Alternate**	**Avoid**
• Color	• Dishes:	• Monotony
• Flavor	–Light to heavy and then	• Elaboration
• Texture	light	• Complex flavors
• Shape and size	–Soft to crisp foods	
• Main ingredient	–Bland to strong-flavored	
• Food groups	foods	
• Cooking methods	–Dry to moist foods	

Consider

- Type of operation
- The occasion—type of menu
- Seasonal effects:
  - Weather (hot/cold)
  - Calendar events
  - Feast or fast
- Availability of supplies/commodities
- Dish selection and balance
- Capability of staff (front/back of house)
- Equipment and plant required

- Location (nationality, religion, customs)
- Customer profile, boredom factor, length of time available for eating
- Type of menu (e.g., à la carte, lunch, dinner)
- Language used: legibility, descriptive, informative, honest; avoid mixing languages unless for national dishes
- Menu cost
- Overall cost framework
- Size of food preparation area
- Dining area, service style, implements

Summary

- Achieve a balance of food constituents, food groups, color, etc.
- Cost framework; dishes per staff ability; equipment and food availability.
- Know your customers likes and dislikes; respect their customs and traditions.

Exhibit 16–2 Example Menu Chart Analyzer

Occasion			Date		
	Cold Starter	Soup	Fish Course	Meat With Garnish	Dessert (or Cheese)
Name of dish					
Main ingredients and color					
Dominant flavor/ component					
Cooking method					
Garnish or sauce (color)					
Herbs, spices, seasoning					
Light or heavy starch, protein, carbohydrate					
Texture					

sequence but with some courses amalgamated while others are omitted. Depending on the nature of the function or occasion, canapés (or Amuse Gueule) might also be served with a reception drink at the start and petit fours at the end with tea or coffee.

The way in which chefs design and present meals largely depends on the image and restrictions of their establishment (e.g., number of chefs, service staff, and covers or the number of diners); type of clientele; selection and availability of foods; and style of cooking. The limitations of a chef singlehandedly producing food for 20 customers are greater than those of a large brigade producing food for a few hundred.

Exhibit 16–1 shows a collation of general rules (from textbook sources designed for students and practitioners) for use when designing meals (Cracknell & Nobis, 1985; Duch, 1961; Fuller, 1974; Herring, 1969; Kinton, Ceserani & Foskett, 1995; Klinger, 1978; Pauli, 1979). Menu design is a gastronomic equation of what the caterer has to consider and what the consumer expects. The first three physiological points for consideration (vary, alternate, avoid) affect the senses and customer perception. These elements should be heightened by variation so that the senses are constantly sharpened and do not become overloaded, diminished, or dulled. The second level of points (i.e., under "Consider") relate to culture, business, environment, enterprise, cost, profit, marketing, viability, and customer care. The three

Exhibit 16–3 Questions Asked of the Chefs

1. Background and Training:
   In what ways have your background and training affected your understanding and interpretation of cooking/style/repertoire of dishes/ menu types?
2. Menu Development and Design:
   Have you kept any old menus from your early career which could be compared to current menus? Has this style changed over the course of your career? What are the cornerstones of your meal design?
3. Menu Examples:
   Do you adopt a different food style for special occasions or menu types? Can you give several examples.
4. Current Influences:
   In what ways, if any, is your menu design influenced by current trends? (e.g., Italy/Mediterranean, California, Pacific Rim.)
5. The Main Elements Considered when Composing a Menu:
   When composing a menu what elements do you consider (in your order of priority)? (e.g., the occasion/client, cost, menu structure, choice, number of courses, nutrition, seasonality of foods, food combinations [classical French/modern], availability of staff, etc.). Please elaborate on the importance of each element in your own menu plan.
6. Service Style:
   In what ways does your menu reflect the style of service (e.g., silver service, plate service).
7. Other Influences:
   A. When planning a menu, do you follow the example of others (e.g., Escoffier, your peers, yourself?) Do you use any books for reference or rely on tried and tested menus from your own repertoire?
   B. Do you have an instinctive/intuitive knowledge of what foods/dishes will complement each other? Was this developed through socialization or learned as part of your craft (nature or nurture)?
   C. Describe what elements you bear in mind when you put together different combinations of foods and dishes to create a 'rounded' eating sensation. How do you personally organize the menu to take account of different menu choices which, in varying combinations, still produce a balanced meal?
   D. Text books refer to the classical structure of 13–14 courses. In what ways do you follow the classical sequence when serving only 4–6 courses? Is the classical structure still, in your opinion, of value? If not, what do you suggest as an alternative?
   E. In eastern cultures it is possible to have a varied choice of food items, cooked by different methods, which may also include a soup, served together. Would you be horrified if, in the future, Europeans went back to the *à la Francaise* days when everything was laid out on the table and diners picked what they wanted?
   F. How often do you change your menu? (Table d'hôte, à la Carte, special menus)?
   G. Do you believe that beverages are an integral part of any meal or should the balance of foods be sufficient?
8. The Future:
   Do you foresee a time when your current style may have to change or adapt? Can you predict or speculate on the future?
9. What menu design advice would you give to a newcomer?

elements of the "Summary" combine the physiological, psychological, and sociological aspects of dining in a business environment.

The skill of the chef lies in cooking a satisfying meal, but his or her creative talent is required to balance the food components so that the flavors, textures, colors and temperatures mix and mingle happily together and meet his customers' expectations. This is where the art of the chef lies. Interpretation of the guidelines is fairly personal to each chef, so to give a broad spectrum of ideas, four world-renowned chefs (Albert Roux, Shaun Hill, Anton Mosimann, and Anton Edelmann) of different nationalities, background, and training have each been asked for his perspective. Set within the context of their own successful business environments, each chef works in a different type of establishment and each discusses the emphasis he places on specific factors when designing and planning a meal at the highest level of gastronomic experience.

For those without the chefs' technical expertise, use of a plan, such as Exhibit 16–2, may aid analysis of the composition of different courses and identify areas of repetition.

The questions set out in Exhibit 16–3 provided the framework on which the four chefs gave their responses. Profiles of the four chefs appear in italics, and their replies to the questions appear in normal print.

ALBERT ROUX, CMA, MCF

Le Gavroche, London

Profile

Albert Roux opened Le Gavroche in London with his brother Michel in 1967, followed by Le Poulbot, Le Gamin, The Waterside Inn (Bray), and Gavvers. The two brothers were awarded the Chevalier du Merite Agricole by the French government in 1975 and both Le Gavroche and The Waterside Inn have achieved many awards for their cuisine, including 3-star Michelin status. The Roux brothers' book, New Classic Cuisine, *published in 1984, typifies the lighter style of classical French cooking for which they have become renowned.*

Albert Roux and Michel Bourdin (Executive Chef at The Connaught, London) were the principals behind the establishment of the British branch of the Académie Culinaire de France formed in 1980, now renamed the Academy of Culinary Arts. As vice-president of the Academy, Albert is actively involved in the management and education aspects of this select group of chefs, patissiers, restaurant managers, sommeliers, and suppliers. Education in the appreciation of food for its taste is a fundamental element of Albert's philosophy. He is looked up to as a Godfather, with reverence, love, and affection, by his peers and by those young people he meets through the Académie Culinaire's 'Adopt a School' program.

The Roux brothers' kitchens have long been a "hot house" for young and enthusiastic chefs, many of whom have received help—financial, business, and culinary—to establish businesses of their own, which in turn have become famous for their food and hospitality. The Roux brothers sponsor a competition for young chefs, with the winner working in top-class establishments in Europe.

Responsibility for the daily cooking at Le Gavroche has been passed from Albert to his son Michel, leaving more time for Albert to develop his extensive international consultancy and business interests.

Background and Training

I never cooked at home but moved to Paris to do an apprenticeship at 14. The training was very traditional, learning the basics of taste. From pastry onto kitchen, my *(traditional)* apprenticeship *(learning from Master chefs)* was purely in private homes. Those meals were very balanced so that the flavor of the menu items did not clash (e.g., if you had a creamy starter you didn't want another creamy course after that). The importance of menu planning with the wine was also studied—they knew they wanted to complement the wine, and I think that was understood right from the beginning.

Menu Development and Design

Things have changed immensely, and the audience participating in good eating is far larger. In the old days it was only in private service, for the very few privileged people; now it is far more open.

For me, the cornerstone of good meal design is that the food has to be seasonal, and I attach a lot of importance to that. Over the years it has unfortunately changed immensely; you are eating strawberries all year round and some of them are totally tasteless. At the end of the day the ultimate of taste is to be blind and to be able to tell what you are going to eat by smelling it; if you cannot define what you are going to eat by the smell, let alone the touch, then you should not eat it.

You have to respond, within reason, to what people are expecting, but you will never see asparagus in the winter, never. We are firmly anchored into the seasons of very many things at Le Gavroche, not as much as I would like, but I would say 70% of the things that we serve are seasonal. Winter is a wonderful season—salsify, all the roots. Vegetables come to their best and that is the time to use them.

Menus for the private dining room at 47 Park Street (*above Le Gavroche*) are more like "a home from home" very much like private service—so its menus become more pure than at Gavroche. The season is observed far more and flavor comes into its own. Exhibits 16–4 and 16–5 display a sample lunch menu and menu exceptionnel from Le Gavroche.

Current Influences

I think we all have a heritage in the world, be it Chinese cuisine, Thai cuisine, English, French, Italian, but I think you should stick to your own tastes. Fusion cuisine and world cuisine are very different from each other. I am open to other cuisines but would never dream of putting lemongrass into a steak and kidney pie; that is fusion cuisine, but it totally loses identity from where it comes and it is the recipe for disaster. Very few people can mix complex flavors—only craftsmen—and few people can solve the meal problem because they do not understand the intimate soul of flavor; they either underdo it or overdo it. Few people can mix flavor beautifully.

There is a lot in our cultural heritage to be looked at, enlightened, and modified without losing its identity to be lightened up. At Le Gavroche we lightened up French cuisine and said goodbye to sauce Espagnole, goodbye to demi-glaze; Troigros did the same and so did Bocuse, but still a lot remains to be done. This was Cuisine Legère or the new Classic Cuisine that we published some 15 years ago in England; Cuisine Nouvelle was very different from this.

As you can see from comparison of styles and interpretations in the menus in Exhibit 16–6, the one I composed for Le Gavroche's 25th anniversary is more traditional than that which Michel created for the 30th anniversary in 1997. His menu is much lighter, with fewer courses, while retaining the balance, marriage with wine, and sense of occasion.

Exhibit 16–4 Le Garoche Lunch Menu

LES VINS

RULLY 1996
Joseph Drouhin

PINOT "BLANCS DE BLANCS" 1997
Leon Beyer

-------0-------

CHATEAU VIEUX SARPE 1993
St Emilion

DOMAINE GAVOTY - CUVEE CLARENDON 1988
Cote de Provence

Lunch menu @ £37.00 per person
inclusive of vat & 1/2 bottle of wine per person
there will be a 12.5% discretionary service charge

Courtesy of Albert Roux, LeGavroche, London, England.

continues

Exhibit 16–4 Continued

LE   MENU

RAVIOLE DE RAIE AU CELERI ET CAPRES

OU

OEUF POCHE CAREME

OU

*RILLETTES DE PORC ET SA SALADE DE PETITS LEGUMES,
VINAIGRETTE DE TRUFFES*

******

*QUEUE DE LOTTE ROTIE AUX ASPERGES ET VENTRECHE,
BEURRE BLANC A LA MOUTARDE*

OU

FRICASSEE DE VOLAILLE FERMIER A L'ESTRAGON ET CRETES DE COQ

OU

QUASI D'AGNEAU ROTI A LA NICOISE

******

PLATEAU DE FROMAGES

OU

LES GLACES ET SORBETS MAISON

OU

MILLEFEUILLE AUX FRAMBOISES ET CREME CHANTILLY A LA VANILLE

******

CAFE ET PETITS FOURS

RELAIS &
CHATEAUX
Relais Gourmands

Exhibit 16–5 Le Garoche Menu Exceptionnel

Menu Exceptionnel

(Pour l'Ensemble de la Table)

Pointes d'Asperges et Saumon Fumé à l'Oeuf Cassé et Caviar
Warm Asparagus and Smoked Salmon with Egg and Caviar Sauce

— ✧ —

Filet de Rouget sur Compote d'Aubergines au Cumin
Grilled Fillet of Red Mullet on Spicy Aubergine and Fennel Salad

— ✧ —

Foie Gras Chaud et Pastilla de Canard à la Cannelle
Hot "Foie Gras" and Crispy Pancake of Duck Flavoured with Cinnamon

— ✧ —

Canon d'Agneau sur Pommes Boulangère et Petits Pois à la Française
Loin of Lamb with French Style Braised Peas

— ✧ —

Barattes de la Saône-et-Loire, Salade Mesclum au Vieux Vinaigre
Fresh and Aged Goats Cheese with an Herb Salad, Red Wine Vinegar Dressing

— ✧ —

Parfait au Chocolat Blanc et Gelée de Framboises
White Chocolate Mousse filled with Poached Raspberries

— ✧ —

Petites Poires d'Api au Caramel et Sauce Chocolat
Toffee Pears and Caramel Mousse, Bitter Chocolate Sauce

— ✧ —

Café et Petits Fours

For the entire Party @ £78 per person, inclusive of VAT
Discretionary service charge 12.5%

Chef Patron: Michel A. Roux   Manager: Silvano Giraldin

Courtesy of Albert Roux, Le Gavroche, London, England.

continues

Exhibit 16–5 continued

Minimum Charge £60.00

VAT Included
Discretionary service
charge 12.5%

Hors-d'Oeuvre

Soufflé Suissesse... ... ... 18.10
Cheese Souffle Cooked on Double Cream

Mousseline de Homard au Champagne 30.20
Lobster Mousse with Caviar and Champagne Butter Sauce

Foie Gras Chaud et Pastilla de Canard à la Cannelle... ... 25.60
Hot "Foie Gras" and Crispy Pancake of Duck Flavoured with Cinnamon

Cassolettes d'Escargots et Cuisses de Grenouilles
    aux Herbes... ... ... 18.90
Snails and Frogs Legs served in little pots flavoured with Herbs

Ragoût de Langoustines Parfumé au Gingembre 23.20
Warm Langoustines in a light Ginger Butter Sauce

Coquilles St Jacques Sautées aux Épices 24.20
Pan Fried Scallops with a Five Spice Sauce

Terrine de Foie Gras d'Oie Truffée à l'Ancienne 26.80
Classic Goose Liver and Truffle Terrine

Filet de Rouget sur Compote d'Aubergines au Cumin ... 15.80
Grilled Fillet of Red Mullet on Spicy Aubergine and Fennel Salad

Pointes d'Asperges et Saumon Fumé à l'Oeuf Cassé
    et Caviar 28.60
Warm Asparagus and Smoked Salmon with Egg and Caviar Sauce

Fricassée de Homard et Pied de Cochon Farci 17.90
Lobster and Stuffed Pigs Trotter, Onion, Caper and Gherkin Sauce

*Although a cigar can be satisfying after a meal, it can be disturbing to other diners. We would therefore respectfully
suggest that you retire to the lounge or the bar to enjoy your coffee and cigar; (no pipe smoking).*

Our Private Dining Room is available from 8-20 people.

continues

Exhibit 16–5 continued

Poissons et Viandes

Tronçonnette de Turbot et Caneloni de Crabe 28.40
Roast Turbot and Crab Caneloni, Light Crab and Carrot Sauce

St. Pierre Rôti Entier, Artichauts Barigoule
  et Pommes Parmentier *(2 pers)* 52.40
Whole Roast John Dory with Artichokes and Smooth Olive Oil, Mashed Potatoes

Pavé de Bar Sur Poêlée de Champignons et Lentilles ...   29.20
Fillet of Sea Bass on a Bed of Pan Fried Mushrooms, Garlic and Lentils

La Marmite Bretonne de Homard 35.60
Hot Pot of Lobster and Selection of Fish

Petites Noix de Ris de Veau et ses Croustillants,
  Jus Aigre Doux 29.10
Crispy Rissoles of Veal Sweetbreads with a Sweet and Sour "Jus"

Mignonette de Boeuf aux Echalotes Confites
  et Queue Farcie 33.40
Fillet of Beef and Stuffed Oxtail with Red Wine Glazed Shallots

Râble de Lapin et Galette au Parmesan 27.80
Roast Saddle of Rabbit with Crispy Potatoes and Parmesan

Poulet Fermier Rôti au Jus, Tagliatelles aux Truffes
  et Foie Gras *(2 pers)* 48.60
Whole Roast Free Range Chicken, Truffled Tagliatelle and Foie Gras

Côte de Veau sur Mousseline d'Haricots Blancs
  et Sauce Verte 31.60
Rib of Veal on a Bed of "Girolles", smooth white Bean "purée", Fresh Herb and Lemon Sauce

Canon d'Agneau sur Pommes Boulangère et Petits Pois
  à la Française 27.40
Loin of Lamb with French Style Braised Peas

The temperature of the dishes depends on their content and varies from tepid to hot, never very hot.
MICHEL A. ROUX

continues

Exhibit 16–5 continued

Fromages

Le Plateau de Fromages Affinés ... ... - par Maître Jacques Vernier, Paris	12.20
Barattes de la Saône-et-Loire, Salade Mesclum au Vieux Vinaigre ... ... *Fresh and Aged Goats Cheese with an Herb Salad, Red Wine Vinegar Dressing*	10.40

Desserts

Les Glaces et Sorbets Maison... ... ... *A selection of home made Sorbets and Ice Creams*	11.80
L'Assiette du Chef ... ... *An assortment of the Chef's favourite Desserts*	23.20
Omelette Rothschild ... ... *Apricot & Cointreau Souffle*	17.10
Le Palet au Chocolat Amer et Praliné Croustillant ... ... *Bitter Chocolate and Praline Indulgence*	18.60
Parfait au Chocolat Blanc et Gelée de Framboises... ... *White Chocolate Mousse filled with Poached Raspberries*	15.80
Petites Poires d'Api au Caramel et Sauce Chocolat... ... *Toffee Pears and Caramel Mousse, Bitter Chocolate Sauce*	16.40
Soufflé aux Fruits de la Passion et son Coulis ... ... *Hot Passion Fruit Souffle*	15.20
Compote de Rhubarbe et Jus de Fraises, Tuiles aux Poivres ... ... *Poached Rhubarb and Strawberry Cordial, Whipped Cream and Pepper Biscuits*	13.10
Délice au Pain d'Épices et Fruits Secs au Tokaji Aszu ... *Ginger Bread and Dry Fruit Surprise served with a Glass of Royal Tokaji "5 Puttonyos"*	19.60
Café et Petits Fours	6.20

The Main Elements Considered When Composing a Menu

Cost is immaterial; it should be the last item in the queue. When you have a dish, you have to say "can the public buy it, will there be enough people to buy it?" Value for money is what is required. I would advise you first on season—what is available, what has good flavor. As far as nutritional value and calories are concerned, I am afraid it is not my scene; I am there to provide joy, not to control your diet. I am only there to propose, you are there to dispose.

At Le Gavroche portions aren't too large but large enough and well balanced, and you should always get up from the table not bloated but thinking "I could do that again." I love cassoulet, and I know that when I have it, I will enjoy it, but I also know I am going to be bloated. That's the beauty of life—diversity.

Exhibit 16–6 Comparison of Styles and Interpretations in Two Menus

Albert Roux's Special Celebration Dinner Menu 25th Anniversary Le Gavroche - London		Michel Roux's Special Celebration Dinner Menu 30th Anniversary Le Gavroche - London	
Les Vins	*Le Menu*	*Les Vins*	*Le Menu*
Baron de "L" 1986 *Domaine de Ladoucette*	Roulade de Saumon au Caviar	Mumm "Grand Cordon" 1985 - Magnum	Canapés
Riesling Vendange Tardive 1981 *Hugel*	Ballotine de Foie Gras à l'Ancienne	Corton Charlemagne 1990 *Bonneau du Martray*	Consommé de Canard en Gelée au Foie Gras et Truffles
Montrachet "Marquis de	Fricasée de Homard aux Girolles	Grand Vin de Château Latour 1967 - Magnum	Ragoût de Homard Parfumé au Gingembre
Laguiche" 1985 *Joseph Droubin*	Sorbet Citron	Château Climens 1967	Carré de Veau aux Trésors des Bois et Gousse d'Ail Confite
Château Latour 1982	Mignonette de Boeuf Gratinée aux Poivres	Cuvée Dom Perignon, Rosé 1985 - Magnum	Feuillantine aux Fruits Rouges et Sorbet au Fromage Blanc
Croft 1963	Séléction de Fromages		
Château d'Yquem 1982	Soufflé Framboises	Bas Armagnac - Castarède 1967	Café et Petits Fours
Remy Martin X.O.	Café et Petit Fours		

Courtesy of Albert Roux, Le Gavroche, London, England.

Service Style

Service is an extraordinary part of the meal, and so when we compile the menu we always try to do something that includes carving at the table so that people can enjoy seeing a maitre d' carving a chicken or duck; it strengthens the bond between the kitchen and our dining room. You can't have waiters who are simply asked to vacuum the room, take the crumbs off of the table, carry plates, and put the cloche out in unison; if it came to that, customers could come in the kitchen and help themselves.

Other Influences

From the age of 18 or 19, Escoffier was my Bible, but my interpretation was not his; I have avidly read Escoffier, Mignon, Careme, and have the highest respect for their work. I read it for inspiration and a very strong guideline of what was and was not permitted. Seventy-five to 80% of my food is classical, and the rest is my own interpretation of what those illustrious people gave me to read. Edouard Nignon was my benchmark, but I read every cookbook that comes into my hands.

I have never believed in six courses; I would rather have one course, either a starter or preferably a main course and a bit of cheese, perfect! My ideal meal is a good starter, main course, cheese, and dessert, but I would be quite happy with only one main course with beautiful flavor and good wine.

Le Gavroche has never been a restaurant that bothers about cost—customer satisfaction is far more important. Most of the time I have made money out of what I have done, but money has never been a driving factor for me.

Food costs of 35% to 40% or 28% are a moveable feast; at the end of the day it is bottoms on seats in the restaurant that makes you money. Would you prefer to be the owner of a restaurant that does 28% food costs and 50% occupancy or a restaurant who does 35% to 40% on food costs and 125% occupancy? I tell you I don't need to be a chartered accountant to know that I will make more money and have a better motivated staff at the end than places that are not that busy.

In relation to high and low cost mix, for watercress soup, the best soup in the world, you could charge £10 and your profit is about 400%, then you bring the cheese, which you are going to lose money on, so it is a balance. I have always looked at the bottom line in restaurants rather than the food and beverage people who look at every item.

At Le Gavroche, customers are king. From the moment they enter our door we have an unwritten contract to satisfy them, provided that they are polite—and 99.9% are nice people. Some of them are awkward and most probably you find them awkward because they are demanding things that should be provided when we are not providing them. We have a very mixed clientele that includes local people, taxi drivers, and people celebrating birthdays.

Menu language at Le Gavroche is unashamedly French, and it will remain like that. I have always believed it should be straightforward. You are not reading a book; pot au feu has no elaboration, and waiters have a lot to do with it because they can explain. How many times have I been deceived by reading something on a menu—it is flowery and expectations are raised?

The Future

In the next 30 years I will be gone, but I think we have broken a great barrier as far as spices and ingredients are concerned. Through transportation we have become globalized. If you have enough money, you could, for example, give a buzz to someone in New Zealand and they could put fresh cherries on a plane. I could send somebody out to the airport and give you the best cherries just picked from the tree 24 hours ago. I am not talking about being able to buy that in a shop; unfortunately, we are not there yet but we will be coming to that, and I think that is when the season will truly disappear as far as taste is concerned. I have just returned from Israel, where they are picking strawberries in the field; they have a fantastic flavor and 24 hours later they are on the shelves in British shops.

You have to live with your time and be wide open about these things. Who would have thought that I would say asparagus from Spain? I am saying that today because my palate is telling me that it is good stuff, quickly picked when it is ready, put into refrigeration, transported, and cooked 48 hours after.

Mise en place—advance preparation—is important, but in a modern kitchen chefs should not have to peel or turn potatoes and carrots. As we evolve, reliable suppliers will deliver turned carrots and potatoes, leaving the chef to do the cooking.

There is nothing new in the styles of cooking; it is simply reinvented, but nothing has to be invented about the way we eat. We are living today in an exciting time, when flavors are mixed and they will find their own way, they will find a true road of identification of the region. We have to preserve the differences; if they are not preserved, people long to go back to them. For instance, with cheese and sausage in this country, people are longing to go back to the regional specialties and in 30 years will yearn to rediscover our heritage.

My Menu Design Advice to a Newcomer

When compiling a menu, keep it simple and honest; cook the way you feel and for your customer; bear in mind the seasons and traditions; and be innovative. French cooking is not the straightjacket it is made out to be. Follow your instinct and common sense and what you cook well, serve well.

SHAUN HILL

The Merchant House, Ludlow, Shropshire

Profile

Chef/patron of The Merchant House, Ludlow, Shropshire, a Jacobean building seating a maximum of 24 diners on five evenings and two lunchtimes per week. Irishman Shaun also acts as a Consultant Chef, author, and translator, and his work includes Archestratus The Life of Luxury, *which purports to be Europe's oldest cookery book. He was formerly Executive Chef at Gidleigh Park Hotel, Chagford, Devon, where under the management of Paul and Kay Henderson, the American owners, he built a reputation for serving lively, individual, and eclectic dishes that embraced classic and modern cooking style as shown in the menu example in Exhibit 16–7, in which Shaun chose the food and Paul selected the wine. A collection of his recipes was published in 1990 under the title* Shaun Hill's Gidleigh Park Cookbook. *In matters of menu planning, he recommends that "taste is everything."*

Background and Training

My start in the catering industry as a chef was reasonably unconventional at the time, but I suppose it is standard now. It was in the late 60s: I was interested in food, and it was almost hobby style. I had read Elizabeth David books and pieces by Robert Carrier. We lived not far from Soho so I used to be able to go to Charlotte Street and see Schmits and the delicatessens that sold interesting things, but it wasn't a great passion, it was an interest.

My background is much more in formal academic work. I went to grammar school and university but didn't finish the degree course, although I still have a Fellowship at a university in classics, not in anything to do with food. I did classics, Greek, Latin, Ancient History, and I still do odds and ends; I worked on Aristophanes, the food references. Again, it is the interest in food but looked at from a slightly different angle. There you have food as a social occasion, the comic aspects of being a chef. The chef is a great comic character in ancient Greek plays, being a rude and boastful man; he was quite a highly paid and prized individual.

Robert Carrier was advertising for staff, and at that time it was very difficult to get staff in London. I asked for the job and got turned down, but only one other person applied and didn't turn up, so I was given a trial and started as commis; I worked hard because I was interested. I was attracted by food and still am, but I am attracted by it as the finished product backwards rather than technique forwards. I like eating an interesting meal that is well constructed, that is stimulating for lots of complex reasons: the company, the surroundings, the whole performance of eating. I like the food and from that stems the interest in how it comes together.

Exhibit 16–7 Gidleigh Park Sample Menu

Gidleigh Park
Lunch, Wednesday, 12 January, 1994

Leek and saffron soup *St Andrews Chardonnay 1990, Napa Valley*	Leek and potato soup based on a rich chicken stock, flavored with saffron and garnished with boneless chicken winglets.
*****	
Grilled red mullet with ginger, garlic and tomato	Tomato concassée stew flavored with fine strips of ginger and crushed garlic and topped with plane grilled red mullet fillets.
*****	
Parsley'd roast rack of lamb *Taltarni Shiraz 1990, Victoria*	Best end of lamb flavored with Dijon mustard and coated in white breadcrumbs and parsley and roasted. Served with Gratin de Jabron potatoes and mange tout and carrots.
*****	
Cheeses	Selection of hand-made cheeses matured in his own cheese cellar.
*****	
Red fruit salad with cinnamon ice cream *Sauternes Baron Philippe 1990*	Fresh pitted cherries, strawberries, raspberries, redcurrants, and Tayberries, sweetened
*****	
Coffee	

Courtesy of Shaun Hill, The Merchant House, Ludlow, Shropshire, England.

When I did eventually get to college when I was working at the Gay Hussar in Greek Street, 4 years later, there were a lot of people who had gone in and had no actual interest in the finished product. They would never have dreamt of spending their money on eating a meal; they were interested in the nuts and bolts of putting it together. There is nothing wrong with that at all, but I found their approach was from a slightly different base than mine.

I worked for Robert Carrier for 4 years. It was an eye opener to start in that sort of place. It was still a time when the formality of eating out extended from the kitchen to the dining room. The kitchen worked on the Repertoire de la Cuisine, and people pretended that they knew what all those names meant. The demarcation of who did what job was fairly strict, and it was largely a grand and nicely but eccentrically dated occupation for people who really weren't concerned with what the food tasted like as long as the fish was served with the head on and all the formalities were accomplished. It was a ritualized business for a certain class of person.

Robert Carrier is an American, and it was different there: it was a different start and it set my outlook fairly dramatically for what followed. He was a TV chef, and at the time he enjoyed that celebrity status with actors and well-known people; Francis Bacon and painters like that ate in his restaurant.

The food was eclectic in its way; we had odds and ends from all over the world, but it was largely provincial France in its outlook. There were aspects of cuisine classique, and among the things we did that were odd that I didn't see afterwards for a time was that we dressed all the salads. This seems obvious now but it wasn't in 1967 and 1968. Salad was tossed in a bowl before each one went out. We dressed it with good dressings made in the kitchen and it had to taste good. Only two vegetables were offered cooked fresh, and only two potatoes, one of which was always gratin Dauphinoise. A lot of aspects were appealing because the end product was so good. We had a real charcoal grill for the little lamb chops. The dishes were interesting for their time, and they used top-quality ingredients.

Dishes were innovative compared with what was served in hotels, and we broke new territory doing it. I am grateful that I started off there because it moved to the end product and you focused more on the customer. It was the other side of the spectrum from ice carving and pulled sugar. There is space for everybody, but this was the meal itself and you could evoke visions of Provence or northern Italy and families eating a wonderful meal from the food there. Good olive oil was used; you didn't see things like that in those days, and dining wasn't as formal there. The waiting staff were theatrical and flamboyant like Carrier himself.

Robert Carrier didn't do any cooking, but he set the menu and set the style. It was up to the chefs to produce the food. It was a small operation, two shifts of three, and it was easy to become enthusiastic about the taste of the food because you didn't spend ages picking spinach because there was no spinach to pick; so you moved on to making things quite quickly and in a small operation you tend to make things from start to finish rather than one aspect in a human production line.

Menu Development and Design

It is difficult to know whether meal design is completely customer led or whether it's cook led. I think it is customer led, and customers have become cooks. The sort of person who ate out in the late 60s formed a narrow band of people; it wasn't something a normal person would think of doing even if they had the money. Now it has moved much more to what people perceive the French outlook to be, where a lorry driver sits next to a prince because they are both interested in food and willing to pay. The other interesting thing to note is the sort of people who do the cooking. The emphasis changed in the 1970s, and I put it down to the beneficial effects of the Nouvelle Cuisine revolution, which is much maligned because it scrapped the Repertoire. Although you could eat the same meal from Lands End to John O'Groats in 1969, it was done very well in one spot and not very well in the other; that is what the difference was.

At the same time the spotlight shifted from the waiter, who filleted your sole in front of you and dusted your chair and made sure everything was all right, to the chef because the chef's taste in compiling the menu became an important factor in whether the restaurant was any good or not. So that change brought people in with an interest in food who a generation earlier wouldn't have dreamt of going into catering. I get applications not from lads in Council houses who couldn't get on the electricity course or the plumbing course but from people from wealthy families who should really have become solicitors who are keen and anxious to learn how to cook. Many will fall by the wayside, but what you have got is people interested in food coming into the industry because they believe it will help them.

The industry has had to respond to a different demand; there weren't enough trained cooks to cope with the burgeoning demand, so young people were promoted beyond their capabilities, and you then found that the difficulties weren't so much in the food but in food costs, administration, and running of a kitchen, things about which an experienced man might have acquired knowledge.

Food Styles for Special Occasions

Special occasions are interesting because they demand specific types of food; if a meal is a celebration, it almost silently demands different food. You want a whole animal so you have a whole duck rather than a chop; you tend to have meat rather than fish. That is interesting to me because when I did Greek, they had fish. They thought that a whole sea bass was the best thing that you could ever offer, and I think it is all part and parcel of the same train of thought as people eating things like peacocks or swans. It has much more to do with what they are like when they are alive than what they taste like when they are dead. There is something celebratory about having a whole beast, so having a celebration meal you will have something like duck or partridge or a big animal like a goose or a suckling pig. It might not go in the order of light to heavy. You will have a centerpiece to the meal, and it can never be a question of expense. Something like sweetbreads cost far more than suckling pig, but you could never have a celebration meal with a kidney or sweetbread because it doesn't fit. What you are looking at is more than the flavor and taste; you are looking at some token to represent success for the occasion.

Current Influences

My menu design is subtly influenced by current trends. On the very first edition of *Masterchef*, the TV program, I was asked what the trends were going to be. I said Moroccan and Tunisian because they are starch based, inexpensive, and different flavors; there's a bit of heat and it's exotic. Well, it didn't come and I felt I got that wrong. But it did come eventually—10 years later.

I think that there are patterns of taste that seem to work together. At the moment the Mediterranean diet is popular for lots of reasons—health is one—and people believe they are living a slightly healthier lifestyle eating lighter foodstuffs.

The Main Elements Considered When Composing a Menu

When composing a menu, the format of the menu is what comes first (i.e., how many dishes you are going to offer), and my format is dictated by purely practical things. The equation is how much time you have, either your own or staff, how many people you are cooking for, which is how many dishes you are going to use and move through. Your framework must be based on technicality; otherwise it will all unravel as soon as you get busy, or the food will be rotten in the refrigerator, or you won't be able to cope. You have to do what will work.

In a place with 80 chefs and 200 people eating you can have 30 choices and each one will turn over at a reasonable rate and it will all work out fine. In a place with one chef and 20 covers it is a very different setup. You can make properly three or four sets of dishes, and the corollary to that is the level of complexity of each dish you can attempt. It has nothing to do with produce, nothing to do with occasion, just the nuts and bolts of getting food out to people that will be acceptable for the price paid and that you are proud of. I can do four dishes at a moderate level of complexity, immediacy, and intricacy and serve them, or I could do two at much more and one at lot more, but you pick what you want. If you have only one thing, you move into the never-ending salmon and chicken area because with people sitting down you can't have anything that offends, you can't ever have offal, you can't have odd tastes because that is all they have to choose at a banquet; it restricts you.

Availability of foods and the joys of menu design and compiling the menu are what come after that. Once you have your size and style of menu set up, then you move to everything else. I then move to produce and markets (i.e., what is available to me). I also see freshness and quality as being things that sadly aren't inevitable with local produce. I select a product not only for its availability but also for its quality. A seasonal product ought to be better, but it isn't always.

I like the idea of seasonality. It has a nice warm feel as a thought and as a marketing ploy because it rings changes; it means the game, the wild salmon, and the asparagus have finished. People often wring their hands and say "Oh, strawberries!" at Christmas, "how disgusting." If strawberries at Christmas tasted like those in May, I would buy them because they would be as good. It might destroy some of the seasonal produce, but I am quite happy to buy leeks year round, which are imported from all over the place, and when the store of shallots has finished, I buy them from whoever is producing them. It tends to be lip service more than reality in most areas where in the winter you would have a quite limited choice for certain periods. Where I think you can do seasonality is in the style of food you offer. In an extreme you wouldn't necessarily want a huge suet pudding on a hot day. Similarly you wouldn't want some light little confection when it's a howling gale outside and people are shivering; they are looking for something heartier, so I think the seasonality is partly in your approach to what you put on the menu.

Service Style

I don't despise silver service; I think it is an amusing and outdated cabaret and is, in fact, a piece of theatre in which you move presentation on a flat *(serving dish)* down to somebody else to present on the plate so that it is presentable and looks appetizing, and that is quite difficult.

Other Influences

I look at cook books from our peers and at Larousse Gastronomique and at magazines. We have *Gourmet* magazine open in the kitchen at the moment because they have some very good pudding and tart recipes, and they stimulate you to try something different. If you want chapter and verse you need a textbook, but the problem with cookery books is that they tend to lose sight of what their function is.

As I said with the Nouvelle Cuisine, it's a matter of taste. If you accept that a good chef ought to have a good level of technical skill that's fine, but once you get beyond that, it only has to be a level of technical skill plus experience of how to organize a service so that food goes out, which is just slightly different than cooking a meal at home. Once you accept these things you then move on, as if you were composing music, to a question of how your mind works, and that is where it starts to get interesting. It is how you can picture the things together. I do tend to eat each new dish in its entirety because a dish can taste fine in a mouthful but starts to pale after a plateful. I try to get down to the simplest part the meal is about, a contrast of two flavors or two textures or whatever, hone in on it, and use that, and I do that with the whole menu.

To a certain extent you tamper with what has gone before at your peril unless you have some serious changes in either availability or technology at hand. I do tend to have fish before meat if I am doing a meal. If someone wants it the other way round, it is not a bother to me, but I think that the progression from lighter to heartier and then lighter again seems to work. I alter the menu every day within a repertoire of dishes that I am able to do for that time of year.

A special menu I did recently was for 12. When there is a large group of people, it is better for many reasons if they have the same menu: they take a long time and it upsets the flow if people have to choose and then we have to remember who has ordered what.

The menu was

- Steamed bass with Chinese spices
- Roast partridge with fresh goat cheese and gnocchi
- Caramel and apple tart with cinnamon ice cream

The fillet of sea bass with Chinese five spice is steamed and served in a fair amount of well-flavored broth. It is quite subtle; there are five spices, a little ginger cooked through in the broth, a few coriander leaves and spring onion pieces, a little soy sauce to provide color and that sort of flavor, but it has to be more flavored than plain soup and has a little bit of chili in it too because it has to act like a sauce for the fish, but the volume of intensity of flavor is halfway between sauce and soup. It forms the function of a soup, but the dish centers around the piece of fish, not the broth.

Moving on, you have got a clean palate from steamed fish and progress to the roast. The goat cheese gnocchi was a semolina gnocchi run through with goat cheese, which has been baked and crispy/crunchy on the top. I served very few vegetables with that, almost certainly Savoy cabbage and something else. I didn't write that on the menu because I think there is a limit to the amount of information you should put on a menu. I am not fond of American menus that are a recipe for each dish because, in my opinion, what you want to know when you look at a menu is what you are going to get. A menu is partly a marketing tool, but it also contains information that needs to be imparted so that basically it tells you what you are going to get.

For the dessert, caramel and apple tart with cinnamon ice cream, there is a bit of pastry, which is a change of texture; the caramel is sweet and ice cream will freshen up the palate. It was a sweet pastry tart with apple puree, creme patissiere, and apples on the top with caramel sauce spooned on top of that. This was a winter meal, and it was appropriate to that time of year.

In relation to the à la carte menu (Exhibit 16–8), you have to offer reasonable choices. The important thing on an a la carte menu is that, looking at the structure of it, it doesn't matter much if you use cheese or garlic or something a lot, as long as it is all on one course because people don't have two first courses. However, if you have a major ingredient, like saffron, among the starters, there will be no main courses that have saffron because what you attempt to do is make it as easy as possible for people to choose a meal that fits together. There should be minimal repetition; I try not to have any. I offer the standard choices, there is a vegetable dish, and I am very fond of risotto. You tend to make a menu and that is the starting point for the next day's menu. First of all you take off what you haven't got any-more or what you are bored with, and then what you replace it with is dictated by what you have taken off. So generally, although I sometimes have three out of four starters that are fish, it probably wouldn't be a good idea to have the fourth one. They are all things I like cooking and eating. I do bend a bit; I always have a butcher's meat, rack of lamb or it could be pork or beef, but I eat very little butcher's meat because I find it quite dull. I am fond of offal, so there is regularly offal on the menu because I am very fond of cooking it and very fond of eating it.

In the Darwinian restaurant-evolving theory, you end up with people who like what you serve them or you end up with no business, but, of course, you have to offer some form of range.

The Future

I can foresee a time when my current style may have to change or adapt because it would be boring to remain forever stuck in any groove; it could change for lots of reasons: if my business got bigger, if I put more tables in, if I got more staff or something. After 5 years, I generally find that I have played out a situation and need to change it a bit, go for something slightly different to keep everybody motivated.

I think that food will start to settle into a slightly less eclectic mix in the future and that the confidence of the punters, of their taste buds—the melting pot—will start to produce new classical dishes. We will take the things that suit the palate and the availability of the ingredients, and everything else will become just like nutmeg did 300 years ago or pepper a thousand years ago: it will become part of the cur-rency of food, and it will settle into something that is more stable.

My Menu Design Advice to a Newcomer

First of all, form a pattern that works and then depart from it; look at what you think is balanced first of all, and have a balance of textures, flavors, and quantities

Exhibit 16–8 Sample à la Carte Menu

£27.50 per person

saffron and artichoke risotto
sautéed monkfish with mustard and cucumber sauce
steamed John dory with juniper sabayon
terrine of foie gras with toasted honey bread

———

grilled brill with vermouth and watercress sauce
rack of lamb with potato and olive cakes
panaché of veal with cassis and red wine sauce
calf's sweetbread kidney and fillet
Aylesbury duckling steamed then crisp fried and served with
celeriac and lentil sauce

———

raspberry crème brûlée
brandysnaps with marmalade ice cream
Somloi
hungarian trifle with walnut rum and apricot
cheese - Rouquefort Munster and Explorateur

———

coffee - tea - tisane   £2.50
cheese as an additional course   £4.00

VAT and service included   Access/Visa cards accepted

Courtesy of Shaun Hill, The Merchant House, Ludlow, Shropshire, England.

and things in the amount of starters, main courses, and puddings that you have. Once you have that, don't be afraid to disturb the balance a little bit to make it individual, to tilt it to what you are after. I like fish and I like the textures of offal, so my menus regularly tilt toward that. I don't think that is a bad thing. We don't all want to be too balanced. Start with the conventional pattern of balance, taste, things like that, and when your confidence is high, feel free to shift it if you want to have an individual mark, but shift it from a base of sound balance.

ANTON MOSIMANN

Mosimann's, London

Profile

Anton Mosimann is one of the world's most celebrated chefs. Born in Switzerland, he trained through a traditional apprenticeship, learning the craft skills from master chefs and attending catering college. He came to Britain in 1975 after spectacular success: he was the youngest chef to receive a Swiss Master Chefs Diploma and worked in many countries— Canada, France, Italy, Japan—before being appointed head chef at London's five star Dorchester Hotel at the age of 28. His food has graced the tables of royalty and heads of state; he has appeared on television, is a consultant chef, and author of many books, including Cuisine Naturelle, *published in 1985, which was a revolutionary trendsetting guide to good food, cooking, and healthy living. He is currently Chef/Patron of Mosimann's Club, a private dining club in Belgrave Square, London, and his business interests include outside catering and Mosimann's Academy, a culinary school. The menu shown in Exhibit 16–9, featured at a gastronomic evening dinner for charity, is typical of the Mosimann style.*

Background and Training

Let's start first with my childhood. I have been very lucky to have had parents who both loved cooking, professionally, because they owned a restaurant. I was their only child and there was nothing nicer for me than going to the markets with my dad early in the morning to smell the produce, touch it, feel it, buy it, and then, maybe on the way home, go to a local restaurant and have a soft drink, passing by the bakery that had just baked freshly made bread. All those memories of course are fantastic and my upbringing from that point of view was very positive. Going home and starting to cook with that lovely produce was exciting. Seeing what you bought just a few hours beforehand on a plate is almost like magic, it just happens, and of course being a very young boy, it was extremely interesting. That is how I became involved in the love of cooking.

Menu Development and Design

In the late 1950s to the early 1960s the choice of menu was very limited, only two or three dishes a day or not even that, maybe just one. People just came and ate what was given to them because they knew it was going to be good. We sometimes had people who were on a diet but not very often; usually people ate what was given to them. The style of cooking was very much bourgeois cuisine; it was honest home cooking. Because you had to go to the market, come home, and cook and serve what you bought in the morning, there was no need for a refrigerator because you sold everything, and the following day you went to the market again. It was a daily trip to the market.

At that time *(early 1960s)* people *(in Switzerland)* ate just two courses, mainly at lunchtime, which was very often a soup or an hors d'oeuvre and then the main course. There was a bit of à la carte in between, cold sausages and cheese and that sort of thing. We didn't have menus at that time, you just sat down and we served you the best possible food of the day.

Of course that time has passed; I have done different things, moved on during my whole career, living in different parts of the world and working all over the world, going to school, and learning from great teachers and wonderful chefs. Eventually, you put all you have learned in one basket, mix it, and you have your own concept.

Menu designing or menu engineering, as I call it, is one of my hobbies; I just love it. It is great creativity and great innovation, and just to make a few decisions from that point of view I think is very, very exciting. For me one of the most important things is to sit down and design something and have the vision of seeing the end result. You have to have the vision of how it might taste automatically.

I love to go to markets and look at, for example, tomatoes: great color, wonderful look, you feel it, you smell it, you can't wait! All you need is a bit of Mozzarella cheese and basil, and you can have a wonderful terrine of tomato, basil, and Mozza-

Exhibit 16–9 Sample Mosimann Menu

Mosimann's
Menu

Symphonie of Scottish Salmon	Cold starter with smoked salmon, marinaded salmon (gravlax) and tartare-style salmon served with poached quails egg and quenelles of seasoned cream cheese.
-oOo-	
Chicken Consommé with Ginger and Coriander	Light, clear chicken soup with a hint of fresh ginger and coriander.
-oOo-	
Fillet of Beef "Nosi Bé" with Four Peppers Market Vegetables	Pan-fried fillets of beef served with whole red and green peppercorns and crushed black and white peppercorns in a reduced veal stock lightly thickened with half whipped cream. Spinach, carrots, and Dauphinoise potatoes.
-oOo-	
Selection of Cheeses	Tête de Moine, shaven in frills on a circular cutter, vermicelli of Stilton, slices of Appenzeller and Roubillac, garnished with grapes and celery.
-oOo-	
Bread and Butter Pudding	A traditional British pudding made to the Mosimann specification. It is one of the dishes for which Anton is renowned—a featherlight creation at the end of the meal.
-oOo-	
Coffee Petit Fours	Almond Tuile biscuits.

Courtesy of Anton Mosimann, Mosimann's, London, England.

---

rella. You have the vision for that. The colors are already on the plate, and you have to taste because you know how tomatoes should taste, but this particular tomato is more exciting than the one you bought last week because it tastes better. You have to have the vision for presentation, taste, looks, texture. I always say when you create a menu, make sure that you look, if you serve three of four courses, for different methods of cooking—one is poached, one is steamed, one is grilled, whatever, that is a very important point.

Food Styles for Special Occasions

The points I would consider are the number of people, guests, type of customer, nationality very often, gender, age group, and religion because of diets. I know that sometimes you think you may be going too far but in fact you can never go too far. If it is a party for 25-year-old people, you cook differently than for 60-year-old people because they eat differently. I had a German person here the other day who was 80 years of age who had lived in England for 60 years. He said, "You know, Mr. Mosimann, I would love to have some rodkraut, but there is nobody in England who can cook rodkraut for me, nobody. I would love venison and spaetzle but long, hand-made spaetzle."

I said leave it to me, and we cooked him that meal. I can tell you he was the happiest man, he loved it. He had been dreaming for so many years of having that meal and he had it here; it made his birthday and was a dream come true.

If it's a party, you want to feel good, so it's no good saying you can't have that. I ask people for their favorites and work around it, I am very flexible.

The à la carte menu (Exhibit 16–10) is very much based on an international theme. It reflects my travels and food I like to eat, what I enjoy myself, and of course what most other people will enjoy. If you create food for the public, you can't go from one extreme to the other: you have to have a good medium, tartare of salmon, tuna gazpacho, a nice Belgium salad with blue cheese dressing—simple food that looks good, tastes good and presents extremely well—a little bit of Oriental duck, and Cornish lobster. Risotto is one of my favorites because I lived and worked in Italy. With the menu here, now you almost eat around the world. People love to experience something different and new; they are ready for a challenge in the food world.

Current Influences

When I did Cuisine Naturelle in 1985 people thought no cream! no butter! I just generally mention that because I always try to be an inch ahead in my designing, in my way of life, my creativity; I love to be just slightly ahead of things—innovative. Today of course one has the influence of many different parts of the world. I love the Far East where I lived for 1 year, and I love that kind of East/West mix, which I started doing many years ago in London. I was one of the first chefs who cooked nouvelle cuisine in a serious way because I had this firsthand experience. My menu is a mix of 60% European and 40% Eastern.

The Main Elements Considered When Composing a Menu

1. Products are most important because they give you the guidelines for the market for the menu. I often create names like rendezvous of seafood, names where I didn't promise you lobster or scallops or anything else because how could I know in 2 weeks time what is going to happen? I left the menu description open, and it is then based on the products I can buy on the day of service.
2. The way the food is cooked: style of cooking, presentation, obviously service, budget—a lot of things to be taken care of.
3. Cost is obviously important because it is a business, and at the end of the day you either make it or break it. Where people very often make the mistake is that they cook for the guides, for Michelin, Gault Milau, and so forth for the prestige or whatever; very often it doesn't work because they don't take cost into account. I am lucky enough not to cook for the food guides anymore because this is a private dining club, so I can allow a slightly different concept from that point of view, a bit more risky quite often, but I cook for my customers. It is an important point because my customers are the people who pay the rent, the interest, or whatever.

The mix of high and low cost is an important factor, and I would recommend to each chef to take a pound—100p—and break it down into staff costs, food costs,

Exhibit 16–10 Mosimann's À La Carte Menus

TARRAGON

MOSIMANN'S
London

Crisp Romaine Salad
with Sour Dough Croûtons

Maize Fed Breast of Chicken
Wild Mushroom, Tomato and
Tarragon Sauce

Market Vegetables

Bakewell Tart
Orange Scented Ice Cream

Coffee
Petits Fours

Courtesy of Anton Mosimann, Mosimann's, London, England.

continues

Exhibit 16–10 continued

<u>ROSEMARY</u>

MOSIMANN'S
London

Panachée of Grilled and Marinated Baby
Vegetables with Goats Cheese

———————

Roast Scottish Salmon with
Chive Mustard Sauce

Leaf Spinach

———————

Chocolate and Coconut Parfait
flavoured with Rum

———————

Coffee
Petits Fours

overhead in general and just see how much is left once you have taken all those expenses out; very often you will be amazed how little there is left. I do this exercise with my staff from time to time just to make them realize that nothing, but nothing, has to be thrown away in the kitchen; everything can be used somehow, somewhere, and once you have that in people's minds, they know how careful they have to be.

I buy as much as possible locally; I believe in the produce that Great Britain produces so little need exists for me to go outside. I remember when I put the English cheese board on the menu at the Dorchester—only British cheese—it was the first cheese board in that sense produced from Britain and was a great success. I stay as much as possible within the seasons and would not for example necessarily buy strawberries from California because there is no need for them. If customers were to ask for them, yes we could get them, but we would recommend the customer not buy them at that time of the year.

Service Style

It is obviously important that when you create a menu, you always put yourself in the customer's shoes: how is he or she going to eat it? Knives, cutlery, or whatever kind of thing you need, fingers, chopsticks, that is an important point. I could do the most wonderful thing in the kitchen and the customer can't eat it (e.g., prawns hanging over the side of a glass at a black tie dinner where people were expected to peel their own prawns); that is not very creative menu planning, so one has to think of the service, ease of eating, and ease of serving.

Other Influences

I have been extremely lucky. I have worked with the most wonderful chefs around the world; I have chosen the chefs because of their reputation and chosen hotels because of their image, so I have been lucky to have chosen the right people. I learned a great deal from everybody about how to do it but also how not to do it, which is always important; I have been influenced by all those people.

I am a great believer that food should not just be presented for color but also for taste. To me this is an important point, and I will not accept it if it has color but not taste. I was once served a white fish on a white plate with white sauce; everything looked fairly unimpressive, and then the chef put a strawberry on my fish, which was unnecessary and very sad.

I think it is very important to follow the classical sequence of courses. Let's say that you start with a cold first course, then a hot soup, then a fish, then the main course, dessert, cheese, you eat my five or six courses and you leave the table saying, "I could just have another chocolate." You have to design the menu to be light; my menus are very light because there is not much cream or butter, or none at all, and you feel that looks good and tastes very good.

Looking at the practicalities, I start with the main course then work backwards (e.g., let's say it's lamb for the main course, you go to the first course so would serve something fishy, then a soup, a fairly light clear soup because you wouldn't want a cream soup after smoked salmon, then you may have a risotto as a third course, and then go into the main course).

If you have a special occasion—like my 50th birthday—I love to give people a choice, to let people choose for themselves. The occasion was a perfect opportunity for people who hadn't seen each other for 20 to 30 years to have a little chat and move to and from the buffet. There was a relaxed feeling, and you could eat the most wonderful and exciting food from all the parts of the world I lived in, I worked in, whatever. It was just such a wonderful opportunity to bring back some of the memories I had. I started planning the party when I was 48, and I mean that seriously. My philosophy has always been to be different, to be an inch ahead, to be innovative. I couldn't have had 600 people from all over the world come and celebrate my birthday and just serve them something very basic. Having been invited, they would wonder what is going to happen tonight, and the anticipation is already there building up, and that, of course, is the way to a successful party. I had great fun creating the menus.

My table d'hôte menu is changed weekly, the à la carte monthly, and the menu surprise daily.

(Note: Escoffier featured a menu surprise, inviting customers to let him know how much they wished to spend and he would create dishes within that budget to "surprise" them. Mosimann's interpretation is slightly different but meets the same basic criteria, a surprise menu at a fixed price.)

I think it is important that food and beverages go hand in hand, that it is a good marriage—the closer the better—because you can have good food and the wrong wine with it and spoil both the wine and the food. At the end of the day it is my cooking that people come for. You don't take risks in serving people calf's head or something; I love it, but in that sense with the public you have to be more careful. If I cook a banquet for 500 people, I will make the final decision about how much salt to put in the soup, it is me—I cannot ask 500 people, so I have to assume with my instinct and experience that that is the right amount of salt.

When designing a menu, it is important to bear in mind how large the kitchen is, the equipment, and whether it can cope. The steamer is probably the most used equipment in my kitchen and the oven for baking, grilling, or whatever.

I always believe in simplicity and write the menus mainly in English except for certain dishes you can't explain, so leave them in their original name. I have just created a menu for a German client and wrote the menu in three different languages, which is against the classical tradition that we were taught. But why should I translate bread and butter pudding into something else? Why should I translate a Chinese-influenced meal into something else in English or Chinese? I am taking the risk and am quite happy with the responsibility for doing that, but I always say it is no good trying to be more Catholic than the Pope. You can explain to people as well, being to the point, in simple and honest terms. The menu language must reflect my honest cooking, taking the best possible produce and doing as little as possible with it.

The Future

When I came to London many years ago, I was actually the first chef who ordered precut meat from the butchers because I came from America and was used to it. People just thought I was crazy and impossible, but today everybody buys precut meat, and I think it is right because it is controlled. It helps to improve the productivity of the chef, and you can concentrate on something else; it is also a cost factor and I think it is a way forward.

I think the healthy aspect is here to stay because we all like to eat more healthily, eat less but better quality food, be fitter. I believe that food will be designed not only for its pleasurable sense but also balanced with the health-giving properties; good balance and a well-cooked meal go a long way, but I always believe in eating the best possible produce and sometimes, lately, less.

My Menu Design Advice to a Newcomer

I think my best advice is choosing the right chef, the right property to learn about the profession, eagerness to win, willingness to work hard; it is a difficult profession, long hours, but it is very exciting. The basics never change and you have to build on that. It is like the foundation of a house: you just build on this foundation; if the house is colored red or yellow, it doesn't matter, but the foundation has to be right.

It is extremely satisfying to see any amount of people you have cooked for, smiling, happy, enjoying themselves; there is nothing that can take that joy away. Whether it's 2 people or 500 and they are smiling and happy, you feel happy too and it is a great feeling. Do they pay? Yes, of course, but cost doesn't necessarily come first; it is the satisfaction, the pleasure, the motivation, and you go home and say that was a good day. So any young person should keep that in mind. It doesn't happen tomorrow; it will take time, but not many professions can offer such satisfaction as cooking.

ANTON EDELMANN

The Savoy London

Profile

Maitre Chef des Cuisines of The Savoy London, a hotel renowned the world over for its opulence and quality. Brought up in Germany and trained in Germany, France, and Switzerland, Anton has a personal liking for Eastern and Mediterranean cuisines. A Savoy Director, he has been head chef since 1982, and his kitchen is acknowledged as being one of the busiest and most creative in London. The River Restaurant, Banqueting, and Private Dining Rooms continue to pride themselves on serving consistently fine food to large numbers of discerning customers. His philosophy of cooking is that true excellence is to be found in the simplest things done well.

Anton Edelmann is the author of several books (e.g., The Savoy Food & Drink Book; Anton Edelmann Creative Cuisine: Chef's Secrets from The Savoy; Canapés and Frivolities; Perfect Pastries, Puddings & Desserts; Fast Feast) *and a regular contributor to food-related articles in the press. He also acts as a consultant and makes guest appearances on television. Official travels representing The Savoy have taken him to the United States, Thailand, and Japan giving cookery demonstrations and lectures. His repertoire of recipes is drawn from the varied culinary cultures of the world, but this melting pot of flavors is firmly rooted in French and British cooking styles. Anton combines the traditional skills of the chef to create food as an art and a science in the knowledge that taste is the ultimate goal.*

The Savoy can boast almost 100 years of tradition, which started under the guiding influence of Caesar Ritz and Auguste Escoffier, the most influential exponents of hospital-

ity of their time. It has an enviable reputation and history that attract a broad mix of customers, with traditional British and international tastes. It endeavors to meet their expectations and hopes to better them at all times, in line with its Mission Statement: "For excellence we strive."

Background and Training

I had a traditional chef's apprenticeship in which the traditions of French and German cuisine formed the foundations of my knowledge. I worked in a good middle-class establishment (Bürgerliche Küche), where we prepared German regional food dishes. During the week a great deal of functional eating occurred, and on Saturdays and at Sunday lunch, we cooked celebratory meals for weddings and families. The menu reflected the occasion, and in general the meals were substantial and heavy in the German style typical of the late 1960s.

Menu Development and Design

I have kept some of my old menus; they have changed quite considerably over the years. As my career progressed, right to the top end, my menus have become more sophisticated, changing from the previous traditional French-based style to an internationally recognizable one, with traditional English dishes featured on a daily basis (e.g., roast beef and Yorkshire pudding, steak and kidney pudding, leg of lamb with mint sauce). These are not the main feature of the menu, they simply acknowledge local taste preferences.

The River Restaurant menu (see Exhibit 16–11) is distinct and reflects my developed understanding of meal design to please our customers. The Savoy is famous for its banqueting (up to 500) for which I design menus with the client. Some sample banqueting menus are available, but I prefer to design the menu to the customer's specifications and needs. In many ways, meal design is like a woman applying makeup: for everyday use, it is kept quick and simple, but for special occasions more time and trouble are taken in its application and a much larger range of products is used.

When composing a menu, I consider the occasion, client, cost, menu structure, choice, number of courses, nutrition, seasonality of foods, and food combinations; however, I no longer use French menu terminology. Menus are in English except on specific occasions (e.g., gourmand dinners).

In essence, I design the menus for banqueting, where customers want the meal to last no longer than 2 hours, and for lunch, where it has to be completed within 1 hour. Time has become more important to our customers, and they no longer wish to spend hours over a meal.

Four times a year I change the menus, working as much as possible within the seasons, constantly trying to be innovative and changing dishes to stimulate consumer demand and keep the interest of the chefs. I also adapt traditional recipes and modernize them to make them lighter. I constantly experiment with new flavors and combinations and test and adjust these with my chefs as part of the ongoing development of recipes. Italian/Mediterranean cooking has the strongest influence on my menu style at present, manifested in my book *Fast Feast* (1996), which reflects my thoughts on food and the selection of dishes.

When designing a meal, I look at the main course first, then the dessert, and cheese (whether served in the order dessert–cheese *or* cheese–dessert), then the fish

Exhibit 16–11 River Restaurant Menu

Starters

Asparagus salad with summer truffles £14.00
San Daniele Parma ham with marinated grilled vegetables £17.00
Wild Scottish salmon carved at your table £18.00
Home-made goose liver pâté with duck confit and truffle salad £19.00
Lobster salad in curry dressing with toasted pinenuts and mustard leaves £22.50
Beluga caviar with buckwheat blinis (30 gr) £57.50
Gull's eggs £4.50 each
Ricotta cheese cannelloni on grilled peppers, anchovies and caper berries £11.50
Glazed scallops on saffron potatoes £14.00
Twice-baked smoked haddock soufflé £14.50

Soups

Chilled essence of organic tomatoes £7.50
Aubergine cream with cumin and coriander croûtons £8.00
Guinea fowl consommé with wild mushrooms dumplings £8.50

Fish and Shellfish

Grilled wild salmon on organic leek fondue and potatoes £24.00
Mix-grill of Atlantic fish with smoked paprika aioli £25.00
Grilled fillet of turbot with roasted Jerusalem artichokes £26.50
Grilled or pan-fried Dover sole £26.50
Spicy roasted lobster with seafood and ginger potato gnocchi £38.50

Meat and Poultry

Roast Gressingham duck with young spring vegetables £23.00
Veal cutlets in Parmesan crust with asparagus £24.00
Grilled sirloin of steak with crisp fried caper sauce and basil tomatoes £24.00
Grilled calf's liver on olive oil and parsnip mash with a herb glaze £25.00
Organic fillet of Welsh lamb on braised fennel and tamarillo chutney £26.00
Tournedos of Buccleuch beef on Port wine shallots £26.50
Chateaubriand, sauce bearnaise and soufflé potatoes £60.00 for 2 persons
Vegetarian collation with tofu and nuts £15.50

All our main courses are fully garnished

Your choice of dessert £7.50
Selection of Continental and British Cheese £7.50
Coffee, cappuccino, espresso, tea and petits-four £3.80

Prices are inclusive of Value Added Tax ~ No Service Charge is added

Courtesy of Anton Edelmann, The Savoy, London, England.

continues

Exhibit 16–11 continued

River Restaurant Luncheon
3 Courses £28.50 ~ 2 Courses £24.00

Rosette of smoked fish with horseradish cream and mustard leaves
Asparagus crown with minted chicken salad
Tian of aubergine, courgette and buffalo mozzarella with Parmesan tuile
Artichokes filled with smoked haddock and poached egg
Seafood cannelloni with chive oil
Cream of onion soup with smoked hock croutons

***

Daily roast carved at your table
Fillet of lemon sole Véronique
Grilled brill with asparagus, tomato and herb dressing
Calf's sweetbread and kidney on spinach in herb crust
Grilled fillet of beef on tallegio polenta topped with pesto
Suckling pig filled with sage stuffing on pickled cabbage and dumplings
Vegetable lasagna with truffle cappuccino

***

Desserts from the Trolley
Selection of Continental and British cheese
Chocolate pavé
Feuilleté of raspberries
Snow eggs

Carver's Trolley
Monday to Thursday
Scottish roast rib of beef with Yorkshire pudding or Roast saddle and best end of lamb
Friday
Savoy chicken pie or Roast saddle and best end of lamb
Saturday
Roast best end and leg of lamb in a herb crust
Sunday
Roast sirloin of beef with Yorkshire pudding

Prices are inclusive of Value Added Tax ~ No Service Charge is added

continues

Exhibit 16–11 continued

River Restaurant ~ Dinner £42.00

Charentais melon
Chilled cream of potato and tomato with basil oil
Grilled asparagus with dried Bayonne ham and frisée salad
Linguini with peas and fresh mint
Seared scallops on black pudding
Risotto with apple and smoked bacon topped with poached gull's eggs

***

Daily roast carved at your table
Grilled brill on lentil cream and tempura vegetables
Halibut wrapped in potato roesti with red wine sauce
Chicken breast filled with langoustines on broad beans with goat's cheese gnocchi
Grilled peppered rumpsteak with pommes frités and herb salad
Filet of veal in Parmesan crust on creamed spinach topped with quail's egg
Suckling pig filled with sage stuffing on pickled cabbage and cumin flavoured sauce
Courgette flower filled with ricotta on Mediterranean vegetables

Desserts

Selection of Continental and British cheese
Chocolate tart with coconut sorbet
Marinated red fruits with its sorbet and biscotti
Amaretto parfait with blackcurrant sauce
Symphony of lemon with a blackcurrant compote
Mille-feuilles of chocolate with praline ice cream
Red fruit parfait with raspberry sorbet and marinated berries
Painter's palette of sorbets
Spicy pineapple pain perdu with banana ice cream
Dundee marmalade soufflé with an orange macaroon and milk ice cream
(Please allow 20 minutes for preparation)

Prices are inclusive of Value Added Tax ~ No Service Charge is added dish or starter. If the main course is light (e.g., fish), I tend to choose something more substantial as a dessert or in the first course. The things that guide me when designing a menu include:

- Avoid butter or cream sauces for the starter if you have a rich sauce for the main course.
- Balance a cold starter with a hot main course, followed by a cold dessert; serve a hot dessert if it is to be followed by cheese.
- Consider the different textures for the meal and don't duplicate them—balance and variety are required.
- Use different cooking methods for each course because each one brings different qualities to the dish.
- Consider color and temperature and avoid repetition of pastry, spices, garnishes, and so on.

Each element of the meal has a distinctive character and individuality that complements the previous and following courses. I also tailor menus to the customer's requirements, particularly in respect to age, gender, and so on, and believe that customers should have the ability to digest their meal without being stuffed or overburdened.

Service Style

Everything on the menu is designed for ease of service. What goes on the plate is managed by the chefs, the waiter only presents the dish to the customer. There are not sufficient professional waiters and they are expensive, so it helps to speed up the service of food if it is plated in the kitchen, and the visual appeal remains with the chef.

Current Influences

I believe that Escoffier was and has been a great chef, but times have changed and moved on. In relation to structure, I still follow a sequence that looks at the balance of the courses and their interaction but make up my own rules depending on the occasion. I can recall having made only one 12-course meal and believe that the ideal should not exceed four courses, excluding canapés and petit fours. This fulfills the time frame and the ability to digest and appreciate the food. The limited number of courses is also driven by the range of food within the dish.

The move is away from formalization in the structure of meals. Big buffets, in a traditional sense, with large centerpieces—garnished hams, salmon, and so on—are becoming something of the past. Customers today like smaller food items and a range of foods.

The longer I am a chef, the more I am drawn to the conclusion that a wine is not always necessary to enhance the meal and I would prefer, to really taste the food, a glass of water.

Other Influences

The Savoy buys the best possible quality food first, but not at any price, and then achieves through it the desired profit margins. In relation to the mix of high-cost and low-cost menu items, fish represents high food cost, whereas soups, salads, and vegetable-based dishes are lowest, and a mix in relation to the cost is engineered to give balance so that some dishes subsidize others.

Strong emphasis is placed on buying British food and staying within seasonal constraints. Certain products are bought in from overseas (e.g., beans, mange tout, courgettes), but in the main, poultry comes from France and ducks from England.

In Great Britain it seems that the only dictation by customers as to what is featured on the menu is in relation to Christmas. Other countries might have customs and traditions where certain food dishes would appear (e.g., lamb at Easter, turkey at Thanksgiving), but the only time the menu features change is because of the availability of food in season (e.g., asparagus, strawberries, salmon, grouse).

The Savoy provides a wide range of menus, breakfast, luncheon, afternoon tea, dinner, supper, function, banquets, and on any given day 1000 high-quality meals may be served on the premises. The menu in Exhibit 16–12 was created specifically for the 1996 Chef's Conference dinner attended by chefs, restaurateurs, and chefs de service. It embodies my meal design style, bringing together mainly traditional British specialties to a group of knowledgeable diners.

The Future

I am working toward lighter meals, more variation of flavors and textures, and believe in greater concentration of flavors in sauces but less quantity (e.g., sea bass, where the parsley and beetroot sauces not only enhance the color of the dish but also give a contrast of flavors), but everything remains centered on the needs of the customers.

We already score menu items to gauge their popularity and contribution/margin to high or low profit. On a daily basis I know the sales analysis of dishes, what is popular and so on. It helps to guide my menu planning (e.g., at the beginning to midweek those who eat regularly from the menu tend to choose fish dishes mainly, whereas, nearer the weekend, meat dishes are chosen). The computer is very much involved in showing trends, and reliance on such equipment will, I believe, increase in the future.

Because of the shortage of qualified trained chefs, I can foresee a time when mise en place of raw prepared food prepared to The Savoy's specification will have to be bought in. At present, chefs use 60% of their time to gather all the various ingredients from the stores and prepare them in a mise en place; only 40% of their time is spent on creative cooking. In the future it will become increasingly important for all quality establishments to use their chefs most productively.

My Menu Design Advice to a Newcomer

London restaurants provide the best examples of several different styles of cuisine, dishes, flavor combinations, more than any other city, except New York. Look around, eat, make notes, adapt. See why they are successful or not.

I believe that eating is one of the greatest pleasures of life and that the food on the plate should be interesting and varied. Cooks should never close their minds to new flavors and textures but take risks, explore and experiment, and create the right atmosphere in which to enjoy good food and good company.

Exhibit 16–12 Menu Created by Anton Edelmann

26th February 1996
Chef's Conference Dinner

Miniature Shepherd's pie and   bite-size appetizers
black pudding in brioche
***

Scallops wrapped in bacon   Specially selected hand-dived Scottish scallops without roe,
                                   with smoked bacon and fresh sage leaf wrapped around the
                                   outside. Served on a bed of ratatouille, surrounded by sun-
***   dried tomato sauce.

Cream of celery with black   Cappuccino of light frothy creamed celeriac soup with shav-
truffles   ings of black truffles
***

Filet and kidney of veal with   Somerset veal tenderloin with a daring combination of veal
cepes and balsamic jus   kidney, surrounded with a garnish of vegetables and cepe
                                   mushrooms and meat juice flavoured with balsamic vinegar.
***

Irish Mileens with celery and   Traditional hand made soft creamy Irish cheese served with
walnut bread   a garnish of lettuces and celery and sliced walnut bread.
***

Warm pear mille feuille   Layers of puff pastry with poached pears

Sweetmeats   traditional petit fours
Savoy Filtered Coffee

Courtesy of Anton Edelmann, The Savoy, London, England.

CONCLUSION

From the comments of the four chefs interviewed, it is apparent that customers' needs are paramount and that it is the responsibility of the chefs to know the tastes and expectations of their customers and their own markets. Although there was no consensus regarding cost, by setting very high standards and charging an appropriate price for the food and service provided, each chef has gained a reputation, which has ensured the viability of each establishment.

The four featured chefs were influenced by enjoyment of eating from an early age; their palates were developed and attuned to the merits of food for its taste and the pleasure it can bring; and their early socialization and subsequent training in culinary skills eventually made them Masters of their art. All four prefer moving from lighter to heavier to lighter foods within a meal and providing a variety of foods and cooking styles. They know their strengths and acknowledge that they must change and develop to stay ahead of the competition. Although they are prepared to consider purchasing some ready-prepared foods (e.g., butchery items and peeled potatoes), they are not prepared to use ingredients that might jeopardize the quality of their dishes. In their own way, they help to shape tastes and fashion in food dishes in their own unique style.

The experience of the four featured chefs indicates that, as a minimum prerequisite, one needs to know (from previous experience) how the finished dishes will look and taste, the technical skills required in their preparation and service, and how combinations of dishes will complement or give a contrast to each other. Technical expertise is required when designing a meal, but this is also interwoven with food as art (feeling and appreciation), which constantly changes depending on values, fashion, and trends.

The art of meal design is learned through experience (Senn, 1954), which in time a caterer may develop, but this alone is not enough; common sense and intuition are also needed because what is food for one man may be poison for another. When considering the fine quality points of a perfect meal, it is important that the natural flavors of the food remain identifiable while at the same time the contrasting colors, textures, shapes, and cooking techniques should generate excitement and provide depth and interest (Culinary Institute of America, 1997). This academic point was confirmed by a food critic after a meal at a 3-star Michelin restaurant in London:

> "The food was a perfect pleasure." On a technical note "the meal was beautifully balanced, so that each course rose and fell naturally into the next... nothing was terribly complicated, often what you ask from a good chef is that he be great enough, and confident enough, to leave the ingredients be. The art of cooking is knowing what not to add." (Gill, 1998).

Meal design is a juggling act involving the availability of ingredients, the capability of the staff, the commercial complexities of business, and consumer motivation and satisfaction; no one element can stand alone or determine what is correct: customers are the final arbiters of taste and satisfaction.

REFERENCES

Aron, J.P. (1975). *The art of eating in France*. London: Peter Owen, Ltd.

Blake, A., & Crewe, Q. (1978). *Great chefs of France*. New York: Harry N. Abrams, Inc.

Cracknell, H.L., & Nobis, G. (1985). *Practical professional gastronomy*. New York: Macmillan Publishing USA.

Duch, K. (1961). *Hand-Lexikon der Koch-kunst*. Linz/Donau: Rudolf Trauner Verlag.

Edelmann, A. (1993). *Anton Edelmann creative cuisine: Chef's serets from the Savoy*. London: Pavilion Books Ltd.

Edelmann, A. (1995). *Fast feast*. London: Harper Collins Publishers.

Edelmann, A. (1988). *The Savoy food and drink book*. London: Octopus Publishing Group.

Edelmann, A. *Perfect pastries, puddings, and desserts*. New York: Harper Collins.

Edelmann, A., & Sutnerling, J. (1991). *Canapés and frivolities*. London: Pavillion.

Escoffier, G.A. (1965). *A guide to modern cookery*. London: William Heinemann, Ltd.

Fuller, J. (1974). *Chefs manual of kitchen management*. London: BT Batsford, Ltd.

Gill, A.A. (1998, 26 April). Table talk. *The Sunday Times Style*, 24–25.

Herring, R. (1969). *Lexikon der Küche*. Gießen: Fachbuchverlag Dr. Pfanneberg & Co.

Hill, S. (1990). *Shaun Hill's Gidleigh Park cookbook*. London: Century.

Kinton, R., Ceserani, V., & Foskett, D. (1995). *The theory of catering*. London: Hodder & Stoughton.

Klinger, H. (1978). *Die Hotel und Restaurations-Küche*. Gießen: Fachbuchverlag Dr. Pfanneberg & Co.

Mennell, S. (1985). *All manners of food*. Oxford, England: Blackwell Publishers.

Mosimann, A. (1985). *Cuisine naturelle*. London: Macmillan London Limited.

Olney, R. (1985). *The French menu cookbook*. New York: Dorling Kindersley Publishing, Inc.

Page, E.B., & Kingsford, P.W. (1971). *The master chefs*. London: Edward Arnold.

Pauli, E. (1979). *Classical cooking the modern way*. Boston: CBI Publishing Company, Inc.

Roux, A., & Roux, M. (1984). *New classic cuisine*. London: Macdonald & Co (Publishers) Ltd.

Saulnier, L. (1976). *Le repertoire de la cuisine*. London: Leon Jaeggi & Sons Ltd.

Senn, C.H. (1954). *The menu book*. London: Ward Lock & Co, Ltd.

The Culinary Institute of America. (1997). *Cooking essentials for the new professional chef*. New York: Van Nostrand Reinhold.

Visser, M. (1991). *The rituals of dinner*. New York: Viking Penguin.

Suggested Readings

Aresty, E.B. (1980). *The exquisite table: A history of French cuisine*. Indianapolis: Bobbs-Merrill Company, Inc.

Brillat-Savarin, J.A. (1972). *The physiology of taste* (M.F.K. Fisher, Trans.). New York: Alfred A. Knopf, Inc.

Kivela, J. (1994). *Menu planning for the hospitality industry*. Victoria, Australia: Hospitality Press.

Paston-Williams, S. (1993). *The art of dining*. London: National Trust Enterprises Limited.

Tannahill, R. (1983). *Food in history*. New York: Stein and Day Publishers.

The Meal:
An Integrative Summary

Herbert L. Meiselman

This volume on "Dimensions of the Meal" has tried to fill gaps in prior volumes dealing with food. We have tried to accomplish this with respect to food products and food development, food science and technology, the psychology and sociology of food habits, the biological and physiological bases of food consumption, the meal habits of a number of different Western and non-Western cultures, and the business and creative aspects of food products and food service. The book has probably raised as many issues and questions as it has resolved. I sincerely hope so, because two of our goals were to appreciate the complexity of the meal and to appreciate the need for more interdisciplinary approaches. This book is aimed at the researcher and scientist, the technologist, the business person, and the creative food professional; anyone interested in food should be interested in meals, and anyone interested in meals has something to share with people who approach food from different perspectives.

I will now attempt to assemble, organize, and integrate some of the major points raised throughout the book. This process of summary and integration must be selective because so many divergent points were raised from all the different disciplines and perspectives represented in this volume. In this summary, I will first cover the "History of the Meal," followed by the basic sciences dealing with the "Biology and Physiology of the Meal." These chapters will be followed by the "Psychology of the Meal" and the "Sociology of the Meal." Four examples of "Meals in Different Cultures" are presented. The final two sections of the summary deal with "Food Service/Catering of Meals" and "Creative Meal Design."

HISTORY OF THE MEAL

A number of chapters present a broad historical perspective on the meal, especially those of John Edwards and Joachim Schafheitle, both of whom work in The School of Service Industries at Bournemouth University and both of whom have worked in the hospitality industry. Food researchers and product developers more often seem unaware of the history of meals. A historical perspective is important because our current conception of meals is quite recent, given man's long history of eating. Marshall also presents interesting historical material in his chapter on the British meal. Edwards begins with a definition, or definitions, of the word *meal*. He notes that *menu* and *meal* are used interchangeably in catering/food service and that the word *menu* reportedly first appeared in 1541. *Meal* might be defined as "a dish or dishes consumed at any one time," whereas *menu* is defined as "a list of

311

dishes that are to be served for a meal" and "the specific order in which they are served in succession." Order is important in catering because many meals at home consist of one course, often served on one plate, whereas meals eaten outside of the home might have multiple courses, most often a starter, main meal, and dessert. Joachim Schafheitle notes in his chapter on meal design that the menu provides information to both the diner and to the kitchen. To the diner the menu should provide "a balance of flavors and textures," whereas to the kitchen the menu provides "a blueprint for the kitchen operation to develop." Order is important to both the diner and to the kitchen, because the diner should expect that "...courses ascend progressively from light, delicate, and more complex flavors to rich, more full bodied and simple flavors." Schafheitle notes that, although the number of courses may vary from one occasion to another, the sequence is always maintained. These basic concepts involved in the definitions of *meal* and *menu* appear and reappear throughout this book and this chapter. We cover menu design in more detail later in this chapter, including more information on course sequence.

Edwards suggests (based on Brillat-Savarin) that meals began when man moved from eating fruits to hunting. Meat brought the family together to eat. The early hospitality industry grew out of providing meals to travelers, either in homes or more likely in monasteries, followed in Europe by inns and taverns. Keep in mind, travel as we know it developed only after the introduction of trains in the mid-nineteenth century. Before that, catering would have been more common for single young men who lived and worked away from their family homes and who ate "at a pie or cook shop or took an 'ordinary' at a chop-house or public house." The meal might have been a slice of meat with vegetables, cheese, and beer. The rich traveled more and ate in hotels and inns, not out of choice but out of necessity. Restaurants in Europe were not established until the nineteenth century, serving mainly French food; and, before that, middle class people ate in a chop-house. Restaurants in France were established earlier, with Boulanger reportedly beginning the first restaurant in Paris in 1765, and the French Revolution making available the chefs who had worked in the houses of the aristocracy.

Both Schafheitle and Edwards detail how the traditional design of meals evolved from *service à la francaise,* which was an opulent buffet of largely cold food presented on the table as guests arrived to be seated. Up to three separate table settings of different courses might have been laid out in sequence. *Service à la russe* replaced this buffet-style service with a meal of courses served in sequence to guests who sat at the table. Along with Russian Service came the menu because all the food was not placed on the table as in French Service. According to Edwards, Russian Service did not replace French Service in England until the latter half of the nineteenth century. Although food service everywhere increasingly used *service à la russe,* "the food culture of the French, traveling with the French Army and to the colonies, was practiced in all civilized western societies." As we approach the end of this millennium, the dominance of French food culture in Western countries continues to decline as the many chapters in this volume illustrate. Local cuisines are being rediscovered, and global travel has produced a spread of food cultures all over the world. Finally, Edwards attributes the introduction of plated service to America of the 1920s, where a single plate contained the various components of the main meal presented together rather than served from separate serving dishes. *Cuisine nouvelle* has brought plated service back to the top end of the market.

The development of mealtimes also follows an interesting history. It is likely that there used to be two main meals per day, one mid-morning and one late afternoon or early evening. These meals tended to slip later over time, requiring the formation of a proper breakfast in both France and England by the eighteenth century. Food was probably taken early in the morning before this time, but it appears not to have been considered a meal, perhaps because it contained no meat or perhaps because it was not eaten with others. Edwards also presents an interesting discussion on terminology for meal names.

BIOLOGY AND PHYSIOLOGY OF THE MEAL

Probably the most basic variables in considering the meal have to do with our own human biology and physiology. In their chapter, "The Psychology of the Meal," Patricia Pliner and Paul Rozin also review the animal research that has included the psychology and physiology of eating in laboratory rats. Pliner and Rozin observe that in addition to physiological and psychological factors, the environment also plays a large role. Simply increasing a laboratory rat's cage size reduces the number of daily meals, probably by increasing the number of activities that compete with eating. Pliner and Rozin argue that the basis for meals is probably ecological; by focusing eating into meals, all animals can reduce the time spent eating. However, meals also stress the body by challenging the homeostasis or balance that the body tries to maintain.

Homeostasis is one of the phenomena that Harry Kissileff addresses in his treatment of the biological bases of meals. Kissileff begins by defining many things in physiological terms. "Meals" become "eating episodes," which end in satiation ('satisfying conclusion') and satiety ('state of inhibition')." Kissileff begins his chapter with a basic review of physiology and physiological methods to acquaint the beginner with how physiological studies are conducted in animals and in man.

Harry Kissileff goes on to describe the different types of studies that have cast light on how physiology controls meal initiation and termination. His ultimate goal is to "show [how] this information could be useful for guiding the development, marketing, and presentation of food products" and to "help providers develop procedures and for determining optimal size for portions." To do this, Kissileff draws on four different types of studies: descriptive studies, construct-driven studies that relate descriptive results to theory, mechanistic studies that relate behavior to anatomy, and practical studies.

One of the more interesting issues in the physiology of eating is the role of homeostasis noted previously. Kissileff traces this notion back to nineteenth century physiology, which led to the thesis that relative constancy of the internal environment is maintained by balancing energy input and energy output. This in turn has lead to "the hypothesis that intake at each meal, which is the foundation of energy homeostasis, must also be equivalent to the energy used since the last meal or anticipates energy used until the next meal." Homeostasis experiments provide a good example of one of Kissileff's main points about physiological methods. Homeostasis studies have tried to use test foods that are more nutrient dense but identical in other ways, but that has turned out to be a challenge because changing one attribute usually brings with it other unintended changes. Kissileff reviews the research literature that has produced evidence both for and against the maintenance of energy intake constancy within meals.

Kissileff also deals, physiologically, with a phenomenon that we see throughout the book and also in the next chapter by Harry Lawless on the senses. Kissileff reviews the research literature showing that people eat more of what they like, but he goes further, looking for the basic physiological mechanisms that might underlie this observation. He poses two possibilities: foods that are liked more might lead to greater insulin release or might lead to greater oxidation, either of which would increase how much is eaten.

Harry Lawless treats the most basic element of human biological reaction to foods in his chapter, "Sensory Combinations in the Meal." Sensory responses to food form the basis of biological responses to foods and the basis of food aesthetics, so we return to this sensory level when we discuss menu and meal design with the four chefs in Chapter 16 of the book. Lawless begins with the omnivore's paradox: the struggle between neophilia (new experiences with food) and neophobia (fear of new and possibly dangerous items). Lawless notes that flavor principles, as discussed in this book by Elizabeth Rozin, are one mechanism to provide familiarity and overcome neophobia.

Lawless sees the meal as "...a combination of diverse sensory experiences in which we have sensory 'struggles.'" The key struggle is in achieving simultaneous sensory balance while maintaining some sensory contrast. Lawless notes that "Probably no other common experience assaults all of the senses as does the consumption of foods"; because "In the simple act of eating we manage to stimulate all the senses at roughly the same time." The senses are much more than taste and smell and include appearance (vision), aroma (olfaction), texture (tactile), temperature (thermal), hearing (audition), and others.

Sensory research is often done with simple pure chemical compounds and with interactions among simple compounds. Lawless notes that "...they attempt to get back to the complexity of real life by examining more complex situations after the simpler more isolated effects are understood." Remember that foods are "...complex combinations of thousands of compounds...." Studies of taste mixtures have generally shown masking or suppression; the taste of each is less in mixture than it would be alone. Thus, sugar in coffee makes the coffee less bitter as do salty pretzels mixed with bitter beer. Lawless discusses the exceptions to mixture suppression rules and the more complex case of blended flavors. Furthermore, Lawless extends mixture rules to "...the entire enterprise of food preparation [which] can be seen as an attempt to make harmonious mixtures." The discussion with the chefs in a later chapter confirms the important role of understanding mixtures to achieve balance and contrast. Lawless contrasts this rather clear situation with mixtures of tastes and smells. Laboratory research with simple chemicals shows additivity (tastes and smells add together), but research with more complex foods is more complicated.

Sensory adaptation, the disappearance of sensation with constant stimulation, is important to consider in foods and in meals. Furthermore, sensory adaptation and mixture suppression can be combined to understand, for example, how a sweet doughnut followed by orange juice leads to less sweet and more sour juice (sweet adaptation plus the removal of sour suppression). Lawless describes examples for both taste and smell. He concludes that, "Part of the adventure of creating a good meal is to arrange possible sequences that may create some contrast effects but not unpleasantly so." The chefs discuss and demonstrate these effects in their discussions and in their sample menus. Lawless also treats the senses of touch, texture, temperature, and pain, and he includes a discussion of texture contrast theory. He emphasizes how the senses work together to produce the complex reactions to foods in combination (i.e., meals). Lawless concludes by arguing that the senses are the basis for the development of sensory-specific satiety and therefore the basis for the desire for variety.

A discussion of sensory-specific satiety and its role in supporting a varied diet follows.

Barbara Rolls defines sensory-specific satiety as the experience of rating foods as less pleasant at the end of the course than at the start of the course in contrast to foods that are sampled but not eaten and maintain their acceptability. This "satiety may be experienced for the foods already eaten, but not for other foods, particularly those with very different sensory properties. Thus satiety need not be for all foods but rather can be sensory-specific. To maintain your enjoyment of a meal, as one food declines in pleasantness, you will switch to another food that remains appealing."

Rolls has determined that sensory-specific satiety does not depend on replacing calories because the phenomenon does not vary by changing the caloric content of the foods but seems to depend on the amount, volume, or weight of food. Because people appear to eat a given amount before satiety, Rolls suggests that "...one way to reduce caloric intake in a meal is to reduce the energy density of the available foods." The sensory nature of sensory-specific satiety is suggested by its time course, which shows the biggest decrease in pleasantness 2 minutes after eating stops. Food pleasantness gradually increases over the next hour. Rolls demonstrated that just chewing (but not swallowing) food or just smelling (but not eating) food also produced reduced pleasantness. Sensory exposure is the key because the effects occur with or without eating, as long as sensory exposure occurs. Furthermore, foods of similar tastes (sweet or salty) interact to produce sensory-specific satiety; sweet foods affect other sweet foods but do not affect salty/savory foods and vice versa. Finally, we learn to associate foods with the satiety that they produce. Rolls notes, "Understanding the basis for these interactions among foods is important for meal planning because the goal of a meal should be to maintain palatability at a high level throughout." This notion of providing balance and contrast throughout a meal recurs throughout this book and is the focus of discussion by the chefs.

Rolls relates sensory-specific satiety to variety as follows: "If satiety is specific to particular properties of foods, then more should be consumed during a meal if a variety of foods are available instead of just one food." Rolls proceeds to document from her own research that variety in shape and flavor can enhance how much is eaten.

Finally, Rolls distinguishes the short-term effects of sensory-specific satiety from the longer term reductions in palatability from monotony. Both sensory-specific satiety and monotony have a cognitive component and may be related to each other. Rolls reviews ways of dealing with monotony by using staple foods, varied diets, foods with initially high palatability, and self-selected foods when possible. Rolls notes, "All companies want consumers to continue buying and eating their products. If they can determine why people tire of foods, potentially they can design foods that are more resistant to changes in palatability with repeated consumption." Rolls' recommendation is to "...include foods that show little (sensory) interaction...." You will note that the chefs try to accomplish this through contrast and through avoidance of the same flavors.

Cees de Graaf approaches the meal from the perspective of a nutritionist, although de Graaf is trained in sensory science as well. De Graaf uses the term "meal" to refer to breakfast, lunch, and dinner, and the term "snack" to refer to eating between these; he analyzes "...the frequency, distribution, and variability of energy and nutrient intake across the day" and across countries. De Graaf cautions that adequate data are not available on enough people for enough time in enough countries, and he further cautions that wide variation exists in eating from one day to the next and from one person to the next. This is reflected in between-subject variability in intake (20% to 30%) and within-subject variability in intake (also 20% to 30%). Snack variability is probably higher than meal variability.

De Graaf begins by examining eating frequency, including snacks, and notes higher eating frequency in Northern and Western European countries and lower eating frequency in Eastern and Southern Europe. Eating frequency in the United States appears to approximate the lower range of frequency of the European countries. Younger people eat more frequently than do older people. When meals in Central Europe and the United States are assessed for how much they contribute to daily intake, breakfast is the smallest, lunch is middle, and dinner the largest, contributing about one third of daily nutrition. Energy intake from snacks follows the same pattern of increasing intake from morning to evening. Younger children appear to eat relatively more at breakfast and relatively less after dinner. De Graaf confirms, however, that breakfast is the smallest meal, although it is also the least variable meal in nutritional content from day to day, and it is the meal most likely to be skipped. The lunch meal in Northern, Southern, and Eastern Europe may constitute the largest meal.

Analysis of the macronutrient composition of meals and snacks shows that dinner is higher in fat and protein in many countries, with breakfast containing more carbohydrate and less protein and fat. Interestingly, the micronutrient composition indicates that snacks do contribute to overall nutrition and are not just empty calories. The percentage of daily nutrition that comes from snacks rather than from meals ranges from 20% to 45% in different countries and different studies.

THE PSYCHOLOGY OF THE MEAL

After considering the biological and physiological influences, we next turn to the level of individual influences, those relating to the psychology of the individual eater. After this, we turn to the sociological and cultural influences of considering people eating in groups. In their chapter, "The Psychology of the Meal," Patricia Pliner and Paul Rozin claim that the meal "is a very real psychological entity." They follow this theme throughout the chapter and conclude: "The meal is a special unit of eating at the psychological level." This conclusion cannot be overstated. Pliner and Rozin are not arguing that the meal is an accident of food provision or an accident of history; rather the meal is a part of our psychology. Pliner and Rozin also provide a simple and minimal definition of what a meal is: "humans...do most of their eating in relatively short periods of time, separated by periods of minimal if any consumption." Within this context of meals, Pliner and Rozin discuss the meal determinants of food palatability and food amount.

Although Pliner and Rozin argue that the three-meal-a-day pattern is a worldwide phenomenon, other chapters in the book present both historical and cultural information that makes this three-meal pattern less certain. In other chapters in this volume, David Marshall and John Edwards inform us that two meals per day were the norm in earlier times. And other chapters indicate that although three hot meals per day were the norm not that long ago, today's meals have changed. The three-meal pattern is very dependent on what one calls "a meal." Also, the definition of the meal yields different estimates of how many daily meals there are. If meals are defined as Pliner and Rozin have defined them above, then English teatime is also a meal, and mid-morning coffee and a sweet might also be a meal. Evening snacking could be another meal. Pliner and Rozin report on their own research, in which they conducted interviews in English with native speakers of 18 different languages. Seventeen of the 18 languages have a word for meal, 16 have unique words for breakfast and lunch, all have a word for the main meal eaten later in the day, and 16 have a word for snack. The French have "the largest meal- (and food-) related vocabulary."

Not only are meals universal linguistic entities, but meals also appear to be important in organizing the day's events, which are planned and remembered as being "after breakfast," "before lunch," and so forth. Pliner and Rozin discuss the memory of meals in some detail, probing the research literature to determine whether different parts of meals might be remembered differentially. They relate this to studies that have reported meal modeling of entire meals and meal components, which have shown that the main dish in a meal is the most predictive of the acceptability of the entire meal. Pliner and Rozin point out that the main dish in a meal is composed of the largest amount of food, which is frequently also the most palatable and most expensive food, and is located in the temporal center of the meal. We return again to the role of the main dish when David Marshall discusses the role of meat in the British diet and when the chefs talk about menu design.

Pliner and Rozin present an interesting discussion of breakfast, entitled "The Anomalies of Breakfast." Indeed, breakfast appears to be unlike other meals. It is the smallest meal, following the longest daily period of not eating. Its culinary makeup differs from other meals, and the difference in breakfast among cultures is vast. Pliner and Rozin argue—and I agree—that many people eat the same breakfast every day in many cultures. If this is true, the mechanisms that promote menu variety and reduce boredom operate differently for breakfast. Certainly breakfast does not follow all the meal rules that apply to other meals. Breakfast probably deserves a lot more special attention and research. Marshall notes that breakfast is the British meal that has changed the most from its traditional form. And de Graaf notes that breakfast is the meal most often skipped, as well as the least variable in nutrition.

CULTURE AND THE SOCIOLOGY OF MEALS

A number of chapters in *Dimensions of the Meal* deal with the sociology of meals and the contribution of culture to meals. Johanna Mäkelä provides a good introduction to the sociology of meals by attempting to define meals from the sociological/cultural perspective. Mäkelä acknowledges that the meal "is a self-evident and common word," and David Marshall and Elizabeth Rozin similarly note that you know a meal when you see one. Marshall notes that only when the rules of meals are broken is it obvious what a meal is. Mäkelä observes that "scholars try to pinpoint some features of our eating habits and our essence as social animals and members of a certain culture." The sociological perspective even changes the meaning of food.

From the perspective of food product developers, a food is the product of agriculture or, more likely, of technology. From the sociological perspective, food is the product of culture because culture defines what is acceptable food and what is desirable food. Paul Rozin has written extensively on what makes food unacceptable, and Patricia Pliner has written on fear of novel foods (food neophobia), which is introduced to the reader by Jeffrey Sobal in his chapter. Culture also defines what a meal is. Mäkelä discusses how meals are constituted when she deals with three different topics: meal format, eating pattern, and the social organization of eating. Marshall uses a similar outline for his chapter.

Meal format refers to both the composition of the main course and the sequence of courses in a multicourse meal (starter or appetizer, main course, dessert, or cheese). Meal format deals with the content and order of meals. Mäkelä and, later, Marshall draw on the work of British anthropologist, Mary Douglas, and British sociologist, Anne Murcott, both of whom examined what constitutes a proper/main meal in Britain, namely meat, potato, vegetable(s), and gravy or dressing. The discussion of what used to fit within a proper British meal and what might fit the definition of a proper British meal today forms an interesting discussion within this book because the topic is addressed in many chapters (see Edwards, Marshall, Schafheitle). Marshall makes the important observation that what distinguishes the British meal is the emphasis on meat. Mäkelä compares the British pattern with studies of meal format in Sweden, Finland, and France. In the chapter on northern European meals, Prättälä presents much more detail on the development and current status of meal structures in Nordic countries.

If eating together is so much a part of the definition of *meal* for sociologists, what occurs when people eat alone? Jeffrey Sobal claims that eating alone is "generally seen as abnormal, undesired, and even unhealthy." It would be interesting to compare views on eating alone longitudinally, comparing attitudes in the last century with attitudes before 1950 and through the end of the twentieth century. The difficulties of obtaining service when eating alone are well established within the restaurant and hospitality folklore, and a recent travel magazine presented a feature story on establishments that accommodate single diners. Sobal discusses the increase in living alone in Western cultures and the increase in eating alone, although he notes that the two do not always go together. Mäkelä also notes the changing pattern of household units and the accompanying changes in meals. It is interesting to question whether sociologists will have to revise their view of sociability as being part of the essence of a meal if single eating continues to increase. Sobal also discusses the special and growing problem of eating alone among the elderly.

Eating pattern, according to Mäkelä, is defined by time, by the number of meals and snacks, and by the alternation of hot and cold meals and snacks. Recall how Pliner and Rozin noted that meals provide a daily time framework (e.g., "I'll be home before dinner."). It appears that at least in Western societies the number of meals and the number of hot meals are declining, so that only one hot meal per day is the norm now. However, one of the distinctions among cultures, and even regions within cultures, is whether the hot meal of the day is midday or in the evening. Prättälä presents an especially interesting presentation of how local customs and government policies have affected the pattern of eating the midday meal in Scandinavian countries.

Previously, some of the peculiar characteristics of breakfast were noted. Breakfast used to be a hot meal in many cultures but has become largely a cold meal of cereal, bread (with butter, jams, or cheese), fruit, and coffee or tea. The substantial, hot cooked breakfast has become a special meal for holidays and weekends. Sobal also notes that breakfast is the meal most likely to be eaten alone, and dinners the least likely. Holiday meals like those at Christmas and Thanksgiving are designed to include other people, with the company being as or more important as the meal. Lucy Long deals extensively with holiday meals in her chapter.

Marshall addresses meal patterns in Britain, both historical and contemporary. Marshall presents interesting historical notes on how meal patterns have changed over the centuries. Eighteenth century Britain saw two meals per day among the upper classes, which was eventually extended to three meals. Conversely, the working classes enjoyed four meals per day into this century. And three hot meals per day were common in Britain until the middle of this century. Prättälä discusses the changing pattern of total meals and hot meals in her chapter. The main meal of the day did not move to dinner until the 1960s to 1970s. Marshall notes that family meals (breakfast, dinner) have moved in the direction of the two-meal pattern of the eighteenth century.

The crux of meal definition to sociologists is sociability. Does the definition of meal require interaction with people? Mäkelä's own research demonstrated that for working mothers in metropolitan Finland, the essential ingredients of a proper meal consisted both of food components and of context: a hot dish, a salad, and company. Jeffery Sobal, in his chapter on sociability and meals, begins with the distinction between just eating, which represents "food events," and eating "in a structured format that is strongly bound by social rules," which represents "meals." Has breakfast become a food event rather than a meal because more people eat breakfast alone? The importance of sociability in meals is probably determined by where you sit. To the food technologist or product developer, a meal is a food or a collection of foods; to the food service provider, a meal is an opportunity to provide an occasion for profit; to the sociologist, a meal is a social event.

Mäkelä also notes the change in eating habits as one's roles change from a single individual eating according to one's own wishes, to a married person sharing meals, to a parent who sees mealtime as a period of socialization for the children. From the social perspective, gender differences also play a role in meals. Women probably continue to engage in more food shopping and preparation, although the role of men in this regard is increasing. It is probable that men hold different perspectives on meals and on what constitutes meals; therefore the increasing role of men in meal preparation might contribute to future changes in meals. The importance of gender differences in discussing and understanding meals also comes from a number of other chapters in this volume. Pliner and Rozin, as well as Sobal, discuss differences between male and female perceptions of how much to eat; Sobal also discusses differences between the roles of mothers and fathers at meals; and Edwards reports differences in how women act and how they are treated in the hospitality industry. Clearly, increasing attention needs to be paid to gender differences as families change, meals change, and the relationship between the two also changes.

Mäkelä deals with the impact of social structure/social hierarchy on meals and vice versa. Although the sharper distinctions between foods and meals in different classes are weakening, differences still exist. Generations ago, different classes ate different foods. Today, the differences are not so much in the foods themselves but in the manners and forms with which foods are consumed. One major point of difference among social classes might be the healthiness of eating. The healthiness of eating is replacing food safety and basic nutrition in much of the developed world as the major concern about eating. Safety concerns have also shifted to long-term concerns (such as genetically modified foods) rather than immediate risks. Several other authors (David Marshall, Anton Mosimann) comment on meal healthfulness being a concern in the future, and Cees de Graaf presents a detailed discussion of some of the health consequences of meal patterns.

Marshall also addresses class differences in eating. The mainstays of the working class diet even into this century were bread (served with a scraping of butter, jam, or margarine) and potatoes at dinner. These items were key for the working classes but merely adjuncts to the meal for the upper classes. Marshall notes, however, that the "breadwinner" still expected a proper meal in the evening. Ritva Prättälä also discusses the important role of bread and potatoes in Nordic meals, where some meals are still based on bread. Marshall also references recent British studies showing the differential purchasing of food among the classes.

The growing concern about healthy eating is one of the contemporary trends that Mäkelä lists at the conclusion of her chapter. She also notes the decline in family meals, although she cautions that family meals are not as dead as some argue. Marshall shares this view and presents data that a large percentage of British adults report eating together as a household unit most days. Marshall further notes that eating together is more prevalent at dinner in persons older than 55, in two-person households, and in lower socioeconomic groups. Marshall questions where the family meal is heading and notes that other writers see the family meal becoming an empty ritual.

In looking toward the future, Mäkelä also notes that the division of labor between the genders and among the family is part of the basis for the trend toward eating outside the home or bringing in prepared food. In the United States, we are just passing the 50% point, where more than half the food dollar is spent out of the home and less than half in the home. The percentage spent out of the home continues to increase. John Edwards presents a lengthy discussion, with supporting studies, of the growing pattern of eating outside the home.

In Jeffrey Sobal's chapter, he deals with three dimensions of meal sociability: facilitation, commensality, and interaction. Social facilitation of eating refers to the enhancement of eating when others are present. This phenomenon is most associated with the scientific research of John de Castro, who has used dietary diaries to obtain data. He has concluded that people eat more with increasing numbers of other people for all meals, for all days, for all settings studied, both with and without alcohol. Sobal reviews the mechanisms that have been proposed to deal with social facilitation of eating. Perhaps the other person(s) serves as a model(s) for eating more; or perhaps it is because increasing the number of companions increases the duration of meals and, hence, the opportunity to eat. And finally, perhaps the presence of others is disinhibiting, as might be seen with alcohol consumption. The disinhibition is probably related to the number of people present and who is present. Interestingly, both men and women tend to eat less when they are with the opposite sex, although men produce social facilitation of eating in women but not in other men. And women who eat less are perceived as more feminine even when eating alone.

Although social facilitation of eating is a powerful effect, Pliner and Rozin note that the effect of the presence of other people while eating is much more complex than simple facilitation. The concept of "eating lightly" is important for some individuals when making a good impression in many Western societies; therefore, people eat less when they are with strangers or coworkers, with superiors in a military organization, when eating with the opposite sex, when under observation, in the presence of a dieter, or with someone who eats very little. People who are sensitive about their weight especially show these effects. Pliner and Rozin note that culture defines what is an appropriate meal size and meal content, noting that two desserts would appear excessive, although they might contain no more nutrients than two other meal components.

Sobal deals with how eating partners are selected and excluded, which is referred to as commensality. Commensality extends the notion of "you are what you eat" to "you are who you eat with." Sobal notes that the concept of commensal circles can be used for classifying actual and potential eating partners, and he uses the caste system of India as a classic example of how commensal circles work. Commensal units exist for particular meals, the most common unit being the family. Pliner and Rozin also include this notion of rules for meal sharing in their chapter. They note that rules for food sharing are more rigid in cultures like Hindu India and Papua New Guinea, but they are also observable in Western societies, where sharing food in the home or especially in a restaurant implies a certain closeness. Commensal units also expand to include the broad range of people involved in celebrations and holiday meals. Sobal notes that the decisions about who is invited are as important as what is eaten, again underscoring the importance of sociability in meals in general and especially in celebratory meals. Sobal further points out that people not only eat together, they diet together in health clubs and other groupings. Marshall notes that commensal rules and rituals may be more marked in eating outside the home than within the home. Outside the home people who are equals tend to eat together, whereas within the home people are not equals. Marshall reminds us of the traditional view that "children should be seen and not heard," and Sobal notes that at home meals, parents communicate more with each other than with children.

Sobal also examines the interactions at meals, the social exchanges that occur at meals. At family meals, the mother tends to have responsibility for both the food and the interaction among family members. She tends to be the arbiter of what is appropriate and what is desirable. A great deal of the interaction at meals consists of narratives, probably of the day's events and related topics. The interaction at family meals permits the transmission of cultural and family ideals to children from parents. Social control is also taught at mealtimes by enforcing the rules of politeness. Finally, Sobal notes the importance of nonverbal communication at mealtimes. Just being present is a form of communication, as is physical body language, how close people sit to one another, and with whom people sit. An interesting point about sociability and eating is that sociability might be downgraded in institutional eating, where time may be as important as eating. Speeding up eating might be accomplished at the cost of increased social and verbal interaction.

A number of chapters throughout this volume have mentioned celebratory meals, perhaps because these occasions highlight many of the elements of meals that are included in this book. Lucy Long focuses on these meals in her chapter on holiday meals, which she subtitles "Rituals of Family Traditions." From the outset, Long positions holiday meals as "multivalent events," "...because such meals represent the intersection of food with two other symbolically rich domains—those of 'family' and those of 'holidays'." Long takes a "folkloristic approach," including the personal and aesthetic aspect of culture and the social construction of meaningfulness. "Holidays and rituals, foodways, and family are all subdisciplines within the field of folklore. The concept of meal is both subject and scholarly construct in each one." Long's chapter uses the holiday of Thanksgiving as a model holiday. In the United States, Thanksgiving is celebrated at the end of November to commemorate the original celebration of the Pilgrim forefathers in the 1620s in the Plimouth Colony. To Long, Thanksgiving "provides an excellent example of the multiple layers of activities and meanings involved in such meals." And further, Long notes that the American Thanksgiving is similar to harvest festivals held around the world. However, Long is quick to point out that virtually every aspect of Thanksgiving, including the authenticity of date, place, food, participants, and the degree of propriety, is subject to claim and counterclaim.

Long divides her treatment of holiday meals into foodways, family, holidays and rituals, and family politics. The meal is "...the visible focal point of a range of activities," and the "...tangible expression or enactment of the beliefs, customs, history, and aesthetics surrounding a culture's eating habits." This extended network of activities is referred to as foodways, which encompasses "the full meaning of food in our lives." Long goes on to discuss each of the activities involved in foodways, the procurement, preservation, preparation, presentation, performance, and consumption of food. Many of these activities involve advance planning and work days before the meal. Long notes that "Most holiday meals...are a mixture of traditional components, variations on those components, and complete innovations." This affects all the activities just noted. The centerpiece of the meal is the whole turkey, using an entire animal for a celebration, as we have seen mentioned elsewhere in this chapter. Because the meal is rather formal and often has many diners, children are often placed at a separate table. Long includes the cleanup as part of the holiday meal tradition.

Lucy Long goes on to discuss the role of family as "folk group," which she defines as "a group in which common experiences are shared and expressive traditions have emerged...." Recall the discussion within this chapter about narrative stories at mealtimes. Long discusses family folklore in terms of family history, family characters, family relationships, and the theme of identity. Identity comprises the characteristics that define a family and "...include: ethnicity, region, socioeconomic class, occupation, religion and ethos, and recreational interests." Family identity is represented by the family holiday meal.

Lucy Long discusses holidays as coming from the words "holy days," thus distinguishing the "...ordinary, secular, and mundane..." from the "...sacred or special...." Long points out that many holidays are not religious. Holidays involve rituals, "...recurring activities with a symbolic reference." Rituals include the rites of spectacle, rites of season, rites of passage, rites of affirmation, rites of unity, and rites of reversal. Long discusses the various roles that rituals perform.

Finally, Long discusses "the politics of family holiday meals" at several different levels. These include class struggles ("gastropolitics"), social relationships ("commensal politics"), and tradition struggles ("cultural politics"). The obvious conclusion is "...eating together is not always a socially positive or pleasant event." The divalent nature of meals and the contribution of some of these issues appear in other chapters of this book.

The reader should note, while exploring the richness of Long's chapter, the information provided on meals in America. Although the chapter focuses on special meals, the material nevertheless tells us about everyday meals in America. There is little or no reference to French food traditions. There is relatively less use of sauces, even at a festive meal. Less focus is on consecutive meal courses and proper sequence of foods (although a sequence does exist).

MEALS IN DIFFERENT CULTURES

Four chapters in the book present information on characteristic meals within cultures, two in Europe and two in Asia.

The two Asian countries, China and Japan, represent two cultures whose food is known in the West, along with the food of India. Increasingly, the food of other Asian cultures, such as Thailand, Vietnam, Cambodia, and Malaysia is also becoming known in the West. Jacqueline Newman presents a detailed chapter on Chinese meals, beginning with the important role of food in Chinese culture and the influence of Chinese culture on other Asian nations, including Japan, Thailand, and Vietnam. Newman claims that food is as important to the Chinese as to the French, and the role of the French in Western food is repeatedly seen in this book. However, Newman sees the role of food in Chinese culture to be different than in the West and of much older heritage, going back as far as 3000 years. Newman presents interesting historical material indicating the ancient roots of many current Chinese foods and the importance of food in traditional China.

Newman presents the Chinese cultural meal traditions that play an important role in Chinese meals. First, "...Chinese children are taught to leave the table 70% full and to eat tsai foods in lesser quantity than fan or grain foods and to listen and learn from adult table conversation." These notions fit in well with the discussions by Sobal and others of the socializing role of meals in the West and the lower status of children at mealtime. Whereas people in the West are just beginning to appreciate the role of food in disease and health, the Chinese view is that "Diseases are most often referred to as 'conditions' that are either *yin* or *yang*. Various foods, also considered *yin* or *yang*, are prescribed to treat them...Meals are made up of lots of grains because they are neutral and a smaller set of hot and cold foods to maintain a person's *yin-yang* balance."

The Chinese eat three meals per day, as we have also observed in Western cultures and in Rozin's observations of linguistic patterns of meals. Under certain conditions the Chinese eat two meals or even one meal plus snacks. The bulk of the diet is carbohydrate (more than 60%), but the most popular grain is wheat, with more rice being consumed in the southern part of the country. Interestingly, "the main carbohydrate is called *fan,* a word that translates as both *rice* and *a meal*. The dishes that accompany the *fan* are referred to as *tsai*. That word translates as 'vegetable' but really means all of the (accompanying) dishes...." A family meal is composed of a large bowl of *fan* and several side dishes of *tsai* to flavor the fan. Snacks of dumplings or noodles are increasing in frequency.

Traditional breakfast consists of rice *congee* or noodle soup in the south or deep-fried crullers in the north. Small amounts of other items accompany the main food in the north and south. The main meal of the day is taken at either midday or evening. The midday meal might be similar to breakfast, or it might be the main meal composed of a vegetable dish, rice or noodles, one or more meat and/or vegetable dishes, or soup, with fruit at the end. The most popular beverage is soup. Tea may be served between meals, at the morning meal, or before or after other meals but rarely during them, except at dim sum. There is a serious protocol of how food is served and consumed, using chopsticks and serving spoons, as well as on other issues of eating and finishing the meal. As Newman notes, "Eating a Chinese meal is a ritual, a social event, and an opportunity for pure enjoyment."

Jacqueline Newman also presents information on both formal family meals and festival meals. At banquets, many courses are served in sequence, with soups or sweet items interspersed and with a sweet soup or fruit as the ending.

For festive occasions, everyday foods such as pork are not used, although whole fish may be used. The use of whole animals/fish for celebratory meals has been noted elsewhere. Certain Chinese celebrations call for special items, just as we saw in the American Thanksgiving presented by Lucy Long.

Regional differences persist in China. Southeastern Chinese prefer mild and sweet foods, cook with very little oil, use different varieties of rice, and originated the tea lunch or dim sum. Southwestern Chinese prefer twice-cooked, hot and piquant foods and brine-preserved vegetables. Northeastern Chinese prefer saltier foods, noodles (mein) made from wheat and other flours, and also grilled and hot pot meals. Eastern Chinese prefer sour and sweet foods. Newman notes that certain seasonings indicate that a meal is Chinese, supporting the flavor principles of Elizabeth Rozin.

Shigeru Otsuka presents Japanese meals, emphasizing both traditional meals and more contemporary changes. The greatest influence on Japanese meals, according to Otsuka, was the imperial ban on eating meat, which began in the seventh century and ended in the nineteenth century. "The prohibition on eating meat led the Japanese to depend on seafood and vegetables... both [of which] change from season to season, and this made the Japanese sensitive to the seasonal nature of food...." Whereas in the West we also see seasonal variation in meals, this topic only receives special attention in this volume in the discussion by the four chefs and is probably less important in the West than in Japan, where "meals constitute a veritable pageant of the seasons." Another significant influence is the importance of ceremony and "other elements having nothing to do with taste and nutrition."

Many Western dishes create flavor from meat, oil, dairy products, and spices, whereas the Japanese use *shoyu* (soy sauce) and *miso* (fermented bean paste), which probably originated from the Asian mainland. Note Newman's discussion of the influence of Chinese food on other Asian foods. Otsuka exclaims that "Shoyu defines the 'flavor of Japan,' is the common denominator of home cooking, and is 'the all-purpose seasoning.'" Otsuka argues that shoyu formed the basis for introducing Western foods into Japanese cuisine, such as the reintroduction in the nineteenth century of beef as sukiyaki flavored with shoyu. Other flavor enhancers are *dashi*, made from dried fish and dried kelp, and *umami*, which is monosodium glutamate. Otsuka presents detailed information on the main components of Japanese meals, including rice, sushi, sashimi and tempura, tofu and other soybean products, and wheat noodles. Otsuka notes that many of these products are based on Asian mainland origins. The many uses of shoya seasoning with the variety of Japanese dishes are outlined in detail. Clearly shoya is at the basis of the flavor principle for Japanese cuisine.

Otsuka ends with a discussion of meal patterns in modern Japan, which underwent rapid change after World War II. This change included the provision of Western-style free lunches in the schools; those free lunch recipients are Japanese adults today. Otsuka presents typical meals for a Japanese family at home, beginning with a breakfast of eggs and a breakfast meat, vegetable or fruit salad, toast and milk, and coffee. Lunch might consist of fried noodles with meat or fish, scallop soup, and fruit. Dinner might be based on seafood or meat, with rice, various pickles, and miso soup. The menus contain an interesting blend of traditional items and Western items such as steak, hamburger, and fried chicken, although Otsuka notes "If not mentioned, most ingredients are chopped and seasoned with shoyu, sugar, and dashi." This demonstrates further how traditional flavor principles have been used to introduce new foods to Japanese meals. Otsuka also presents typical school lunch meals.

It is interesting to consider whether Chinese meals will undergo an intense process of internationalization as occurred in Japan. The two major factors that Otsuka identifies are not present in China today. Japan had an Occupation Army after World War II that ate a different diet, rice was in very short supply after the war, and bread was used as a substitute. It is also worthwhile for the reader to look in depth at these presentations of two different Asian cuisines and to examine the misconceptions of these cuisines as viewed in the West. The Chinese cuisine is not primarily rice based. Fish is far more used in Japan because of the prohibition on meat for a millennium. Traditional flavoring is composed of several rather simple materials.

We turn now to two examples of European cultures and their meals, the broad culture of the Nordic countries in Northern Europe and the other in Great Britain. From the perspective of the National Public Health Institute in Finland, Ritva Prättälä presents her observations and collected research covering Denmark, Finland, Norway, and Sweden. The four countries are bound together by history and geography, and different combinations of the countries were linked politically at different points in history. Prättälä concentrates on the workday pattern of eating and distinguishes the traditional meal pattern from the modern emerging meal pattern, as we saw in Otsuka's presentation on Japanese meals.

The traditional meal pattern included three hot meals and two or three snacks. Prättälä describes the traditional hot meals and snacks in the four different countries. Breakfast was not always a distinctive meal in terms of its components. The common ingredients of all meals were bread, potatoes, fish, porridge or gruel, and some meat. The midday meal was usually the largest meal.

The present meal pattern is similar in all the four Nordic countries. Breakfast is based on bread served with spreads, cheese, milk, fruit juice, and coffee. The main meal of the day is the midday meal in Finland and Sweden, and the evening meal in Norway and Denmark. The main meal in all countries consists of boiled potatoes with meat or fish. Prättälä notes that "Although rice and pasta are often used as substitutes for potatoes, the Nordic people seem to emphasize the role of potatoes as a component of a proper meal. Vegetable salads are common, whereas desserts, with the exception of fresh fruit, rarely belong to weekday meals." Midday meals in school or the workplace are composed of bread and fresh vegetable salads.

Prättälä confirms some observations in other chapters in her discussion of Nordic meal patterns. In Finland, there are three mealtimes, but only 40% of Finns report three meals per day, and most people eat only one hot meal. Breakfast is not a hot meal, and lunch is usually the main meal, accounting for the largest peak in nutrient intake and the largest percentage of people participating. The pattern of the large lunch evolved from government policy to provide lunch to schoolchildren, beginning in 1948, and to many workplace employees in the 1970s. This government policy has also produced less demand for a large, hot evening meal.

Sweden has an official dietary recommendation of three meals and two to three snacks per day, and according to Prättälä "Swedes follow their recommendations." Breakfast is a smaller uncooked meal, and the percentage skipping breakfast is reported to be only 5% to 10%. Both lunch and dinner are cooked meals, and either can be the main meal; but many Swedes only eat one hot meal per day. Schools have provided lunches since 1946, as in Finland. Both Finland and Sweden have large national catering services that exert a great impact on the daily meals of the populations. Nevertheless, in Sweden the cooked family dinner is still a regular occurrence.

Norwegians eat five times per day, including two bread-based meals and one hot meal at dinner when the family eats together. Lunch is cold, and children bring sandwiches to school. Dinner includes boiled potatoes and vegetables, along with meat or fish. Danes also eat about five times per day, with two cold meals and one hot meal, the latter usually taken at dinner in the evening. The Danish school lunch also consists of sandwiches.

Prättälä concludes with the identification of some patterns across the four Nordic countries. Three hot meals per day have evolved into one hot meal, but the other lighter meals occur at mealtimes and do not appear to qualify as grazing. The cooked breakfast has disappeared, as has been observed elsewhere. Meal components have changed with the introduction of some new foods (pizza, hamburgers, fresh vegetables) and the disappearance of some old foods (porridge). In Finland and Sweden the hot lunch is still the norm, whereas in Norway and Denmark the sandwich lunch is the norm. Prättälä discusses this in terms of the social policies and socioeconomic conditions of the different countries.

David Marshall brings together both cultural considerations and eating habit considerations in his chapter on British meals, in which he asks the question, "What makes a meal distinctly British?" Marshall begins by reviewing some of the definitions of meal and by distinguishing meals from snacks, using Mäkelä's criteria of ingredients (varied-simple), quantity (more-less), quality (healthy-unhealthy), food preparation (cooked-uncooked), situation (sit down–not sit down), sociability (company-alone), and planning (planned-unplanned). Mäkelä currently uses three main dimensions in categorizing meals, as was noted previously: meal format, eating pattern, and social organization.

Marshall focuses on the characteristics of British meals throughout his chapter, including a lengthy discussion of the "proper (family) meal." The proper (British) meal consists of one course comprising hot meat, potatoes, vegetables, and gravy. Ann Murcott has noted that the various components of this one plateful require separate preparation, separate cooking techniques, and lengthy preparations. Marshall reports that his own research underscores the widespread appeal of the proper meal in northern England. The proper meal is epitomized in the traditional British Sunday lunch of three courses, including the plated main dish with its accompaniments. The Sunday lunch is the most likely to bring the family together. Marshall cautions that there is an "emerging plurality" of British meals, and he predicts that more variety is in the future.

Marshall proposes a "speculative classification of British meals," which includes celebratory meals, main weekend meals (Sunday lunch), main weekday meals, light meals, and snacks. Marshall notes that as one moves from celebratory meals to snacks, the meals become "more informal, less structured, more frequent, include fewer courses, fewer items/ingredients, are more likely to be eaten alone, and are more likely to be prepared individually. Less time is required for planning, food preparation, cooking, eating...." The importance of this proposed classification according to Marshall is that the distribution of these meals is changing, with an increase in light meals.

Marshall ends his chapter with a thoughtful discussion of meals in Britain, concluding, "British meals have remained relatively resilient to change despite the proliferation of new products, the rise in eating out, and greater exposure, through the media, to new cuisine." Marshall notes the importance of this phenomenon because "The 'proper meal' is symbolic of British family life and central to understanding how meals are defined." Interestingly, Marshall notes that British breakfast has changed in a pronounced way. This provides yet another example of how breakfast differs from other meals. Has breakfast changed more than lunch or dinner in other cultures as well? As we grapple throughout this book with an adequate definition of a meal, Marshall draws the conclusion that "The distinguishing feature of the meal lies in the combination and presentation of the food." The reader should see how the chefs expand on this basic notion. Concern about preparation might further subside as more partially prepared and fully prepared products enter the market.

FOOD SERVICE/CATERING OF MEALS

Edwards presents a detailed discussion of dining out from the caterer's perspective. When combined with Schafheitle's discussion with the four chefs, we get a rich view of restaurant meals. Edwards divides meals into eating out for pleasure, eating out at work or for business, and eating out through necessity. Each of these situations brings with it a host of environmental, contextual, or situational factors that contribute to the enjoyment of the meal. Herbert Meiselman and his colleagues have written extensively on these contextual factors. Edwards discusses producing meals in each of these situations, following different concepts of production.

There appear to be continuing shifts in the rates at which people eat out for pleasure and eat out for work. Interestingly, surveys show that English food is the first choice in Britain, although when all the ethnic choices are combined, they are greater. The popularity of English food is interesting in the context of the preceding discussion on the proper British meal, which remains popular, especially in the form of the Sunday lunch. Fast food restaurants appear to be much more popular in the United States than in the United Kingdom. Eating out for business in the United Kingdom does not appear to be increasing at the same rate as eating out for pleasure, and many more people bring a packed meal to work. These shifts are part of what leads to changing concepts of a midday meal and a meal in general. Generations ago these workers would have had a full hot meal at midday, often at home. Subsidized meals for workers remain important in the United Kingdom and throughout Europe.

Edwards presents an overview of the catering industry, which provides many insights for those readers unfamiliar with the behind-the-scenes realities of the food service industry. For example, in the United Kingdom in 1994, 85% of all catering was provided by small firms doing less than ¼ million pounds of annual business. The image of big business dominating the foodservice industry is not supported by the facts. I assume that the percentage of big industry catering in the United States would be higher. Edwards notes that the U. S. consumer spends the largest amount per year on eating out, with consumers in Australia, Canada, Singapore, Hong Kong, and Japan in a second tier. European spending for eating out is much lower, with France, Germany, and Italy grouping together near the top of European countries. All the meals combined add up to one of the largest industries in the world, employing more than 10% of the global workforce, which produces more than 10% of the world gross domestic product. Edwards details the different techniques that have developed to produce and store catering meals in addition to producing everything fresh.

Edwards ends his chapter with a forecast of future changes in the catering industry, including increased availability of new and modified foods, increased casual and family dining, and increased number of catering outlets. The chapters by Edwards and Schafheitle should make it clear to the reader that catering or food service has changed greatly in the past 100 to 200 years, and even in the past 5 to 10 years, and will continue to change. What the meal or menu will be in 2050 is very difficult to predict now, but no one should assume it will stay the same.

PRODUCT DEVELOPMENT OF MEALS

Howard Moskowitz provides the perspectives of researcher, market researcher, and product developer. Moskowitz begins by comparing the traditional approach to meal preparation with newer approaches that are geared more to mathematical models of market acceptance by the consumer. In the traditional method, a marketing group created a product concept and handed it off to the development group, which produced a physical product. In today's team development, the product developer, marketer, market researcher, package designer, and others all work together. The team develops the concept and brings the concept through product development. The team is responsible for all the steps in designing and producing meal prototypes: "…today's team identifies trends in the market, sets business objectives, musters the relevant resources, creates the prototypes, does the necessary premarket research… creates the packaging, launches the project, and manages the product through its first year or two." This new approach to product development has created new technologies in the past 10 to 15 years, aimed at identifying product opportunities, creating and quantifying concepts, and managing product launch.

Moskowitz then brings the reader through product examples in much the same way that the chefs bring the reader through menu design based on giving the customer what he or she wants. However, Moskowitz uses extremely sophisticated mathematical techniques to handle consumer data and product data. In his first example, the goal is to develop a packaged lunch meal for children for sale in both food service outlets and in supermarkets. The project begins with a list of five problems or objectives: general product concept, number and identification of lunch components, number of different lunches, optimal formulation of the meat item ("the key component"), and package design. Note the reliance on the meat item as the main item, as we have seen previously and will see later in the chef-designed meals. The project team divided this work into six stages and along the way conducted interviews with targeted customer samples. For example, the team sought to link together the various components of the lunch meal into one concept and used interviews with 53 teenagers to accomplish this. The teenagers rated 50 different lunch concepts involving 20 different lunch food items. At the same time, product developers rated the same set of 50 concepts on feasibility. Further tests with consumers refined both the concept and the items that fit the concept from the perspective of the target group of customers. As Moskowitz notes, the developing trend in commercial meal design is to involve consumers earlier in meal development and to use "consumers as guides through a database of alternatives rather simply as evaluators."

Moskowitz provides the reader with a good introduction to meal modeling, which we first saw in the chapter by Pliner and Rozin. Moskowitz presents both his own and other research in assessing the contribution of each meal component to overall meal acceptability (the main dish dominates). More complicated, however, are the time preference relationships that result from repetitive consumption of foods and the decline in acceptance with repetition and exposure. Interestingly, although the main dish (meat) items dominate the overall meal, these same items suffer a rapid decline in acceptability after eating. Staple items such as bread exhibit little effect of exposure. Finally, Moskowitz reviews his own research on mathematical analysis of food combinations in meals. All these approaches demonstrate that the complexity of meals is not beyond quantitative scientific investigation and that new scientific techniques can help both to describe meal events and to aid the product designer and product developer in making decisions based on data.

Recall that David Marshall stated that "The distinguishing feature of the meal lies in the combination and presentation of the food." Elizabeth Rozin approaches meal development challenge by considering how "...people...produce food that is palatable and enjoyable...." "The means by which people accomplish this goal is the behavior we call cooking, the deliberate manipulation or transformation of basic foodstuffs into appropriate edibles." And Elizabeth Rozin points out that although cooking exists in "...almost limitless variety..." across the world, "...much of the enterprise is common and universal." Rozin identifies the universal components of all culinary systems as the basic foods of the system, the culinary techniques, and the flavors associated with the system.

The basic foods of any system are the foods that are readily available and prepared for consumption. These food selections are based on geography, environment (climate, soil, precipitation), ease of production, and other factors. They are maintained even when new foods are introduced. Examples are the use of lamb in the Middle East and corn in Mexico.

Culinary techniques refer to how basic foods are transformed into edible food. Elizabeth Rozin classifies these into three categories: (1) processes that change the physical size, shape, and mass of food, such as chopping; (2) processes that change the water content of foods, such as drying; and (3) processes that change the food chemically, such as cooking or fermentation. The selection of these processes produces large differences in the final food products, such as occurs when three different cultures preserve local fish: "...the natives of the American Northwest smoke their salmon, whereas the Portuguese salt and dry their codfish, and the Scandinavians pickle their herring."

Elizabeth Rozin notes that flavor "...seems to be...that part of culinary practice that is most capable of evoking a particular ethnicity, that is most crucial in providing a sensory and cultural label for the food of any group." Rozin goes on to describe the use of "flavor principles," which are combinations of flavoring ingredients used consistently and pervasively. Flavor principles provide flavor profiles "...that are familiar and pleasing to those within the system, recognizable and replicable to those from without." Rozin contrasts the flavors of two lamb dishes, both cooked over charcoal: the Geek lamb is flavored with lemon and oregano; the Indonesian lamb is flavored with soy sauce, coconut, chiles, and ground peanuts. And she contrasts the flavoring of boiled wheat noodles in China (soy sauce, gingerroot, and sesame oil) and in Italy (olive oil, garlic, tomatoes, basil, and oregano).

Elizabeth Rozin goes on to dissect flavor principles into their essential components, called "flavor systems." These include fats or oils, whether used in cooking, in spreads, or as condiments. Fats and oils possess their own flavor, as well as carrying other flavors. Another flavor system is the liquid component, which can be animal or vegetable, fresh or fermented. Fresh animal-based liquids include chicken stock, and fermented animal-based liquids include fish sauce from Southwest Asia and yogurt. The third flavor system is the wide variety of flavoring ingredients, both fresh and processed, including aromatics (onion), fresh green herbs (basil), spices (seeds), and acidic ingredients (vinegar).

Elizabeth Rozin discusses the role of flavor principles and flavor systems in improving the level of palatability worldwide, where most people cannot choose to eat anything they wish. She notes that "...the heavier the dependence on plant or vegetable foods, the more pronounced the seasonings (Southwest Asia, India, Africa, Mexico); the heavier the consumption of animal foods, the less pronounced the seasonings (Western Europe, Central Asia)." Rozin argues that "...the heavier seasoning of vegetable foods is a way of compensating for the lack of...red meat with its savory juices and heavy load of fat." Recall the central role of meat in Britain, according to David Marshall.

Elizabeth Rozin goes on to discuss the preparation of salads and the use of condiments in different cultures. While focusing on the factors that promote stability in cuisines, Rozin also notes "...the unprecedented swing toward novelty..." in the twentieth century, and she discusses changing food habits. Other authors in this volume also comment on changing food habits, some arguing that certain meals retain their importance (the British Sunday lunch), whereas others argue that foods are changing. Chef Shaun Hill sees less novelty in the future. This issue of where meals are heading is an interesting issue, for which there is no simple answer; understanding all the factors covered in this book makes it clear why predicting the future of meals is so difficult. But Elizabeth Rozin ends on a simpler note: "...we all know a meal when we see one. And we all know what makes it taste good." We have seen the first opinion expressed earlier in this volume, and on the basis of Rozin's descriptions of flavor principles, we now have a better idea of how to accomplish the second point. The last section of the book deals with how four chefs create good food and combine that into good meals.

CREATIVE MEAL DESIGN

Joachim Schafheitle introduces the reader to menu design before embarking on the interviews with the four chefs. Of special interest is the material that he has organized in a figure covering "a collation of general rules... for use when designing meals." Schafheitle notes that the first three points to consider (vary, alternate, avoid) affect the senses and customer perception. The next level of consideration (consider) affects culture and business/marketing issues, and the final three elements (summary) combine social and biological aspects within a business environment. "The skill of the chef lies in cooking a satisfying meal, but his creative talent is required to balance the food components so that the flavors, textures, colors, and temperatures mix and mingle happily together and meet his customers' expectation." We will next turn to the chefs and hear from them directly about how to achieve this skill.

Schafheitle's dialogue with the four chefs moves this book across the full range from science through application to creative art. The chefs represent European training and backgrounds, both traditional and nontraditional. Albert Roux and Anton Mosimann were classically trained by apprenticeship to master chefs, Roux in France and Mosimann in Switzerland; Anton Edelmann was classically trained in Germany, France, and Switzerland; Shaun Hill plunged right into the restaurant business working for an American restaurateur in London. At present, Albert Roux has two restaurants, both of which have attained 3-star Michelin status. Shaun Hill was previously Executive Chef at one of England's premier country house hotels and now runs a country restaurant of his own. Anton Mosimann has also served as Executive Chef at a top English hotel and currently owns a private dining club in London. Anton Edelman is Maitre Chefs des Cuisines at one of the top London Hotels.

Albert Roux's philosophy of menu design is nicely summarized at the end of his interview: "When compiling a menu, keep it simple and honest; cook the way you feel and for your customer; bear in mind the seasons and traditions and be innovative...." Roux emphasizes simplicity (perhaps one should read *elegant* simplicity) in menus, rather than excess. A great emphasis is placed on pleasing the customer and on the use of fresh, seasonal ingredients to provide seasonal variety and excellent quality. In other words use "what is available, what has good flavor," although Roux recognizes the realities and potential of global transportation. Traditions are also important to this classically trained chef, who emphasizes traditional cuisine: "Stick to your own tastes," meaning keep cuisines pure rather than fusion; "Very few people mix flavor beautifully"; "we have to preserve the differences...." Roux uses the sociological aspect of eating by emphasizing the relationship between server and diner through service performed at the table (e.g., carving) because "it strengthens the bond between the kitchen and our dining room." Finally, in keeping with Roux's own background and training and our historical observations on classical cuisine, "Menu language at *Le Gavroche* is unashamedly French."

Throughout his interview, Chef Shaun Hill tries to balance the role of customer and cook in menu design. "It is difficult to know whether meal design is completely customer led or whether it's cook led. I think it is customer led, and customers have become cooks." But he later adds, "...the chef's taste in compiling the menu became an important factor...." Hill advises on a number of important factors in menu design: preparation time considerations, availability, and seasonality ("I like the idea of seasonality. It has a nice warm feel..."), flavors ("I do tend to eat each new dish in its entirety because a dish can taste fine in a mouthful but starts to pale after a plateful. I try to get down to what the simplest part the meal is about, a contrast of two flavors or two textures...," sequence ("...progression from lighter to heartier and then lighter again..."), and minimal or no repetition of flavors from course to course. Throughout this discussion, it is important to keep in mind the material on sensory aspects of meals by Harry Lawless; much of the design of menus comes from the sensory basis of meals. Hill also looks to the future and predicts that "Food will become slightly less eclectic...."

Hill also comments on celebratory meals, which have been covered elsewhere in this book, especially in Lucy Long's chapter. "Special celebrations...demand specific types of food..." and "You want a whole animal...and you tend to have meat rather than fish." In some cultures (Greek, Japanese) you might have fish. "...you are looking for some token to represent success for the occasion."

Anton Mosimann contrasts his earlier experiences in menu design in Switzerland with the pattern today. For menu designing, "you have to have the vision for presentation, taste, looks, texture. I always say when you create a menu, make sure you look; if you serve three or four courses, for different methods of cooking—one is poached, one is steamed, one is grilled...."

Mosimann concentrates on local fresh food and seasonal food: "I buy as much as possible locally..." and "I stay as much as possible within the seasons...." The presentation should combine both color and taste. Mosimann also prefers to follow the classical sequence of courses. Mosimann begins menu design with the main course, emphasizing the importance of the main course that was noted previously when talking about meal modeling. Because Mosimann runs a private dining club, he changes the Table d'hôte menu weekly, the à la carte menu monthly, and the menu surprise daily. Menu changes vary with the type of establishment and the frequency of customers eating at the establishment. Mosimann does not use French on the menu. For the future of menu design, Mosimann sees a continuing focus on health and fitness. In advising a new cook about menu design, Anton Mosimann returns to his own classical training, and he advises the new cook to work at the right establishment with the right chef.

Anton Edelmann also uses English language menus, and in designing menus he emphasizes some of the points that have been seen with the other chefs: "...choice, number of courses, nutrition, seasonality of foods, food combinations...." Like the others, Edelmann likes local (British) foods. Edelmann sums it up this way: "In many ways, meal design is like a woman applying makeup: for everyday use, it is kept quick and simple but for special occasions more time and trouble are taken in its application and a much larger range of products is used." Edelmann notes that time considerations for the customers are also important, with dinner lasting 2 hours and lunch 1 hour, reflecting the faster pace of life today. Edelmann changes the menus four times yearly, working within the seasons.

Edelmann agrees with what we have heard before about the importance of the main course, and he begins meal design with it. Second come the dessert and cheese, and then the fish or appetizer. Edelmann tries to achieve balance and variety in sauces (rich-light), temperature (hot-cold), texture, cooking methods, color, spices, and so forth. Edelmann also notes that no celebratory meal is more fixed by tradition in Britain than Christmas dinner. When looking toward the future, Edelmann sees further decrease in meal formality, with continued preference for smaller food portions and variety.

Do our four chefs agree on menu design? Yes, to a remarkable degree, especially in view of the fact that they were trained in several different countries. In addition, they now run very different types of establishments: a top-end restaurant in the city, a more intimate country restaurant, a private city dining club, and a big city hotel restaurant. Despite these differences, the chefs all emphasize the roles of variety and balance in maintaining interest throughout the meal. This is accomplished by the following:

- Seasonal variation
- Proper food combinations in sequence, avoiding duplication (e.g., fish-fish)
- Variety in food preparation techniques from course to course
- Variety in sauces from course to course
- Variety in texture, taste, color, etc.

How can we possibly summarize all the material in this book? Clearly, this selective summary proves that we cannot because meals are so complex and require many disciplines to understand them. Marshall concludes his chapter with what to me is the theme for the entire volume: "Understanding eating patterns, meal formats, and the changing social nature of eating is the key not only to new product development and product acceptability but also to the adoption of healthful dietary advice, good nutrition, and health promotion." I hope that this lively presentation of what meals are about will expand how we look at meals and lead us to include considerations of meals in our research, writing, product developing, and creative catering.

Index